STATA QUICK REFERENCE AND INDEX

RELEASE 12

A Stata Press Publication
StataCorp LP
College Station, Texas

Table of contents

Combined subject table of contents

This is the complete contents for all the Reference manuals.

Every estimation command has a postestimation entry; however, the postestimation entries are not listed in the subject table of contents.

Getting Started

[GSM]	*Getting Started with Stata for Mac*	....
[GSU]	*Getting Started with Stata for Unix*	....
[GSW]	*Getting Started with Stata for Windows*	....
[U]	Chapter 3	.... Resources for learning and using Stata
[U]	Chapter 4	.... Stata's help and search facilities
[R]	help	.... Display online help
[R]	search	.... Search Stata documentation

Data manipulation and management

Basic data commands

[D]	codebook	....,,...... Describe data contents
[D]	data management	.... Introduction to data management commands
[D]	data types	....,,,,,..... Quick reference for data types
[D]	datetime	.... Date and time values and variables
[D]	describe	.... Describe data in memory or in file
[D]	edit	.... Browse or edit data with Data Editor
[D]	format	.... Set variables' output format
[D]	inspect	.... Display simple summary of data's attributes
[D]	label	.... Manipulate labels
[D]	list	.... List values of variables
[D]	missing values	.... Quick reference for missing values
[D]	rename	.... Rename variable
[D]	save	.... Save Stata dataset
[D]	sort	.... Sort data
[D]	use	.... Load Stata dataset
[D]	varmanage	,................ Manage variable labels, formats, and other properties

Creating and dropping variables

[D]	clear	.... Clear memory
[D]	compress	.... Compress data in memory
[D]	drop	.... Eliminate variables or observations
[D]	egen	.... Extensions to generate
[D]	functions	.... Functions
[D]	generate	.... Create or change contents of variable

Functions and expressions

[U]	Chapter 13	.... Functions and expressions
[D]	egen	.... Extensions to generate
[D]	functions	.... Functions

2

Dates and times

[U]   Section 12.5.3 .... Date and time formats
[U]   Chapter 24 .... Working with dates and times
[D]   bcal .... Business calendar file manipulation
[D]   datetime .... Date and time values and variables
[D]   datetime business calendars .... Business calendars
[D]   datetime business calendars creation .... Business calendars creation
[D]   datetime display formats .... Display formats for dates and times
[D]   datetime translation .... String to numeric date translation functions

Loading, saving, importing, and exporting data

[GS]   Chapter 6 (GSM, GSU, GSW) .... Using the Data Editor
[U]   Chapter 21 .... Inputting and importing data
[D]   edit .... Browse or edit data with Data Editor
[D]   export .... Overview of exporting data from Stata
[TS]   haver .... Load data from Haver Analytics database
[D]   import .... Overview of importing data into Stata
[D]   import excel .... Import and export Excel files
[D]   import sasxport .... Import and export datasets in SAS XPORT format
[D]   infile (fixed format) .... Read text data in fixed format with a dictionary
[D]   infile (free format) .... Read unformatted text data
[D]   infix (fixed format) .... Read text data in fixed format
[D]   input .... Enter data from keyboard
[D]   insheet .... Read text data created by a spreadsheet
[D]   odbc .... Load, write, or view data from ODBC sources
[D]   outfile .... Export dataset in text format
[D]   outsheet .... Write spreadsheet-style dataset
[D]   save .... Save Stata dataset
[D]   sysuse .... Use shipped dataset
[D]   use .... Load Stata dataset
[D]   webuse .... Use dataset from Stata website
[D]   xmlsave .... Export or import dataset in XML format

Combining data

[U]   Chapter 22 .... Combining datasets
[D]   append .... Append datasets
[MI]   mi append .... Append mi data
[D]   cross .... Form every pairwise combination of two datasets
[D]   joinby .... Form all pairwise combinations within groups
[D]   merge .... Merge datasets
[MI]   mi merge .... Merge mi data

Reshaping datasets

[D]   collapse .... Make dataset of summary statistics
[D]   contract .... Make dataset of frequencies and percentages
[D]   expand .... Duplicate observations
[D]   expandcl .... Duplicate clustered observations
[D]   fillin .... Rectangularize dataset
[D]   obs .... Increase the number of observations in a dataset
[D]   reshape .... Convert data from wide to long form and vice versa

[MI]   mi reshape .... Reshape mi data
[TS]   rolling .... Rolling-window and recursive estimation
[D]   separate .... Create separate variables
[SEM]   ssd .... Making summary statistics data
[D]   stack .... Stack data
[D]   statsby .... Collect statistics for a command across a by list
[D]   xpose .... Interchange observations and variables

Labeling, display formats, and notes

[GS]   Chapter 7 (GSM, GSU, GSW) .... Using the Variables Manager
[U]   Section 12.5 .... Formats: Controlling how data are displayed
[U]   Section 12.6 .... Dataset, variable, and value labels
[D]   format .... Set variables' output format
[D]   label .... Manipulate labels
[D]   label language .... Labels for variables and values in multiple languages
[D]   labelbook .... Label utilities
[D]   notes .... Place notes in data
[D]   varmanage .... Manage variable labels, formats, and other properties

Changing and renaming variables

[GS]   Chapter 7 (GSM, GSU, GSW) .... Using the Variables Manager
[U]   Chapter 25 .... Working with categorical data and factor variables
[D]   clonevar .... Clone existing variable
[D]   destring .... Convert string variables to numeric variables and vice versa
[D]   encode .... Encode string into numeric and vice versa
[D]   generate .... Create or change contents of variable
[D]   mvencode .... Change missing values to numeric values and vice versa
[D]   order .... Reorder variables in dataset
[D]   recode .... Recode categorical variables
[D]   rename .... Rename variable
[D]   rename group .... Rename groups of variables
[D]   split .... Split string variables into parts
[D]   varmanage .... Manage variable labels, formats, and other properties

Examining data

[GS]   Chapter 6 (GSM, GSU, GSW) .... Using the Data Editor
[D]   cf .... Compare two datasets
[D]   codebook .... Describe data contents
[D]   compare .... Compare two variables
[D]   count .... Count observations satisfying specified conditions
[D]   describe .... Describe data in memory or in file
[D]   ds .... List variables matching name patterns or other characteristics
[D]   duplicates .... Report, tag, or drop duplicate observations
[D]   edit .... Browse or edit data with Data Editor
[D]   gsort .... Ascending and descending sort
[D]   inspect .... Display simple summary of data's attributes
[D]   isid .... Check for unique identifiers
[D]   lookfor .... Search for string in variable names and labels
[R]   misstable .... Tabulate missing values
[MI]   mi describe .... Describe mi data
[MI]   mi misstable .... Tabulate pattern of missing values

[D]   pctile . . . . Create variable containing percentiles
[ST]   stdescribe . . . . Describe survival-time data
[R]   summarize . . . . Summary statistics
[SVY]   svy: tabulate oneway . . . . One-way tables for survey data
[SVY]   svy: tabulate twoway . . . . Two-way tables for survey data
[P]   tabdisp . . . . Display tables
[R]   table . . . . Tables of summary statistics
[R]   tabstat . . . . Display table of summary statistics
[R]   tabulate oneway . . . . One-way tables of frequencies
[R]   tabulate twoway . . . . Two-way tables of frequencies
[R]   tabulate, summarize() . . . . One- and two-way tables of summary statistics
[XT]   xtdescribe . . . . Describe pattern of xt data

File manipulation

[D]   cd . . . . Change directory
[D]   cf . . . . Compare two datasets
[D]   changeeol . . . . Convert end-of-line characters of text file
[D]   checksum . . . . Calculate checksum of file
[D]   copy . . . . Copy file from disk or URL
[D]   dir . . . . Display filenames
[D]   erase . . . . Erase a disk file
[D]   filefilter . . . . Convert text or binary patterns in a file
[D]   mkdir . . . . Create directory
[D]   rmdir . . . . Remove directory
[D]   type . . . . Display contents of a file
[D]   zipfile . . . . Compress and uncompress files and directories in zip archive format

Miscellaneous data commands

[D]   corr2data . . . . Create dataset with specified correlation structure
[D]   drawnorm . . . . Draw sample from multivariate normal distribution
[R]   dydx . . . . Calculate numeric derivatives and integrals
[D]   icd9 . . . . ICD-9-CM diagnostic and procedure codes
[D]   ipolate . . . . Linearly interpolate (extrapolate) values
[D]   range . . . . Generate numerical range
[D]   sample . . . . Draw random sample

Multiple imputation

[MI]   mi add . . . . Add imputations from another mi dataset
[MI]   mi append . . . . Append mi data
[MI]   mi convert . . . . Change style of mi data
[MI]   mi copy . . . . Copy mi flongsep data
[MI]   mi describe . . . . Describe mi data
[MI]   mi erase . . . . Erase mi datasets
[MI]   mi expand . . . . Expand mi data
[MI]   mi export . . . . Export mi data
[MI]   mi export ice . . . . Export mi data to ice format
[MI]   mi export nhanes1 . . . . Export mi data to NHANES format
[MI]   mi extract . . . . Extract original or imputed data from mi data
[MI]   mi import . . . . Import data into mi
[MI]   mi import flong . . . . Import flong-like data into mi
[MI]   mi import flongsep . . . . Import flongsep-like data into mi

[MI]	mi import ice	Import ice-format data into mi
[MI]	mi import nhanes1	Import NHANES-format data into mi
[MI]	mi import wide	Import wide-like data into mi
[MI]	mi merge	Merge mi data
[MI]	mi misstable	Tabulate pattern of missing values
[MI]	mi passive	Generate/replace and register passive variables
[MI]	mi ptrace	Load parameter-trace file into Stata
[MI]	mi rename	Rename variable
[MI]	mi replace0	Replace original data
[MI]	mi reset	Reset imputed or passive variables
[MI]	mi reshape	Reshape mi data
[MI]	mi set	Declare multiple-imputation data
[MI]	mi stsplit	Stsplit and stjoin mi data
[MI]	mi update	Ensure that mi data are consistent
[MI]	mi varying	Identify variables that vary across imputations
[MI]	mi xeq	Execute command(s) on individual imputations
[MI]	mi XXXset	Declare mi data to be svy, st, ts, xt, etc.
[MI]	noupdate option	The noupdate option
[MI]	styles	Dataset styles
[MI]	workflow	Suggested workflow

Utilities

Basic utilities

[GS]	Chapter 13 (GSM, GSU, GSW)	Using the Do-file Editor—automating Stata
[U]	Chapter 4	Stata's help and search facilities
[U]	Chapter 15	Saving and printing output—log files
[U]	Chapter 16	Do-files
[R]	about	Display information about your Stata
[D]	by	Repeat Stata command on subsets of the data
[R]	copyright	Display copyright information
[R]	do	Execute commands from a file
[R]	doedit	Edit do-files and other text files
[R]	exit	Exit Stata
[R]	help	Display online help
[R]	hsearch	Search help files
[R]	level	Set default confidence level
[R]	log	Echo copy of session to file
[D]	obs	Increase the number of observations in a dataset
[R]	#review	Review previous commands
[R]	search	Search Stata documentation
[R]	translate	Print and translate logs
[R]	view	View files and logs
[D]	zipfile	Compress and uncompress files and directories in zip archive format

Error messages

[U]	Chapter 8	Error messages and return codes
[P]	error	Display generic error message and exit
[R]	error messages	Error messages and return codes
[P]	rmsg	Return messages

Saved results

[U]	Section 13.5	Accessing coefficients and standard errors
[U]	Section 18.8	Accessing results calculated by other programs
[U]	Section 18.9	Accessing results calculated by estimation commands
[U]	Section 18.10	Saving results
[P]	creturn	Return c-class values
[P]	ereturn	Post the estimation results
[R]	estimates	Save and manipulate estimation results
[R]	estimates describe	Describe estimation results
[R]	estimates for	Repeat postestimation command across models
[R]	estimates notes	Add notes to estimation results
[R]	estimates replay	Redisplay estimation results
[R]	estimates save	Save and use estimation results
[R]	estimates stats	Model statistics
[R]	estimates store	Store and restore estimation results
[R]	estimates table	Compare estimation results
[R]	estimates title	Set title for estimation results
[P]	_return	Preserve saved results
[P]	return	Return saved results
[R]	saved results	Saved results

Internet

[U]	Chapter 28	Using the Internet to keep up to date
[R]	adoupdate	Update user-written ado-files
[D]	checksum	Calculate checksum of file
[D]	copy	Copy file from disk or URL
[R]	net	Install and manage user-written additions from the Internet
[R]	net search	Search the Internet for installable packages
[R]	netio	Control Internet connections
[R]	news	Report Stata news
[R]	sj	Stata Journal and STB installation instructions
[R]	ssc	Install and uninstall packages from SSC
[R]	update	Update Stata
[D]	use	Load Stata dataset

Data types and memory

[U]	Chapter 6	Managing memory
[U]	Section 12.2.2	Numeric storage types
[U]	Section 12.4.4	String storage types
[U]	Section 13.11	Precision and problems therein
[U]	Chapter 23	Working with strings
[D]	compress	Compress data in memory
[D]	data types	Quick reference for data types
[R]	matsize	Set the maximum number of variables in a model
[D]	memory	Memory management
[D]	missing values	Quick reference for missing values
[D]	recast	Change storage type of variable

Advanced utilities

[D]	assert	Verify truth of claim
[D]	cd	Change directory
[D]	changeeol	Convert end-of-line characters of text file
[D]	checksum	Calculate checksum of file
[D]	copy	Copy file from disk or URL
[P]	_datasignature	Determine whether data have changed
[D]	datasignature	Determine whether data have changed
[R]	db	Launch dialog
[P]	dialog programming	Dialog programming
[D]	dir	Display filenames
[P]	discard	Drop automatically loaded programs
[D]	erase	Erase a disk file
[P]	file	Read and write ASCII text and binary files
[D]	filefilter	Convert text or binary patterns in a file
[D]	hexdump	Display hexadecimal report on file
[D]	mkdir	Create directory
[R]	more	The —more— message
[R]	query	Display system parameters
[P]	quietly	Quietly and noisily perform Stata command
[D]	rmdir	Remove directory
[R]	set	Overview of system parameters
[R]	set cformat	Format settings for coefficient tables
[R]	set_defaults	Reset system parameters to original Stata defaults
[R]	set emptycells	Set what to do with empty cells in interactions
[R]	set seed	Specify initial value of random-number seed
[R]	set showbaselevels	Display settings for coefficient tables
[D]	shell	Temporarily invoke operating system
[P]	signestimationsample	Determine whether the estimation sample has changed
[P]	smcl	Stata Markup and Control Language
[P]	sysdir	Query and set system directories
[D]	type	Display contents of a file
[R]	which	Display location and version for an ado-file

Graphics

Common graphs

[G-2]	graph	The graph command
[G-2]	graph bar	Bar charts
[G-2]	graph box	Box plots
[G-2]	graph combine	Combine multiple graphs
[G-2]	graph copy	Copy graph in memory
[G-2]	graph describe	Describe contents of graph in memory or on disk
[G-2]	graph dir	List names of graphs in memory and on disk
[G-2]	graph display	Display graph stored in memory
[G-2]	graph dot	Dot charts (summary statistics)
[G-2]	graph drop	Drop graphs from memory
[G-2]	graph export	Export current graph
[G-2]	graph manipulation	Graph manipulation commands
[G-2]	graph matrix	Matrix graphs

8

[R]   marginsplot .... Graph results from margins (profile plots, etc.)
[G-2]   palette .... Display palettes of available selections

Distributional graphs

[R]   cumul .... Cumulative distribution
[R]   diagnostic plots .... Distributional diagnostic plots
[R]   ladder .... Ladder of powers
[R]   spikeplot .... Spike plots and rootograms

Multivariate graphs

[MV]   biplot .... Biplots
[MV]   ca postestimation .... Postestimation tools for ca and camat
[MV]   cluster dendrogram .... Dendrograms for hierarchical cluster analysis
[MV]   mca postestimation .... Postestimation tools for mca
[MV]   mds postestimation .... Postestimation tools for mds, mdsmat, and mdslong
[MV]   scoreplot .... Score and loading plots
[MV]   screeplot .... Scree plot

Quality control

[R]   cusum .... Graph cumulative spectral distribution
[R]   qc ,,, Quality control charts
[R]   serrbar ....,,,,   ,,,, Graph standard error bar chart

Regression diagnostic plots

[R]   regress postestimation .... Postestimation tools for regress

ROC analysis

[R]   logistic postestimation .... Postestimation tools for logistic
[R]   roccomp .... Tests of equality of ROC areas
[R]   rocfit postestimation .... Postestimation tools for rocfit
[R]   rocregplot .... Plot marginal and covariate-specific ROC curves after rocreg
[R]   roctab .... Nonparametric ROC analysis

Smoothing and densities

[R]   kdensity .... Univariate kernel density estimation
[R]   lowess .... Lowess smoothing
[R]   lpoly .... Kernel-weighted local polynomial smoothing
[R]   sunflower .... Density-distribution sunflower plots

Survival-analysis graphs

[ST]   ltable .... Life tables for survival data
[ST]   stci .... Confidence intervals for means and percentiles of survival time
[ST]   stcox PH-assumption tests .... Tests of proportional-hazards assumption
[ST]   stcurve ... Plot survivor, hazard, cumulative hazard, or cumulative incidence function
[ST]   strate .... Tabulate failure rates and rate ratios
[ST]   sts graph .... Graph the survivor and cumulative hazard functions

Time-series graphs

[TS]   corrgram .... Tabulate and graph autocorrelations
[TS]   cumsp .... Cumulative spectral distribution
[TS]   fcast graph .... Graph forecasts of variables computed by fcast compute
[TS]   irf cgraph ..... Combine graphs of IRFs, dynamic-multiplier functions, and FEVDs
[TS]   irf graph .... Graph IRFs, dynamic-multiplier functions, and FEVDs
[TS]   irf ograph .... Graph overlaid IRFs, dynamic-multiplier functions, and FEVDs
[TS]   pergram .... Periodogram
[TS]   tsline .... Plot time-series data
[TS]   varstable .... Check the stability condition of VAR or SVAR estimates
[TS]   vecstable .... Check the stability condition of VECM estimates
[TS]   wntestb .... Bartlett's periodogram-based test for white noise
[TS]   xcorr .... Cross-correlogram for bivariate time series

More statistical graphs

[R]   dotplot .... Comparative scatterplots
[ST]   epitab .... Tables for epidemiologists
[R]   fracpoly postestimation .... Postestimation tools for fracpoly
[R]   grmeanby .... Graph means and medians by categorical variables
[R]   pkexamine .... Calculate pharmacokinetic measures
[R]   pksumm .... Summarize pharmacokinetic data
[R]   stem .... Stem-and-leaf displays
[XT]   xtline .... Panel-data line plots

Editing

[G-1]   graph editor .... Graph Editor

Graph utilities

[G-2]   set graphics .... Set whether graphs are displayed
[G-2]   set printcolor .... Set how colors are treated when graphs are printed
[G-2]   set scheme .... Set default scheme

Graph schemes

[G-4]   schemes intro .... Introduction to schemes
[G-4]   scheme economist .... Scheme description: economist
[G-4]   scheme s1 .... Scheme description: s1 family
[G-4]   scheme s2 .... Scheme description: s2 family
[G-4]   scheme sj .... Scheme description: sj

Graph concepts

[G-4]   concept: gph files .... Using gph files
[G-4]   concept: lines .... Using lines
[G-4]   concept: repeated options .... Interpretation of repeated options
[G-4]   text .... Text in graphs

Statistics

ANOVA and related

[U]	Chapter 26	.... Overview of Stata estimation commands
[R]	anova	.... Analysis of variance and covariance
[R]	contrast	.... Contrasts and linear hypothesis tests after estimation
[R]	loneway	.... Large one-way ANOVA, random effects, and reliability
[MV]	manova	.... Multivariate analysis of variance and covariance
[R]	oneway	.... One-way analysis of variance
[R]	pkcross	.... Analyze crossover experiments
[R]	pkshape	.... Reshape (pharmacokinetic) Latin-square data
[R]	pwcompare	.... Pairwise comparisons
[R]	regress	.... Linear regression
[XT]	xtmixed	.... Multilevel mixed-effects linear regression
[XT]	xtreg	.. Fixed-, between-, and random-effects and population-averaged linear models

Basic statistics

[R]	anova	.... Analysis of variance and covariance
[R]	bitest	.... Binomial probability test
[R]	ci	.... Confidence intervals for means, proportions, and counts
[R]	correlate	.... Correlations (covariances) of variables or coefficients
[D]	egen	.... Extensions to generate
[R]	mean	.... Estimate means
[R]	misstable	.... Tabulate missing values
[MV]	mvtest	.... Multivariate tests
[R]	oneway	.... One-way analysis of variance
[R]	proportion	.... Estimate proportions
[R]	prtest	.... One- and two-sample tests of proportions
[R]	pwmean	.... Pairwise comparisons of means
[R]	ranksum	.... Equality tests on unmatched data
[R]	ratio	.... Estimate ratios
[R]	regress	.... Linear regression
[R]	sampsi	.... Sample size and power for means and proportions
[R]	sdtest	.... Variance-comparison tests
[R]	signrank	.... Equality tests on matched data
[D]	statsby	.... Collect statistics for a command across a by list
[R]	summarize	.... Summary statistics
[R]	table	.... Tables of summary statistics
[R]	tabstat	.... Display table of summary statistics
[R]	tabulate oneway	.... One-way tables of frequencies
[R]	tabulate twoway	.... Two-way tables of frequencies
[R]	tabulate, summarize()	.... One- and two-way tables of summary statistics
[R]	total	.... Estimate totals
[R]	ttest	.... Mean-comparison tests

Binary outcomes

[U]	Chapter 20	.... Estimation and postestimation commands
[U]	Section 26.6	.... Binary-outcome qualitative dependent-variable models
[R]	binreg	.... Generalized linear models: Extensions to the binomial family
[R]	biprobit	.... Bivariate probit regression

[R] cloglog .... Complementary log-log regression
[R] exlogistic .... Exact logistic regression
[R] glm .... Generalized linear models
[R] glogit .... Logit and probit regression for grouped data
[R] heckprob .... Probit model with sample selection
[R] hetprob .... Heteroskedastic probit model
[R] ivprobit .... Probit model with continuous endogenous regressors
[R] logistic .... Logistic regression, reporting odds ratios
[R] logit .... Logistic regression, reporting coefficients
[R] probit .... Probit regression
[R] rocfit .... Parametric ROC models
[R] rocreg .... Receiver operating characteristic (ROC) regression
[R] scobit .... Skewed logistic regression
[XT] xtcloglog .... Random-effects and population-averaged cloglog models
[XT] xtlogit .... Fixed-effects, random-effects, and population-averaged logit models
[XT] xtmelogit .... Multilevel mixed-effects logistic regression
[XT] xtprobit .... Random-effects and population-averaged probit models

Categorical outcomes

[U] Chapter 20 .... Estimation and postestimation commands
[U] Section 26.9 .... Multiple-outcome qualitative dependent-variable models
[R] asclogit .... Alternative-specific conditional logit (McFadden's choice) model
[R] asmprobit .... Alternative-specific multinomial probit regression
[R] clogit .... Conditional (fixed-effects) logistic regression
[R] mlogit .... Multinomial (polytomous) logistic regression
[R] mprobit .... Multinomial probit regression
[R] nlogit .... Nested logit regression
[R] slogit .... Stereotype logistic regression

Cluster analysis

[U] Section 26.23 .... Multivariate and cluster analysis
[MV] cluster .... Introduction to cluster-analysis commands
[MV] cluster dendrogram .... Dendrograms for hierarchical cluster analysis
[MV] cluster generate   ... Generate summary or grouping variables from a cluster analysis
[MV] cluster kmeans and kmedians .... Kmeans and kmedians cluster analysis
[MV] cluster linkage .... Hierarchical cluster analysis
[MV] cluster notes .... Place notes in cluster analysis
[MV] cluster programming subroutines .... Add cluster-analysis routines
[MV] cluster programming utilities .... Cluster-analysis programming utilities
[MV] cluster stop .... Cluster-analysis stopping rules
[MV] cluster utility .... List, rename, use, and drop cluster analyses
[MV] clustermat .... Introduction to clustermat commands
[MV] matrix dissimilarity .... Compute similarity or dissimilarity measures
[MV] measure_option .... Option for similarity and dissimilarity measures
[MV] multivariate .... Introduction to multivariate commands

Correspondence analysis

[MV] ca .... Simple correspondence analysis
[MV] mca .... Multiple and joint correspondence analysis

Count outcomes

[U]	Chapter 20	Estimation and postestimation commands
[U]	Section 26.10	Count dependent-variable models
[U]	Section 26.17.5	Count dependent-variable models with panel data
[R]	expoisson	Exact Poisson regression
[R]	nbreg	Negative binomial regression
[R]	poisson	Poisson regression
[R]	tnbreg	Truncated negative binomial regression
[R]	tpoisson	Truncated Poisson regression
[XT]	xtmepoisson	Multilevel mixed-effects Poisson regression
[XT]	xtnbreg	Fixed-effects, random-effects, & population-averaged negative binomial models
[XT]	xtpoisson	Fixed-effects, random-effects, and population-averaged Poisson models
[R]	zinb	Zero-inflated negative binomial regression
[R]	zip	Zero-inflated Poisson regression

Discriminant analysis

[MV]	candisc	Canonical linear discriminant analysis
[MV]	discrim	Discriminant analysis
[MV]	discrim estat	Postestimation tools for discrim
[MV]	discrim knn	kth-nearest-neighbor discriminant analysis
[MV]	discrim lda	Linear discriminant analysis
[MV]	discrim logistic	Logistic discriminant analysis
[MV]	discrim qda	Quadratic discriminant analysis
[MV]	scoreplot	Score and loading plots
[MV]	screeplot	Scree plot

Do-it-yourself generalized method of moments

[U]	Section 26.19	Generalized method of moments (GMM)
[R]	gmm	Generalized method of moments estimation
[P]	matrix	Introduction to matrix commands

Do-it-yourself maximum likelihood estimation

[P]	matrix	Introduction to matrix commands
[R]	ml	Maximum likelihood estimation

Endogenous covariates

[U]	Chapter 20	Estimation and postestimation commands
[U]	Chapter 26	Overview of Stata estimation commands
[R]	gmm	Generalized method of moments estimation
[R]	ivprobit	Probit model with continuous endogenous regressors
[R]	ivregress	Single-equation instrumental-variables regression
[R]	ivtobit	Tobit model with continuous endogenous regressors
[R]	reg3	Three-stage estimation for systems of simultaneous equations
[R]	treatreg	Treatment-effects model
[XT]	xtdpd	Linear dynamic panel-data estimation
[XT]	xtdpdsys	Arellano–Bover/Blundell–Bond linear dynamic panel-data estimation
[XT]	xthtaylor	Hausman–Taylor estimator for error-components models
[XT]	xtivreg	Instrumental variables and two-stage least squares for panel-data models

14

Epidemiology and related

[R]	binreg	.... Generalized linear models: Extensions to the binomial family
[R]	brier	.... Brier score decomposition
[R]	dstdize	.... Direct and indirect standardization
[ST]	epitab	.... Tables for epidemiologists
[R]	exlogistic	.... Exact logistic regression
[D]	icd9	.... ICD-9-CM diagnostic and procedure codes
[R]	kappa	.... Interrater agreement
[R]	logistic	.... Logistic regression, reporting odds ratios
[R]	pk	.... Pharmacokinetic (biopharmaceutical) data
[R]	pkcollapse	.... Generate pharmacokinetic measurement dataset
[R]	pkcross	.... Analyze crossover experiments
[R]	pkequiv	.... Perform bioequivalence tests
[R]	pkexamine	.... Calculate pharmacokinetic measures
[R]	pkshape	.... Reshape (pharmacokinetic) Latin-square data
[R]	pksumm	.... Summarize pharmacokinetic data
[R]	roc	.... Receiver operating characteristic (ROC) analysis
[R]	roccomp	.... Tests of equality of ROC areas
[R]	rocfit	.... Parametric ROC models
[R]	rocreg	.... Receiver operating characteristic (ROC) regression
[R]	roctab	.... Nonparametric ROC analysis
[R]	symmetry	.... Symmetry and marginal homogeneity tests
[R]	tabulate twoway	.... Two-way tables of frequencies

Estimation related

[R]	BIC note	.... Calculating and interpreting BIC
[R]	constraint	.... Define and list constraints
[R]	*eform_option*	.... Displaying exponentiated coefficients
[R]	estimation options	.... Estimation options
[R]	fracpoly	.... Fractional polynomial regression
[R]	maximize	.... Details of iterative maximization
[R]	mfp	.... Multivariable fractional polynomial models
[R]	mkspline	.... Linear and restricted cubic spline construction
[R]	stepwise	.... Stepwise estimation
[R]	*vce_option*	.... Variance estimators
[XT]	*vce_options*	.... Variance estimators

Exact statistics

[U]	Section 26.11	.... Exact estimators
[R]	bitest	.... Binomial probability test
[R]	centile	.... Report centile and confidence interval
[R]	ci	.... Confidence intervals for means, proportions, and counts
[R]	dstdize	.... Direct and indirect standardization
[ST]	epitab	.... Tables for epidemiologists
[R]	exlogistic	.... Exact logistic regression
[R]	expoisson	.... Exact Poisson regression
[R]	ksmirnov	.... Kolmogorov–Smirnov equality-of-distributions test
[R]	loneway	.... Large one-way ANOVA, random effects, and reliability
[R]	ranksum	.... Equality tests on unmatched data
[R]	roctab	.... Nonparametric ROC analysis

[R]   symmetry .... Symmetry and marginal homogeneity tests
[R]   tabulate twoway .... Two-way tables of frequencies
[R]   tetrachoric .... Tetrachoric correlations for binary variables

Factor analysis and principal components

[R]   alpha .... Compute interitem correlations (covariances) and Cronbach's alpha
[MV] canon .... Canonical correlations
[MV] factor .... Factor analysis
[MV] pca .... Principal component analysis
[MV] rotate .... Orthogonal and oblique rotations after factor and pca
[MV] rotatemat .... Orthogonal and oblique rotations of a Stata matrix
[MV] scoreplot .... Score and loading plots
[MV] screeplot .... Scree plot
[R]   tctrachoric .... Tetrachoric correlations for binary variables

Generalized linear models

[U]   Chapter 20 .... Estimation and postestimation commands
[U]   Section 26.5 .... Generalized linear models
[R]   binreg .... Generalized linear models: Extensions to the binomial family
[R]   glm .... Generalized linear models
[XT] xtgee .... Fit population-averaged panel-data models by using GEE

Indicator and categorical variables

[U]   Section 11.4.3 .... Factor variables
[U]   Chapter 25 .... Working with categorical data and factor variables
[R]   fvset .... Declare factor-variable settings

Linear regression and related

[U]   Chapter 20 .... Estimation and postestimation commands
[U]   Chapter 26 .... Overview of Stata estimation commands
[R]   areg .... Linear regression with a large dummy variable set
[R]   cnsreg .... Constrained linear regression
[R]   constraint .... Define and list constraints
[R]   eivreg .... Errors-in-variables regression
[I]   estimation commands .... Quick reference for estimation commands
[R]   fracpoly .... Fractional polynomial regression
[R]   frontier .... Stochastic frontier models
[R]   glm .... Generalized linear models
[R]   gmm .... Generalized method of moments estimation
[R]   heckman .... Heckman selection model
[R]   intreg .... Interval regression
[R]   ivregress .... Single equation instrumental-variables regression
[R]   ivtobit .... Tobit model with continuous endogenous regressors
[R]   logit .... Logistic regression, reporting coefficients
[R]   lpoly .... Kernel-weighted local polynomial smoothing
[R]   mfp .... Multivariable fractional polynomial models
[R]   mvreg .... Multivariate regression
[R]   nbreg .... Negative binomial regression
[R]   nestreg .... Nested model statistics
[TS] newey .... Regression with Newey–West standard errors
[R]   nl .... Nonlinear least-squares estimation

[R]   nlsur . . . . Estimation of nonlinear systems of equations
[R]   orthog . . . . Orthogonalize variables and compute orthogonal polynomials
[R]   poisson . . . . Poisson regression
[TS]   prais . . . . Prais–Winsten and Cochrane–Orcutt regression
[R]   qreg . . . . Quantile regression
[R]   reg3 . . . . Three-stage estimation for systems of simultaneous equations
[R]   regress . . . . Linear regression
[R]   rocfit . . . . Parametric ROC models
[R]   rreg . . . . Robust regression
[ST]   stcox . . . . Cox proportional hazards model
[ST]   stcrreg . . . . Competing-risks regression
[R]   stepwise . . . . Stepwise estimation
[ST]   streg . . . . Parametric survival models
[R]   sureg . . . . Zellner's seemingly unrelated regression
[R]   tnbreg . . . . Truncated negative binomial regression
[R]   tobit . . . . Tobit regression
[R]   tpoisson . . . . Truncated Poisson regression
[R]   treatreg . . . . Treatment-effects model
[R]   truncreg . . . . Truncated regression
[R]   vwls . . . . Variance-weighted least squares
[XT]   xtabond . . . . Arellano–Bond linear dynamic panel-data estimation
[XT]   xtcloglog . . . . Random-effects and population-averaged cloglog models
[XT]   xtdpd . . . . Linear dynamic panel-data estimation
[XT]   xtdpdsys . . . . . . Arellano–Bover/Blundell–Bond linear dynamic panel-data estimation
[XT]   xtfrontier . . . . Stochastic frontier models for panel data
[XT]   xtgee . . . . Fit population-averaged panel-data models by using GEE
[XT]   xtgls . . . . Fit panel-data models by using GLS
[XT]   xthtaylor . . . . Hausman–Taylor estimator for error-components models
[XT]   xtintreg . . . . Random-effects interval-data regression models
[XT]   xtivreg . . . . . Instrumental variables and two-stage least squares for panel-data models
[XT]   xtlogit . . . . Fixed-effects, random-effects, and population-averaged logit models
[XT]   xtmelogit . . . . Multilevel mixed-effects logistic regression
[XT]   xtmepoisson . . . . Multilevel mixed-effects Poisson regression
[XT]   xtmixed . . . . Multilevel mixed-effects linear regression
[XT]   xtnbreg Fixed-effects, random-effects, & population-averaged negative binomial models
[XT]   xtpcse . . . . Linear regression with panel-corrected standard errors
[XT]   xtpoisson . . . . Fixed-effects, random-effects, and population-averaged Poisson models
[XT]   xtprobit . . . . Random-effects and population-averaged probit models
[XT]   xtrc . . . . Random-coefficients model
[XT]   xtreg . . Fixed-, between-, and random-effects and population-averaged linear models
[XT]   xtregar . . . . Fixed- and random-effects linear models with an AR(1) disturbance
[XT]   xttobit . . . . Random-effects tobit models
[R]   zinb . . . . Zero-inflated negative binomial regression
[R]   zip . . . . Zero-inflated Poisson regression

Logistic and probit regression

[U]   Chapter 20 . . . . Estimation and postestimation commands
[U]   Chapter 26 . . . . Overview of Stata estimation commands
[R]   asclogit . . . . Alternative-specific conditional logit (McFadden's choice) model
[R]   asmprobit . . . . Alternative-specific multinomial probit regression
[R]   asroprobit . . . . Alternative-specific rank-ordered probit regression

Longitudinal data/panel data

[XT]   xtrc . . . . Random-coefficients model
[XT]   xtreg . . Fixed-, between-, and random-effects and population-averaged linear models
[XT]   xtregar . . . . Fixed- and random-effects linear models with an AR(1) disturbance
[XT]   xtset . . . . Declare data to be panel data
[XT]   xtsum . . . . Summarize xt data
[XT]   xttab . . . . Tabulate xt data
[XT]   xttobit . . . . Random-effects tobit models
[XT]   xtunitroot . . . . Panel-data unit-root tests

Mixed models

[U]   Chapter 20 . . . . Estimation and postestimation commands
[R]   anova . . . . Analysis of variance and covariance
[MV]   manova . . . . Multivariate analysis of variance and covariance
[XT]   xtcloglog . . . . Random-effects and population-averaged cloglog models
[XT]   xtintreg . . . . Random-effects interval-data regression models
[XT]   xtlogit . . . . Fixed-effects, random-effects, and population-averaged logit models
[XT]   xtmelogit . . . . Multilevel mixed-effects logistic regression
[XT]   xtmepoisson . . . . Multilevel mixed-effects Poisson regression
[XT]   xtmixed . . . . Multilevel mixed-effects linear regression
[XT]   xtprobit . . . . Random-effects and population-averaged probit models
[XT]   xtrc . . . . Random-coefficients model
[XT]   xtreg . . Fixed-, between-, and random-effects and population-averaged linear models
[XT]   xttobit . . . . Random-effects tobit models

Multidimensional scaling and biplots

[MV]   biplot . . . . Biplots
[MV]   mds . . . . Multidimensional scaling for two-way data
[MV]   mdslong . . . . Multidimensional scaling of proximity data in long format
[MV]   mdsmat . . . . Multidimensional scaling of proximity data in a matrix
[MV]   measure_option . . . . Option for similarity and dissimilarity measures

Multilevel/hierarchical models

[XT]   xtmelogit . . . . Multilevel mixed-effects logistic regression
[XT]   xtmepoisson . . . . Multilevel mixed-effects Poisson regression
[XT]   xtmixed . . . . Multilevel mixed-effects linear regression

Multiple imputation

[U]   Section 26.22 . . . . Multiple imputation
[MI]   estimation . . . . Estimation commands for use with mi estimate
[MI]   intro substantive . . . . Introduction to multiple-imputation analysis
[MI]   mi estimate . . . . Estimation using multiple imputations
[MI]   mi estimate using . . . . Estimation using previously saved estimation results
[MI]   mi estimate postestimation . . . . Postestimation tools for mi estimate
[MI]   mi impute . . . . Impute missing values
[MI]   mi impute chained . . . . Impute missing values using chained equations
[MI]   mi impute intreg . . . . Impute using interval regression
[MI]   mi impute logit . . . . Impute using logistic regression
[MI]   mi impute mlogit . . . . Impute using multinomial logistic regression
[MI]   mi impute monotone . . . . Impute missing values in monotone data
[MI]   mi impute mvn . . . . Impute using multivariate normal regression

[MI]   mi impute nbreg .... Impute using negative binomial regression
[MI]   mi impute ologit .... Impute using ordered logistic regression
[MI]   mi impute pmm .... Impute using predictive mean matching
[MI]   mi impute poisson .... Impute using Poisson regression
[MI]   mi impute regress .... Impute using linear regression
[MI]   mi impute truncreg .... Impute using truncated regression
[MI]   mi predict .... Obtain multiple-imputation predictions
[MI]   mi test .... Test hypotheses after mi estimate

Multivariate analysis of variance and related techniques

[U]   Section 26.23 .... Multivariate and cluster analysis
[MV]   canon .... Canonical correlations
[MV]   hotelling .... Hotelling's T-squared generalized means test
[MV]   manova .... Multivariate analysis of variance and covariance
[R]   mvreg .... Multivariate regression
[MV]   mvtest covariances .... Multivariate tests of covariances
[MV]   mvtest means .... Multivariate tests of means

Nonparametric statistics

[R]   bitest .... Binomial probability test
[R]   bootstrap .... Bootstrap sampling and estimation
[R]   bsample .... Sampling with replacement
[R]   bstat .... Report bootstrap results
[R]   centile .... Report centile and confidence interval
[R]   cusum .... Graph cumulative spectral distribution
[R]   kdensity .... Univariate kernel density estimation
[R]   ksmirnov .... Kolmogorov–Smirnov equality-of-distributions test
[R]   kwallis .... Kruskal–Wallis equality-of-populations rank test
[R]   lowess .... Lowess smoothing
[R]   lpoly .... Kernel-weighted local polynomial smoothing
[R]   nptrend .... Test for trend across ordered groups
[R]   prtest .... One- and two-sample tests of proportions
[R]   qreg .... Quantile regression
[R]   ranksum .... Equality tests on unmatched data
[R]   roc .... Receiver operating characteristic (ROC) analysis
[R]   roccomp .... Tests of equality of ROC areas
[R]   rocreg .... Receiver operating characteristic (ROC) regression
[R]   rocregplot .... Plot marginal and covariate-specific ROC curves after rocreg
[R]   roctab .... Nonparametric ROC analysis
[R]   runtest .... Test for random order
[R]   signrank .... Equality tests on matched data
[R]   simulate .... Monte Carlo simulations
[R]   smooth .... Robust nonlinear smoother
[R]   spearman .... Spearman's and Kendall's correlations
[R]   symmetry .... Symmetry and marginal homogeneity tests
[R]   tabulate twoway .... Two-way tables of frequencies

Ordinal outcomes

[U]   Chapter 20 .... Estimation and postestimation commands
[R]   asroprobit .... Alternative-specific rank-ordered probit regression

[R]   ologit . . . . Ordered logistic regression
[R]   oprobit . . . . Ordered probit regression
[R]   rologit . . . . Rank-ordered logistic regression

Other statistics

[R]   alpha . . . . Compute interitem correlations (covariances) and Cronbach's alpha
[R]   ameans . . . . Arithmetic, geometric, and harmonic means
[R]   brier . . . . Brier score decomposition
[R]   centile . . . . Report centile and confidence interval
[R]   kappa . . . . Interrater agreement
[MV]   mvtest correlations . . . . Multivariate tests of correlations
[R]   pcorr . . . . Partial and semipartial correlation coefficients
[D]   pctile . . . . Create variable containing percentiles
[D]   range . . . . Generate numerical range

Pharmacokinetic statistics

[U]   Section 26.24 . . . . Pharmacokinetic data
[R]   pk . . . . Pharmacokinetic (biopharmaceutical) data
[R]   pkcollapse . . . . Generate pharmacokinetic measurement dataset
[R]   pkcross . . . . Analyze crossover experiments
[R]   pkequiv . . . . Perform bioequivalence tests
[R]   pkexamine . . . . Calculate pharmacokinetic measures
[R]   pkshape . . . . Reshape (pharmacokinetic) Latin-square data
[R]   pksumm . . . . Summarize pharmacokinetic data

Power and sample size

[R]   sampsi . . . . Sample size and power for means and proportions
[ST]   stpower . . . . . Sample-size, power, and effect-size determination for survival analysis
[ST]   stpower cox Sample size, power, and effect size for the Cox proportional hazards model
[ST]   stpower exponential . . . . Sample size and power for the exponential test
[ST]   stpower logrank . . . . Sample size, power, and effect size for the log-rank test

Quality control

[R]   cusum . . . . Graph cumulative spectral distribution
[R]   qc . . . . Quality control charts
[R]   serrbar . . . . Graph standard error bar chart

ROC analysis

[U]   Section 26.7 . . . . ROC analysis
[R]   roc . . . . Receiver operating characteristic (ROC) analysis
[R]   roccomp . . . . Tests of equality of ROC areas
[R]   rocfit . . . . Parametric ROC models
[R]   rocfit postestimation . . . . Postestimation tools for rocfit
[R]   rocreg . . . . Receiver operating characteristic (ROC) regression
[R]   rocreg postestimation . . . . Postestimation tools for rocreg
[R]   rocregplot . . . . Plot marginal and covariate-specific ROC curves after rocreg
[R]   roctab . . . . Nonparametric ROC analysis

Rotation

[MV]   procrustes .... Procrustes transformation
[MV]   rotate .... Orthogonal and oblique rotations after factor and pca
[MV]   rotatemat .... Orthogonal and oblique rotations of a Stata matrix

Sample selection models

[U]   Chapter 20 .... Estimation and postestimation commands
[U]   Section 26.15 .... Models with endogenous sample selection
[R]   heckman .... Heckman selection model
[R]   heckprob .... Probit model with sample selection
[R]   treatreg .... Treatment-effects model

Simulation/resampling

[R]   bootstrap .... Bootstrap sampling and estimation
[R]   bsample .... Sampling with replacement
[R]   jackknife .... Jackknife estimation
[R]   permute .... Monte Carlo permutation tests
[R]   simulate .... Monte Carlo simulations

Standard postestimation tests, tables, and other analyses

[U]   Section 13.5 .... Accessing coefficients and standard errors
[U]   Chapter 20 .... Estimation and postestimation commands
[R]   contrast .... Contrasts and linear hypothesis tests after estimation
[R]   correlate .... Correlations (covariances) of variables or coefficients
[R]   estat .... Postestimation statistics
[R]   estimates .... Save and manipulate estimation results
[R]   estimates describe .... Describe estimation results
[R]   estimates for .... Repeat postestimation command across models
[R]   estimates notes .... Add notes to estimation results
[R]   estimates replay .... Redisplay estimation results
[R]   estimates save .... Save and use estimation results
[R]   estimates stats .... Model statistics
[R]   estimates store .... Store and restore estimation results
[R]   estimates table .... Compare estimation results
[R]   estimates title .... Set title for estimation results
[R]   hausman .... Hausman specification test
[R]   lincom .... Linear combinations of estimators
[R]   linktest .... Specification link test for single-equation models
[R]   lrtest .... Likelihood-ratio test after estimation
[R]   margins .... Marginal means, predictive margins, and marginal effects
[R]   margins, contrast .... Contrasts of margins
[R]   margins, pwcompare .... Pairwise comparisons of margins
[R]   marginsplot .... Graph results from margins (profile plots, etc.)
[MV]   mvtest .... Multivariate tests
[R]   nlcom .... Nonlinear combinations of estimators
[R]   predict .... Obtain predictions, residuals, etc., after estimation
[R]   predictnl .... Obtain nonlinear predictions, standard errors, etc., after estimation
[R]   pwcompare .... Pairwise comparisons
[R]   suest .... Seemingly unrelated estimation

[R]   test . . . .   Test linear hypotheses after estimation
[R]   testnl . . . .   Test nonlinear hypotheses after estimation

Structural equation modeling

[U]   Section 26.3 . . . .   Structural equation modeling (SEM)
[SEM]   estat eqgof . . . .   Equation-level goodness-of-fit statistics
[SEM]   estat eqtest . . . .   Equation-level test that all coefficients are zero
[SEM]   estat framework . . . .   Display estimation results in modeling framework
[SEM]   estat ggof . . . .   Group-level goodness-of-fit statistics
[SEM]   estat ginvariant . . . .   Tests for invariance of parameters across groups
[SEM]   estat gof . . . .   Goodness-of-fit statistics
[SEM]   estat mindices . . . .   Modification indices
[SEM]   estat residuals . . . .   Display mean and covariance residuals
[SEM]   estat scoretests . . . .   Score tests
[SEM]   estat stable . . . .   Check stability of nonrecursive system
[SEM]   estat stdize . . . .   Test standardized parameters
[SEM]   estat summarize . . . .   Report summary statistics for estimation sample
[SEM]   estat teffects . . . .   Decomposition of effects into total, direct, and indirect
[SEM]   example 1 . . . .   Single-factor measurement model
[SEM]   example 2 . . . .   Creating a dataset from published covariances
[SEM]   example 3 . . . .   Two-factor measurement model
[SEM]   example 4 . . . .   Goodness-of-fit statistics
[SEM]   example 5 . . . .   Modification indices
[SEM]   example 6 . . . .   Linear regression
[SEM]   example 7 . . . .   Nonrecursive structural model
[SEM]   example 8 . . . .   Testing that coefficients are equal, and constraining them
[SEM]   example 9 . . . .   Structural model with measurement component
[SEM]   example 10 . . . .   MIMIC model
[SEM]   example 11 . . . .   estat framework
[SEM]   example 12 . . . .   Seemingly unrelated regression
[SEM]   example 13 . . . .   Equation-level Wald test
[SEM]   example 14 . . . .   Predicted values
[SEM]   example 15 . . . .   Higher-order CFA
[SEM]   example 16 . . . .   Correlation
[SEM]   example 17 . . . .   Correlated uniqueness model
[SEM]   example 18 . . . .   Latent growth model
[SEM]   example 19 . . . .   Creating multiple-group summary statistics data
[SEM]   example 20 . . . .   Two-factor measurement model by group
[SEM]   example 21 . . . .   Group-level goodness of fit
[SEM]   example 22 . . . .   Testing parameter equality across groups
[SEM]   example 23 . . . .   Specifying parameter constraints across groups
[SEM]   example 24 . . . .   Reliability
[SEM]   example 25 . . . .   Creating summary statistics data from raw data
[SEM]   example 26 . . . .   Fitting a model using data missing at random
[SEM]   GUI . . . .   Graphical user interface
[SEM]   intro 1 . . . .   Introduction
[SEM]   intro 2 . . . .   Learning the language: Path diagrams and command language
[SEM]   intro 3 . . . .   Substantive concepts
[SEM]   intro 4 . . . .   Tour of models
[SEM]   intro 5 . . . .   Comparing groups
[SEM]   intro 6 . . . .   Postestimation tests and predictions

[SEM]   intro 7 . . . . Robust and clustered standard errors
[SEM]   intro 8 . . . . Standard errors, the full story
[SEM]   intro 9 . . . . Fitting models using survey data
[SEM]   intro 10 . . . . Fitting models using summary statistics data
[SEM]   lincom . . . . Linear combinations of parameters
[SEM]   lrtest . . . . Likelihood-ratio test of linear hypothesis
[SEM]   methods and formulas . . . . Methods and formulas
[SEM]   nlcom . . . . Nonlinear combinations of parameters
[SEM]   predict . . . . Factor scores, linear predictions, etc.
[SEM]   sem . . . . Structural equation model estimation command
[SEM]   sem estimation options . . . . Options affecting estimation
[SEM]   sem group options . . . . Fitting models on different groups
[SEM]   sem model description options . . . . Model description options
[SEM]   sem option constraints( ) . . . . Specifying constraints
[SEM]   sem option covstructure( ) . . . . Specifying covariance restrictions
[SEM]   sem option from( ) . . . . Specifying starting values
[SEM]   sem option method( ) . . . . Specifying method and calculation of VCE
[SEM]   sem option noxconditional . Computing means, etc. of observed exogenous variables
[SEM]   sem option reliability( ) . . . . Fraction of variance not due to measurement error
[SEM]   sem option select( ) . . . . Using sem with summary statistics data
[SEM]   sem path notation . . . . Command syntax for path diagrams
[SEM]   sem postestimation . . . . Postestimation tools for sem
[SEM]   sem reporting options . . . . Options affecting reporting of results
[SEM]   sem ssd options . . . . Options for use with summary statistics data
[SEM]   sem syntax options . . . . Options affecting interpretation of syntax
[SEM]   ssd . . . . Making summary statistics data
[SEM]   test . . . . Wald test of linear hypotheses
[SEM]   testnl . . . . Wald test of nonlinear hypotheses

Survey data

[U]   Chapter 20 . . . . Estimation and postestimation commands
[U]   Section 26.21 . . . . Survey data
[SVY]   survey . . . . Introduction to survey commands
[SVY]   *bootstrap_options* . . . . More options for bootstrap variance estimation
[SVY]   *brr_options* . . . . More options for BRR variance estimation
[SVY]   direct standardization . . . . Direct standardization of means, proportions, and ratios
[SVY]   estat . . . . Postestimation statistics for survey data
[SVY]   *jackknife_options* . . . . More options for jackknife variance estimation
[SVY]   ml for svy . . . . Maximum pseudolikelihood estimation for survey data
[SVY]   poststratification . . . . Poststratification for survey data
[P]   _robust . . . . Robust variance estimates
[SVY]   *sdr_options* . . . . More options for SDR variance estimation
[SVY]   subpopulation estimation . . . . Subpopulation estimation for survey data
[SVY]   svy . . . . The survey prefix command
[SVY]   svy bootstrap . . . . Bootstrap for survey data
[SVY]   svy brr . . . . Balanced repeated replication for survey data
[SVY]   svy estimation . . . . Estimation commands for survey data
[SVY]   svy jackknife . . . . Jackknife estimation for survey data
[SVY]   svy postestimation . . . . Postestimation tools for svy
[SVY]   svy sdr . . . . Successive difference replication for survey data
[SVY]   svy: tabulate oneway . . . . One-way tables for survey data

[SVY]   svy: tabulate twoway . . . . Two-way tables for survey data
[SVY]   svydescribe . . . . Describe survey data
[SVY]   svymarkout . . . Mark observations for exclusion on the basis of survey characteristics
[SVY]   svyset . . . . Declare survey design for dataset
[MI]   mi XXXset . . . . Declare mi data to be svy, st, ts, xt, etc.
[SVY]   variance estimation . . . . Variance estimation for survey data

Survival analysis

[U]   Chapter 20 . . . . Estimation and postestimation commands
[U]   Section 26.18 . . . . Survival-time (failure-time) models
[ST]   survival analysis Introduction to survival analysis & epidemiological tables commands
[ST]   ct . . . . Count-time data
[ST]   ctset . . . . Declare data to be count-time data
[ST]   cttost . . . . Convert count-time data to survival-time data
[ST]   discrete . . . . Discrete-time survival analysis
[ST]   ltable . . . . Life tables for survival data
[ST]   snapspan . . . . Convert snapshot data to time-span data
[ST]   st . . . . Survival-time data
[ST]   st_is . . . . Survival analysis subroutines for programmers
[ST]   stbase . . . . Form baseline dataset
[ST]   stci . . . . Confidence intervals for means and percentiles of survival time
[ST]   stcox . . . . Cox proportional hazards model
[ST]   stcox PH-assumption tests . . . . Tests of proportional-hazards assumption
[ST]   stcrreg . . . . Competing-risks regression
[ST]   stcurve . . . Plot survivor, hazard, cumulative hazard, or cumulative incidence function
[ST]   stdescribe . . . . Describe survival-time data
[R]   stepwise . . . . Stepwise estimation
[ST]   stfill . . . . Fill in by carrying forward values of covariates
[ST]   stgen . . . . Generate variables reflecting entire histories
[ST]   stir . . . . Report incidence-rate comparison
[ST]   stpower . . . . . Sample-size, power, and effect-size determination for survival analysis
[ST]   stpower cox Sample size, power, and effect size for the Cox proportional hazards model
[ST]   stpower exponential . . . . Sample size and power for the exponential test
[ST]   stpower logrank . . . . Sample size, power, and effect size for the log-rank test
[ST]   stptime . . . . Calculate person-time, incidence rates, and SMR
[ST]   strate . . . . Tabulate failure rates and rate ratios
[ST]   streg . . . . Parametric survival models
[ST]   sts . . . . . Generate, graph, list, and test the survivor and cumulative hazard functions
[ST]   sts generate . . . . Create variables containing survivor and related functions
[ST]   sts graph . . . . Graph the survivor and cumulative hazard functions
[ST]   sts list . . . . List the survivor or cumulative hazard function
[ST]   sts test . . . . Test equality of survivor functions
[ST]   stset . . . . Declare data to be survival-time data
[MI]   mi XXXset . . . . Declare mi data to be svy, st, ts, xt, etc.
[ST]   stsplit . . . . Split and join time-span records
[MI]   mi stsplit . . . . Stsplit and stjoin mi data
[ST]   stsum . . . . Summarize survival-time data
[ST]   sttocc . . . . Convert survival-time data to case–control data
[ST]   sttoct . . . . Convert survival-time data to count-time data
[ST]   stvary . . . . Report whether variables vary over time

Time series, multivariate

[U]   Section 11.4.4 .... Time-series varlists
[U]   Section 13.9 .... Time-series operators
[U]   Chapter 20 .... Estimation and postestimation commands
[U]   Section 26.16 .... Models with time-series data
[TS]   time series .... Introduction to time-series commands
[TS]   dfactor .... Dynamic-factor models
[TS]   fcast compute Compute dynamic forecasts of dependent variables after var, svar, or vec
[TS]   fcast graph .... Graph forecasts of variables computed by fcast compute
[TS]   haver .... Load data from Haver Analytics database
[TS]   irf .... Create and analyze IRFs, dynamic-multiplier functions, and FEVDs
[TS]   irf add .... Add results from an IRF file to the active IRF file
[TS]   irf cgraph ..... Combine graphs of IRFs, dynamic-multiplier functions, and FEVDs
[TS]   irf create .... Obtain IRFs, dynamic-multiplier functions, and FEVDs
[TS]   irf ctable .... Combine tables of IRFs, dynamic-multiplier functions, and FEVDs
[TS]   irf describe .... Describe an IRF file
[TS]   irf drop .... Drop IRF results from the active IRF file
[TS]   irf graph .... Graph IRFs, dynamic-multiplier functions, and FEVDs
[TS]   irf ograph .... Graph overlaid IRFs, dynamic-multiplier functions, and FEVDs
[TS]   irf rename .... Rename an IRF result in an IRF file
[TS]   irf set .... Set the active IRF file
[TS]   irf table .... Create tables of IRFs, dynamic-multiplier functions, and FEVDs
[TS]   mgarch .... Multivariate GARCH models
[TS]   mgarch ccc .... Constant conditional correlation multivariate GARCH models
[TS]   mgarch dcc .... Dynamic conditional correlation multivariate GARCH models
[TS]   mgarch dvech .... Diagonal vech multivariate GARCH models
[TS]   mgarch vcc .... Varying conditional correlation multivariate GARCH models
[TS]   psdensity .... Parametric spectral density estimation after arima, arfima, and ucm
[TS]   rolling .... Rolling-window and recursive estimation
[TS]   sspace .... State-space models
[TS]   tsappend .... Add observations to a time-series dataset
[TS]   tsfill .... Fill in gaps in time variable
[TS]   tsline .... Plot time-series data
[TS]   tsreport .... Report time-series aspects of a dataset or estimation sample
[TS]   tsrevar .... Time-series operator programming command
[TS]   tsset .... Declare data to be time-series data
[TS]   var intro .... Introduction to vector autoregressive models
[TS]   var svar .... Structural vector autoregressive models
[TS]   var .... Vector autoregressive models
[TS]   varbasic .... Fit a simple VAR and graph IRFs or FEVDs
[TS]   vargranger .... Perform pairwise Granger causality tests after var or svar
[TS]   varlmar ...... Perform LM test for residual autocorrelation after var or svar
[TS]   varnorm .... Test for normally distributed disturbances after var or svar
[TS]   varsoc .... Obtain lag-order selection statistics for VARs and VECMs
[TS]   varstable .... Check the stability condition of VAR or SVAR estimates
[TS]   varwle .... Obtain Wald lag-exclusion statistics after var or svar
[TS]   vec intro .... Introduction to vector error-correction models
[TS]   vec .... Vector error-correction models
[TS]   veclmar .... Perform LM test for residual autocorrelation after vec
[TS]   vecnorm .... Test for normally distributed disturbances after vec
[TS]   vecrank .... Estimate the cointegrating rank of a VECM

[TS]   vecstable .... Check the stability condition of VECM estimates
[TS]   xcorr .... Cross-correlogram for bivariate time series

Time series, univariate

[U]   Section 11.4.4 .... Time-series varlists
[U]   Section 13.9 .... Time-series operators
[U]   Chapter 20 .... Estimation and postestimation commands
[U]   Section 26.16 .... Models with time-series data
[TS]   time series .... Introduction to time-series commands
[TS]   arch ..... Autoregressive conditional heteroskedasticity (ARCH) family of estimators
[TS]   arfima .... Autoregressive fractionally integrated moving-average models
[TS]   arima .... ARIMA, ARMAX, and other dynamic regression models
[TS]   corrgram .... Tabulate and graph autocorrelations
[TS]   cumsp .... Cumulative spectral distribution
[TS]   dfgls .... DF-GLS unit-root test
[TS]   dfuller .... Augmented Dickey–Fuller unit-root test
[TS]   haver .... Load data from Haver Analytics database
[TS]   newey .... Regression with Newey–West standard errors
[TS]   pergram .... Periodogram
[TS]   pperron .... Phillips–Perron unit-root test
[TS]   prais .... Prais–Winsten and Cochrane–Orcutt regression
[TS]   rolling .... Rolling-window and recursive estimation
[TS]   tsappend .... Add observations to a time-series dataset
[TS]   tsfill .... Fill in gaps in time variable
[TS]   tsfilter .... Filter a time-series, keeping only selected periodicities
[TS]   tsfilter bk .... Baxter–King time-series filter
[TS]   tsfilter bw .... Butterworth time-series filter
[TS]   tsfilter cf .... Christiano–Fitzgerald time-series filter
[TS]   tsfilter hp .... Hodrick–Prescott time-series filter
[TS]   tsline .... Plot time-series data
[TS]   tsreport .... Report time-series aspects of a dataset or estimation sample
[TS]   tsrevar .... Time-series operator programming command
[TS]   tsset .... Declare data to be time-series data
[TS]   tssmooth .... Smooth and forecast univariate time-series data
[TS]   tssmooth dexponential .... Double-exponential smoothing
[TS]   tssmooth exponential .... Single-exponential smoothing
[TS]   tssmooth hwinters .... Holt–Winters nonseasonal smoothing
[TS]   tssmooth ma .... Moving-average filter
[TS]   tssmooth nl .... Nonlinear filter
[TS]   tssmooth shwinters .... Holt–Winters seasonal smoothing
[TS]   ucm .... Unobserved-components model
[TS]   wntestb .... Bartlett's periodogram-based test for white noise
[TS]   wntestq .... Portmanteau (Q) test for white noise
[TS]   xcorr .... Cross-correlogram for bivariate time series

Transforms and normality tests

[R]   boxcox .... Box–Cox regression models
[R]   fracpoly .... Fractional polynomial regression
[R]   ladder .... Ladder of powers
[R]   lnskew0 .... Find zero-skewness log or Box–Cox transform
[R]   mfp .... Multivariable fractional polynomial models

[MV]   mvtest normality . . . . Multivariate normality tests
[R]   sktest . . . . Skewness and kurtosis test for normality
[R]   swilk . . . . Shapiro–Wilk and Shapiro–Francia tests for normality

Matrix commands

Basics

[U]   Chapter 14 . . . . Matrix expressions
[P]   matlist . . . . Display a matrix and control its format
[P]   matrix . . . . Introduction to matrix commands
[P]   matrix define . . . . Matrix definition, operators, and functions
[P]   matrix utility . . . . List, rename, and drop matrices

Programming

[P]   ereturn . . . . Post the estimation results
[P]   matrix accum . . . . Form cross-product matrices
[P]   matrix rownames . . . . Name rows and columns
[P]   matrix score . . . . Score data from coefficient vectors
[R]   ml . . . . Maximum likelihood estimation
[M]   *Mata Reference Manual* . . . .

Other

[P]   makecns . . . . Constrained estimation
[P]   matrix dissimilarity . . . . Compute similarity or dissimilarity measures
[P]   matrix eigenvalues . . . . Eigenvalues of nonsymmetric matrices
[P]   matrix get . . . . Access system matrices
[P]   matrix mkmat . . . . Convert variables to matrix and vice versa
[P]   matrix svd . . . . Singular value decomposition
[P]   matrix symeigen . . . . Eigenvalues and eigenvectors of symmetric matrices

Mata

[D]   putmata . . . . Put Stata variables into Mata and vice versa
[M]   *Mata Reference Manual* . . . .

Programming

Basics

[U]   Chapter 18 . . . . Programming Stata
[U]   Section 18.3 . . . . Macros
[U]   Section 18.11 . . . . Ado-files
[P]   comments . . . . Add comments to programs
[P]   fvexpand . . . . Expand factor varlists
[P]   macro . . . . Macro definition and manipulation
[P]   program . . . . Define and manipulate programs
[P]   return . . . . Return saved results

Program control

[U]   Section 18.11.1 . . . . Version
[P]   capture . . . . Capture return code
[P]   continue . . . . Break out of loops
[P]   error . . . . Display generic error message and exit
[P]   foreach . . . . Loop over items
[P]   forvalues . . . . Loop over consecutive values
[P]   if . . . . if programming command
[P]   version . . . . Version control
[P]   while . . . . Looping

Parsing and program arguments

[U]   Section 18.4 . . . . Program arguments
[P]   confirm . . . . Argument verification
[P]   gettoken . . . . Low-level parsing
[P]   levelsof . . . . Levels of variable
[P]   numlist . . . . Parse numeric lists
[P]   syntax . . . . Parse Stata syntax
[P]   tokenize . . . . Divide strings into tokens

Console output

[P]   dialog programming . . . . Dialog programming
[P]   display . . . . Display strings and values of scalar expressions
[P]   smcl . . . . Stata Markup and Control Language
[P]   tabdisp . . . . Display tables

Commonly used programming commands

[P]   byable . . . . Make programs byable
[P]   #delimit . . . . Change delimiter
[P]   exit . . . . Exit from a program or do-file
[R]   fvrevar . . . . Factor-variables operator programming command
[P]   mark . . . . Mark observations for inclusion
[P]   matrix . . . . Introduction to matrix commands
[P]   more . . . . Pause until key is pressed
[P]   nopreserve option . . . . nopreserve option
[P]   preserve . . . . Preserve and restore data
[P]   quietly . . . . Quietly and noisily perform Stata command
[P]   scalar . . . . Scalar variables
[P]   smcl . . . . Stata Markup and Control Language
[P]   sortpreserve . . . . Sort within programs
[P]   timer . . . . Time sections of code by recording and reporting time spent
[TS]   tsrevar . . . . Time-series operator programming command

Debugging

[P]   pause . . . . Program debugging command
[P]   timer . . . . Time sections of code by recording and reporting time spent
[P]   trace . . . . Debug Stata programs

Advanced programming commands

[P]   automation .... Automation
[P]   break .... Suppress Break key
[P]   char .... Characteristics
[M-2]   class .... Object-oriented programming (classes)
[P]   class .... Class programming
[P]   class exit .... Exit class-member program and return result
[P]   classutil .... Class programming utility
[P]   estat programming .... Controlling estat after user-written commands
[P]   _estimates .... Manage estimation results
[P]   file .... Read and write ASCII text and binary files
[P]   findfile .... Find file in path
[P]   include .... Include commands from file
[P]   macro .... Macro definition and manipulation
[P]   macro lists .... Manipulate lists
[R]   ml .... Maximum likelihood estimation
[M-5]   moptimize( ) .... Model optimization
[M-5]   optimize( ) .... Function optimization
[P]   plugin .... Load a plugin
[P]   postfile .... Save results in Stata dataset
[P]   _predict .. Obtain predictions, residuals, etc., after estimation programming command
[P]   program properties .... Properties of user-defined programs
[D]   putmata .... Put Stata variables into Mata and vice versa
[P]   _return .... Preserve saved results
[P]   _rmcoll .... Remove collinear variables
[P]   _robust .... Robust variance estimates
[P]   serset .... Create and manipulate sersets
[D]   snapshot .... Save and restore data snapshots
[P]   unab .... Unabbreviate variable list
[P]   unabcmd .... Unabbreviate command name
[P]   varabbrev .... Control variable abbreviation
[P]   viewsource .... View source code

Special-interest programming commands

[R]   bstat .... Report bootstrap results
[MV]   cluster programming subroutines .... Add cluster-analysis routines
[MV]   cluster programming utilities .... Cluster-analysis programming utilities
[R]   fvrevar .... Factor-variables operator programming command
[P]   matrix dissimilarity .... Compute similarity or dissimilarity measures
[MI]   mi select .... Programmer's alternative to mi extract
[ST]   st_is .... Survival analysis subroutines for programmers
[SVY]   svymarkout ... Mark observations for exclusion on the basis of survey characteristics
[MI]   technical .... Details for programmers
[TS]   tsrevar .... Time-series operator programming command

File formats

[P]   file formats .dta .... Description of .dta file format

Mata

[M]   *Mata Reference Manual* ....

Interface features

[GS]	Chapter 1 (GSM, GSU, GSW)	Introducing Stata—sample session
[GS]	Chapter 2 (GSM, GSU, GSW)	The Stata user interface
[GS]	Chapter 3 (GSM, GSU, GSW)	Using the Viewer
[GS]	Chapter 6 (GSM, GSU, GSW)	Using the Data Editor
[GS]	Chapter 7 (GSM, GSU, GSW)	Using the Variables Manager
[GS]	Chapter 13 (GSM, GSU, GSW)	Using the Do-file Editor—automating Stata
[GS]	Chapter 15 (GSM, GSU, GSW)	Editing graphs
[P]	dialog programming	Dialog programming
[R]	doedit	Edit do-files and other text files
[D]	edit	Browse or edit data with Data Editor
[P]	sleep	Pause for a specified time
[P]	smcl	Stata Markup and Control Language
[D]	varmanage	Manage variable labels, formats, and other properties
[P]	viewsource	View source code
[P]	window programming	Programming menus and windows

Title

data types — Quick reference for data types

Description

This entry provides a quick reference for data types allowed by Stata. See [U] **12 Data** for details.

Storage type	Minimum	Maximum	Closest to 0 without being 0	bytes
byte	-127	100	± 1	1
int	$-32,767$	32,740	± 1	2
long	$-2,147,483,647$	2,147,483,620	± 1	4
float	$-1.70141173319 \times 10^{38}$	$1.70141173319 \times 10^{38}$	$\pm 10^{-38}$	4
double	$-8.9884656743 \times 10^{307}$	$8.9884656743 \times 10^{307}$	$\pm 10^{-323}$	8

Precision for float is 3.795×10^{-8}

Precision for double is 1.414×10^{-16}

String storage type	Maximum length	Bytes
str1	1	1
str2	2	2
...	.	.
...	.	.
...	.	.
str244	244	244

Also see

[D] **compress** — Compress data in memory

[D] **destring** — Convert string variables to numeric variables and vice versa

[D] **encode** — Encode string into numeric and vice versa

[D] **format** — Set variables' output format

[D] **recast** — Change storage type of variable

[U] **12.2.2 Numeric storage types**

[U] **12.4.4 String storage types**

[U] **12.5 Formats: Controlling how data are displayed**

[U] **13.11 Precision and problems therein**

Title

> **estimation commands** — Quick reference for estimation commands

Description

This entry provides a quick reference for Stata's estimation commands. Because enhancements to Stata are continually being made, type `search estimation commands` for possible additions to this list; see [R] **search**.

For a discussion of properties shared by all estimation commands, see [U] **20 Estimation and postestimation commands**.

For a list of prefix commands that can be used with many of these estimation commands, see [U] **11.1.10 Prefix commands**.

Command	Description	See
anova	Analysis of variance and covariance	[R] **anova**
arch	ARCH family of estimators	[TS] **arch**
areg	Linear regression with a large dummy-variable set	[R] **areg**
arfima	Autoregressive fractionally integrated moving-average models	[TS] **arfima**
arima	ARIMA, ARMAX, and other dynamic regression models	[TS] **arima**
asclogit	Alternative-specific conditional logit (McFadden's choice) model	[R] **asclogit**
asmprobit	Alternative-specific multinomial probit regression	[R] **asmprobit**
asroprobit	Alternative-specific rank-ordered probit regression	[R] **asroprobit**
binreg	Generalized linear models: Extensions to the binomial family	[R] **binreg**
biprobit	Bivariate probit regression	[R] **biprobit**
blogit	Logistic regression for grouped data	[R] **glogit**
boxcox	Box–Cox regression models	[R] **boxcox**
bprobit	Probit regression for grouped data	[R] **glogit**
bsqreg	Bootstrapped quantile regression	[R] **qreg**
ca	Simple correspondence analysis	[MV] **ca**
camat	Simple correspondence analysis of a matrix	[MV] **ca**
candisc	Canonical linear discriminant analysis	[MV] **candisc**
canon	Canonical correlations	[MV] **canon**
clogit	Conditional (fixed-effects) logistic regression	[R] **clogit**
cloglog	Complementary log-log regression	[R] **cloglog**
cnsreg	Constrained linear regression	[R] **cnsreg**
contrast, post	Post contrasts as estimation results	[R] **contrast**
dfactor	Dynamic-factor models	[TS] **dfactor**
discrim knn	kth-nearest-neighbor discriminant analysis	[MV] **discrim knn**
discrim lda	Linear discriminant analysis	[MV] **discrim lda**
discrim logistic	Logistic discriminant analysis	[MV] **discrim logistic**
discrim qda	Quadratic discriminant analysis	[MV] **discrim qda**

Command	Description	See
eivreg	Errors-in-variables regression	[R] **eivreg**
exlogistic	Exact logistic regression	[R] **exlogistic**
expoisson	Exact Poisson regression	[R] **exlogistic**
factor	Factor analysis	[MV] **factor**
factormat	Factor analysis of a correlation matrix	[MV] **factor**
frontier	Stochastic frontier models	[R] **frontier**
glm	Generalized linear models	[R] **glm**
glogit	Weighted least-squares logistic regression for grouped data	[R] **glogit**
gmm	Generalized method of moments estimation	[R] **gmm**
gnbreg	Generalized negative binomial model	[R] **nbreg**
gprobit	Weighted least-squares probit regression for grouped data	[R] **glogit**
heckman	Heckman selection model	[R] **heckman**
heckprob	Probit model with selection	[R] **heckprob**
hetprob	Heteroskedastic probit model	[R] **hetprob**
intreg	Interval regression	[R] **intreg**
iqreg	Interquantile range regression	[R] **qreg**
ivprobit	Probit model with endogenous regressors	[R] **ivprobit**
ivregress	Single-equation instrumental-variables estimation	[R] **ivregress**
ivtobit	Tobit model with endogenous regressors	[R] **ivtobit**
logistic	Logistic regression, reporting odds ratios	[R] **logistic**
logit	Logistic regression, reporting coefficients	[R] **logit**
manova	Multivariate analysis of variance and covariance	[MV] **manova**
margins, post	Post margins as estimation results	[R] **margins**
mca	Multiple and joint correspondence analysis	[MV] **mca**
mds	Multidimensional scaling for two-way data	[MV] **mds**
mdslong	Multidimensional scaling of proximity data in long format	[MV] **mdslong**
mdsmat	Multidimensional scaling of proximity data in a matrix	[MV] **mdsmat**
mean	Estimate means	[R] **mean**
mgarch ccc	Constant conditional correlation multivariate GARCH model	[TS] **mgarch ccc**
mgarch dcc	Dynamic conditional correlation multivariate GARCH model	[TS] **mgarch dcc**
mgarch dvech	Diagonal vech multivariate GARCH model	[TS] **mgarch dvech**
mgarch vcc	Varying conditional correlation multivariate GARCH model	[TS] **mgarch vcc**
mlogit	Multinomial (polytomous) logistic regression	[R] **mlogit**
mprobit	Multinomial probit regression	[R] **mprobit**
mvreg	Multivariate regression	[R] **mvreg**
nbreg	Negative binomial regression	[R] **nbreg**
newey	Regression with Newey–West standard errors	[TS] **newey**
nl	Nonlinear least-squares estimation	[R] **nl**
nlogit	Nested logit regression (RUM-consistent and nonnormalized)	[R] **nlogit**
nlsur	Systems of nonlinear equations	[R] **nlsur**
ologit	Ordered logistic regression	[R] **ologit**
oprobit	Ordered probit regression	[R] **oprobit**

Command	Description	See
pca	Principal component analysis	[MV] **pca**
pcamat	Principal component analysis of a correlation or covariance matrix	[MV] **pca**
poisson	Poisson regression	[R] **poisson**
prais	Prais–Winsten and Cochrane–Orcutt regression	[TS] **prais**
probit	Probit regression	[R] **probit**
procrustes	Procrustes transformation	[MV] **procrustes**
proportion	Estimate proportions	[R] **proportion**
pwcompare, post	Post pairwise comparisons as estimation results	[R] **pwcompare**
pwmean	Perform pairwise comparisons of means	[R] **pwmean**
_qreg	Internal estimation command for quantile regression	[R] **qreg**
qreg	Quantile regression	[R] **qreg**
ratio	Estimate ratios	[R] **ratio**
reg3	Three-stage estimation for systems of simultaneous equations	[R] **reg3**
regress	Linear regression	[R] **regress**
rocfit	Parametric ROC models	[R] **rocfit**
rocreg	Parametric and nonparametric ROC regression	[R] **rocreg**
rologit	Rank-ordered logistic regression	[R] **rologit**
rreg	Robust regression	[R] **rreg**
scobit	Skewed logistic regression	[R] **scobit**
sem	Structural equation models	[SEM] **sem**
slogit	Stereotype logistic regression	[R] **slogit**
sqreg	Simultaneous-quantile regression	[R] **qreg**
sspace	State-space models	[TS] **sspace**
stcox	Cox proportional hazards model	[ST] **stcox**
stcrreg	Competing-risks regression	[ST] **stcrreg**
streg	Parametric survival models	[ST] **streg**
sureg	Zellner's seemingly unrelated regression	[R] **sureg**
svy: *command**	Estimation commands for survey data	[SVY] **svy estimation**
svy: tabulate oneway	One-way tables for survey data	[SVY] **svy: tabulate oneway**
svy: tabulate twoway	Two-way tables for survey data	[SVY] **svy: tabulate twoway**
tnbreg	Truncated negative binomial regression	[R] **tnbreg**
tobit	Tobit regression	[R] **tobit**
total	Estimate totals	[R] **total**
tpoisson	Truncated Poisson regression	[R] **tpoisson**
treatreg	Treatment-effects model	[R] **treatreg**
truncreg	Truncated regression	[R] **truncreg**
ucm	Unobserved-components model	[TS] **ucm**

*See the table below for a list of commands that support the svy prefix.

Command	Description	See
var	Vector autoregressive models	[TS] **var**
var svar	Structural vector autoregressive models	[TS] **var svar**
varbasic	Fit a simple VAR and graph IRFs or FEVDs	[TS] **varbasic**
vec	Vector error-correction models	[TS] **vec**
vwls	Variance-weighted least squares	[R] **vwls**
xtabond	Arellano–Bond linear dynamic panel-data estimation	[XT] **xtabond**
xtcloglog	Random-effects and population-averaged cloglog models	[XT] **xtcloglog**
xtdpd	Linear dynamic panel-data estimation	[XT] **xtdpd**
xtdpdsys	Arellano–Bond/Blundell–Bond estimation	[XT] **xtdpdsys**
xtfrontier	Stochastic frontier models for panel data	[XT] **xtfrontier**
xtgee	Fit population-averaged panel-data models using GEE	[XT] **xtgee**
xtgls	Fit panel-data models using GLS	[XT] **xtgls**
xthtaylor	Hausman–Taylor estimator for error-components models	[XT] **xthtaylor**
xtintreg	Random-effects interval data regression models	[XT] **xtintreg**
xtivreg	Instrumental variables and two-stage least squares for panel-data models	[XT] **xtivreg**
xtlogit	Fixed-effects, random-effects, and population-averaged logit models	[XT] **xtlogit**
xtmelogit	Multilevel mixed-effects logistic regression	[XT] **xtmelogit**
xtmepoisson	Multilevel mixed-effects Poisson regression	[XT] **xtmepoisson**
xtmixed	Multilevel mixed-effects linear regression	[XT] **xtmixed**
xtnbreg	Fixed-effects, random-effects, and population-averaged negative binomial models	[XT] **xtnbreg**
xtpcse	OLS or Prais–Winsten models with panel-corrected standard errors	[XT] **xtpcse**
xtpoisson	Fixed-effects, random-effects, and population-averaged Poisson models	[XT] **xtpoisson**
xtprobit	Random-effects and population-averaged probit models	[XT] **xtprobit**
xtrc	Random-coefficients models	[XT] **xtrc**
xtreg	Fixed-, between-, and random-effects, and population-averaged linear models	[XT] **xtreg**
xtregar	Fixed- and random-effects linear models with an AR(1) disturbance	[XT] **xtregar**
xttobit	Random-effects tobit models	[XT] **xttobit**
zinb	Zero-inflated negative binomial regression	[R] **zinb**
zip	Zero-inflated Poisson regression	[R] **zip**

The following estimation commands support the `mi estimate` prefix.

Command	Description	See
Linear regression models		
regress	Linear regression	[R] **regress**
cnsreg	Constrained linear regression	[R] **cnsreg**
mvreg	Multivariate regression	[R] **mvreg**
Binary-response regression models		
logistic	Logistic regression, reporting odds ratios	[R] **logistic**
logit	Logistic regression, reporting coefficients	[R] **logit**
probit	Probit regression	[R] **probit**
cloglog	Complementary log-log regression	[R] **cloglog**
binreg	GLM for the binomial family	[R] **binreg**
Count-response regression models		
poisson	Poisson regression	[R] **poisson**
nbreg	Negative binomial regression	[R] **nbreg**
gnbreg	Generalized negative binomial regression	[R] **nbreg**
Ordinal-response regression models		
ologit	Ordered logistic regression	[R] **ologit**
oprobit	Ordered probit regression	[R] **oprobit**
Categorical-response regression models		
mlogit	Multinomial (polytomous) logistic regression	[R] **mlogit**
mprobit	Multinomial probit regression	[R] **mprobit**
clogit	Conditional (fixed-effects) logistic regression	[R] **clogit**
Quantile regression models		
qreg	Quantile regression	[R] **qreg**
iqreg	Interquantile range regression	[R] **qreg**
sqreg	Simultaneous-quantile regression	[R] **qreg**
bsqreg	quantile regression	[R] **qreg**
Survival regression models		
stcox	Cox proportional hazards model	[ST] **stcox**
streg	Parametric survival models	[ST] **streg**
stcrreg	Competing-risks regression	[ST] **stcrreg**
Other regression models		
glm	Generalized linear models	[R] **glm**
areg	Linear regression with a large dummy-variable set	[R] **areg**
rreg	Robust regression	[R] **rreg**
truncreg	Truncated regression	[R] **truncreg**
Descriptive statistics		
mean	Estimate means	[R] **mean**
proportion	Estimate proportions	[R] **proportion**
ratio	Estimate ratios	[R] **ratio**

Command	Description	See
Panel-data models		
xtreg	Fixed-, between- and random-effects, and population-averaged linear models	[XT] **xtreg**
xtmixed	Multilevel mixed-effects linear regression	
xtrc	Random-coefficients regression	[XT] **xtmixed**
xtlogit	Fixed effects, random-effects, and population-averaged logit models	[XT] **xtlogit**
xtprobit	Random-effects and population-averaged probit models	[XT] **xtprobit**
xtcloglog	Random-effects and population-averaged cloglog models	[XT] **xtcloglog**
xtpoisson	Fixed-effects, random-effects, and population-averaged Poisson models	[XT] **xtcloglog**
xtnbreg	Fixed-effects, random-effects, and population-averaged negative binomial models	[XT] **xtnbreg**
xtmelogit	Multilevel mixed-effects logistic regression	[XT] **xtmelogit**
xtmepoisson	Multilevel mixed-effects Poisson regression	[XT] **xtmepoisson**
xtgee	Fit population-averaged panel-data models by using GEE	[XT] **xtgee**
Survey regression models		
svy:	Estimation commands for survey data (excluding commands that are not listed above)	[SVY] **svy**

The following estimation commands support the svy prefix.

Command	Description	See
Descriptive statistics		
mean	Estimate means	[R] **mean**
proportion	Estimate proportions	[R] **proportion**
ratio	Estimate ratios	[R] **ratio**
total	Estimate totals	[R] **total**
Linear regression models		
cnsreg	Constrained linear regression	[R] **cnsreg**
glm	Generalized linear models	[R] **glm**
intreg	Interval regression	[R] **intreg**
nl	Nonlinear least-squares estimation	[R] **nl**
regress	Linear regression	[R] **regress**
sem	Structural equation models	[SEM] **sem**
tobit	Tobit regression	[R] **tobit**
treatreg	Treatment-effects model	[R] **treatreg**
truncreg	Truncated regression	[R] **truncreg**
Survival-data regression models		
stcox	Cox proportional hazards model	[ST] **stcox**
streg	Parametric survival model	[ST] **streg**

Command	Description	See
Binary-response regression models		
biprobit	Bivariate probit regression	[R] **biprobit**
cloglog	Complementary log-log regression	[R] **cloglog**
hetprob	Heteroskedastic probit model	[R] **hetprob**
logistic	Logistic regression, reporting odds ratios	[R] **logistic**
logit	Logistic regression, reporting coefficients	[R] **logit**
probit	Probit regression	[R] **probit**
scobit	Skewed logistic regression	[R] **scobit**
Discrete-response regression models		
clogit	Conditional (fixed-effects) logistic regression	[R] **clogit**
mlogit	Multinomial (polytomous) logistic regression	[R] **mlogit**
mprobit	Multinomial probit regression	[R] **mprobit**
ologit	Ordered logistic regression	[R] **ologit**
oprobit	Ordered probit regression	[R] **oprobit**
slogit	Stereotype logistic regression	[R] **slogit**
Poisson regression models		
gnbreg	Generalized negative binomial regression	[R] **nbreg**
nbreg	Negative binomial regression	[R] **nbreg**
poisson	Poisson regression	[R] **poisson**
tnbreg	Truncated negative binomial regression	[R] **tnbreg**
tpoisson	Truncated Poisson regression	[R] **tpoisson**
zinb	Zero-inflated negative binomial regression	[R] **zinb**
zip	Zero-inflated Poisson regression	[R] **zip**
Instrumental-variables regression models		
ivprobit	Probit model with continuous endogenous regressors	[R] **ivprobit**
ivregress	Single-equation instrumental-variables regression	[R] **ivregress**
ivtobit	Tobit model with continuous endogenous regressors	[R] **ivtobit**
Regression models with selection		
heckman	Heckman selection model	[R] **heckman**
heckprob	Probit model with sample selection	[R] **heckprob**

* The xtmixed command for fitting linear multilevel models supports survey data, including allowing sampling weights and robust–cluster standard errors. Because sampling weights are handled differently by xtmixed than by other commands, xtmixed does not work with the svy: prefix. See *Survey data* in [XT] **xtmixed** for details.

Also see

[U] **20 Estimation and postestimation commands**

Title

file extensions — Quick reference for default file extensions

Description

This entry provides a quick reference for default file extensions that are used by various commands.

Extension	Reference	Description
.ado	[U] **17 Ado-files**	automatically loaded do-files
.dct	[D] **infile (fixed format)**	ASCII data dictionary
.do	[U] **16 Do-files**	do-file
.dta	[D] **save**, [D] **use**	Stata-format dataset
.dtasig	[D] **datasignature**	datasignature file
.gph	[G-2] **graph save**, [G-2] **graph use**	graph
.grec	[G-1] **graph editor**	Graph Editor recording (ASCII format)
.irf	[TS] **irf set**	impulse–response function datasets
.log	[R] **log**	log file in text format
.mata	[M-1] **source**	Mata source code
.mlib	[M-3] **mata mlib**	Mata library
.mmat	[M-3] **mata matsave**	Mata matrix
.mo	[M-3] **mata mosave**	Mata object file
.out	[D] **outsheet**	file saved by outsheet
.raw	[D] **infile (free format)**, [D] **insheet**	ASCII-format dataset
.smcl	[R] **log**	log file in SMCL format
.ster	[R] **estimates save**	saved estimates
.stbcal	[D] **datetime business calendars**	business calendars
.sthlp	[P] **smcl**	help files
.stptrace	[MI] **mi ptrace**	parameter-trace file
.sum	[D] **checksum**	checksum files to verify network transfer
.zip	[D] **zipfile**	zip file

The following files are of interest only to advanced programmers or are for Stata's internal use.

Extension	Reference	Description
.class	[P] **class**	class file for object-oriented programming
.dlg	[P] **dialog programming**	dialog resource file
.idlg	[P] **dialog programming**	dialog resource include file
.ihlp	[P] **smcl**	help include file
.key	[R] **search**	search's keyword database file
.maint		maintenance file (for Stata's internal use only)
.mnu		menu file (for Stata's internal use only)
.pkg	[R] **net**	user-site package file
.plugin	[P] **plugin**	compiled addition (DLL)
.scheme	[G-4] **schemes intro**	control file for a graph scheme
.style	[G-2] **graph query**	graph style file
.toc	[U] **28.5 Making your own download site**	user-site description file

Title

> **format** — Quick reference for numeric and string display formats

Description

This entry provides a quick reference for display formats.

Remarks

The default formats for each of the numeric variable types are

```
byte     %8.0g
int      %8.0g
long     %12.0g
float    %9.0g
double   %10.0g
```

To change the display format for variable myvar to %9.2f, type

```
format myvar %9.2f
```

or

```
format %9.2f myvar
```

Stata will understand either statement.

Four values displayed in different numeric display formats

%9.0g	%9.0gc	%9.2f	%9.2fc	%-9.0g	%09.2f	%9.2e
12345	12,345	12345.00	12,345.00	12345	012345.00	1.23e+04
37.916	37.916	37.92	37.92	37.916	000037.92	3.79e+01
3567890	3567890	3.57e+06	3.57e+06	3567890	3.57e+06	3.57e+06
.9165	.9165	0.92	0.92	.9165	000000.92	9.16e-01

Left-aligned and right-aligned string display formats

%-17s	%17s
AMC Concord	AMC Concord
AMC Pacer	AMC Pacer
AMC Spirit	AMC Spirit
Buick Century	Buick Century
Buick Opel	Buick Opel

Also see

[U] **12.5 Formats: Controlling how data are displayed**

Title

immediate commands — Quick reference for immediate commands

Description

An *immediate* command is a command that obtains data not from the data stored in memory, but from numbers types as arguments.

Command	Reference	Description
bitesti	[R] **bitest**	Binomial probability test
cci csi iri mcci	[ST] **epitab**	Tables for epidemiologists
cii	[R] **ci**	Confidence intervals for means, proportions, and counts
prtesti	[R] **prtest**	One- and two-sample tests of proportions
sampsi	[R] **sampsi**	Sample size and power for means and proportions
sdtesti	[R] **sdtest**	Variance comparison tests
symmi	[R] **symmetry**	Symmetry and marginal homogeneity tests
tabi	[R] **tabulate twoway**	Two-way tables of frequencies
ttesti	[R] **ttest**	Mean comparison tests
twoway pci	[G-2] **graph twoway pci**	Paired-coordinate plot with spikes or lines
twoway pcarrowi	[G-2] **graph twoway pcarrowi**	Paired-coordinate plot with arrows
twoway scatteri	[G-2] **graph twoway scatteri**	Twoway scatterplot

Also see

[U] **19 Immediate commands**

Title

---
missing values — Quick reference for missing values

Description

This entry provides a quick reference for Stata's missing values.

Remarks

Stata has 27 numeric missing values:

., the default, which is called the *system missing value* or `sysmiss`

and

.a, .b, .c, ..., .z, which are called the *extended missing values*.

Numeric missing values are represented by "large positive values". The ordering is

$$\text{all nonmissing numbers} < . < .a < .b < \cdots < .z$$

Thus the expression

$$\texttt{age} > 60$$

is true if variable `age` is greater than 60 or missing.

To exclude missing values, ask whether the value is less than '.'.

    `. list if age > 60 & age < .`

To specify missing values, ask whether the value is greater than or equal to '.'. For instance,

    `. list if age >=.`

Stata has one string missing value, which is denoted by `""` (blank).

Also see

[U] **12.2.1 Missing values**

Title

> **postestimation commands** — Quick reference for postestimation commands

Description

This entry provides a quick reference for Stata's postestimation commands. Because enhancements to Stata are continually being made, type `search postestimation commands` for possible additions to this list; see [R] **search**.

Available after most estimation commands

Command	Description
contrast	contrasts and ANOVA-style joint tests of estimates
estat ic	AIC and BIC
estat summarize	estimation sample summary
estat vce	VCE
estimates	cataloging estimation results
hausman	Hausman's specification test
lincom	point estimates, standard errors, testing, and inference for linear combinations of coefficients
linktest	link test for model specification for single-equation models
lrtest	likelihood-ratio test
margins	marginal means, predictive margins, marginal effects, and average marginal
marginsplot	graph the results from margins (profile plots, interaction plots, etc.) effects
nlcom	point estimates, standard errors, testing, and inference for nonlinear combinations of coefficients
predict	predictions, residuals, influence statistics, and other diagnostic measures
predictnl	point estimates, standard errors, testing, and inference for generalized predictions
pwcompare	pairwise comparisons of estimates
suest	seemingly unrelated estimation
test	Wald tests of simple and composite linear hypotheses
testnl	Wald tests of nonlinear hypotheses

Special-interest postestimation commands

Command	Description
anova	
dfbeta	DFBETA influence statistics
estat hettest	tests for heteroskedasticity
estat imtest	information matrix test
estat ovtest	Ramsey regression specification-error test for omitted variables
estat szroeter	Szroeter's rank test for heteroskedasticity
estat vif	variance inflation factors for the independent variables
acprplot	augmented component-plus-residual plot
avplot	added-variable plot
avplots	all added-variables plots in one image
cprplot	component-plus-residual plot
lvr2plot	leverage-versus-squared-residual plot
rvfplot	residual-versus-fitted plot
rvpplot	residual-versus-predictor plot
arfima	
psdensity	estimate the spectral density
arima	
psdensity	estimate the spectral density
asclogit	
estat alternatives	alternative summary statistics
estat mfx	marginal effects
asmprobit and asroprobit	
estat alternatives	alternative summary statistics
estat covariance	covariance matrix of the latent-variable errors for the alternatives
estat correlation	correlation matrix of the latent-variable errors for the alternatives
estat facweights	covariance factor weights matrix
estat mfx	marginal effects
bootstrap	
estat bootstrap	percentile-based and bias-corrected CI tables

Command	Description
ca and camat	
cabiplot	biplot of row and column points
caprojection	CA dimension projection plot
estat coordinates	display row and column coordinates
estat distances	display χ^2 distances between row and column profiles
estat inertia	display inertia contributions of the individual cells
estat loadings	display correlations of profiles and axes
estat profiles	display row and column profiles
[†] estat summarize	estimation sample summary
estat table	display fitted correspondence table
screeplot	plot singular values
candisc	
estat anova	ANOVA summaries table
estat canontest	tests of the canonical discriminant functions
estat classfunctions	classification functions
estat classtable	classification table
estat correlations	correlation matrices and p-values
estat covariance	covariance matrices
estat errorrate	classification error-rate estimation
estat grdistances	Mahalanobis and generalized squared distances between the group means
estat grmeans	group means and variously standardized or transformed means
estat grsummarize	group summaries
estat list	classification listing
estat loadings	canonical discriminant-function coefficients (loadings)
estat manova	MANOVA table
estat structure	canonical structure matrix
estat summarize	estimation sample summary
loadingplot	plot standardized discriminant-function loadings
scoreplot	plot discriminant-function scores
screeplot	plot eigenvalues
canon	
estat correlations	show correlation matrices
estat loadings	show loading matrices
estat rotate	rotate raw coefficients, standard coefficients, or loading matrices
estat rotatecompare	compare rotated and unrotated coefficients or loadings
screeplot	plot canonical correlations

[†] estat summarize is not available after camat.

Command	Description
discrim knn and discrim logistic	
estat classtable	classification table
estat errorrate	classification error-rate estimation
estat grsummarize	group summaries
estat list	classification listing
estat summarize	estimation sample summary
discrim lda	
estat anova	ANOVA summaries table
estat canontest	tests of the canonical discriminant functions
estat classfunctions	classification functions
estat classtable	classification table
estat correlations	correlation matrices and p-values
estat covariance	covariance matrices
estat errorrate	classification error-rate estimation
estat grdistances	Mahalanobis and generalized squared distances between the group means
estat grmeans	group means and variously standardized or transformed means
estat grsummarize	group summaries
estat list	classification listing
estat loadings	canonical discriminant-function coefficients (loadings)
estat manova	MANOVA table
estat structure	canonical structure matrix
estat summarize	estimation sample summary
loadingplot	plot standardized discriminant-function loadings
scoreplot	plot discriminant-function scores
screeplot	plot eigenvalues
discrim qda	
estat classtable	classification table
estat correlations	group correlation matrices and p-values
estat covariance	group covariance matrices
estat errorrate	classification error-rate estimation
estat grdistances	Mahalanobis and generalized squared distances between the group means
estat grsummarize	group summaries
estat list	classification listing
estat summarize	estimation sample summary
exlogistic	
estat predict	single-observation prediction
estat se	report ORs or coefficients and their asymptotic standard errors
expoisson	
estat se	report coefficients or IRRs and their asymptotic standard errors

Command	Description
factor and factormat	
estat anti	anti-image correlation and covariance matrices
estat common	correlation matrix of the common factors
estat factors	AIC and BIC model-selection criteria for different numbers of factors
estat kmo	Kaiser–Meyer–Olkin measure of sampling adequacy
estat residuals	matrix of correlation residuals
estat rotatecompare	compare rotated and unrotated loadings
estat smc	squared multiple correlations between each variable and the rest
estat structure	correlations between variables and common factors
† estat summarize	estimation sample summary
loadingplot	plot factor loadings
rotate	rotate factor loadings
scoreplot	plot score variables
screeplot	plot eigenvalues
fracpoly	
fracplot	plot data and fit from most recently fit fractional polynomial model
fracpred	create variable containing prediction, deviance residuals, or SEs of fitted values
gmm	
estat overid	perform test of overidentifying restrictions
ivprobit	
estat classification	report various summary statistics, including the classification table
lroc	compute area under ROC curve and graph the curve
lsens	graph sensitivity and specificity versus probability cutoff
ivregress	
estat endogenous	perform tests of endogeneity
estat firststage	report "first-stage" regression statistics
estat overid	perform tests of overidentifying restrictions
logistic and logit	
estat classification	report various summary statistics, including the classification table
estat gof	Pearson or Hosmer–Lemeshow goodness-of-fit test
lroc	compute area under ROC curve and graph the curve
lsens	graph sensitivity and specificity versus probability cutoff
manova	
manovatest	multivariate tests after manova
screeplot	plot eigenvalues

† estat summarize is not available after factormat.

Command	Description
mca	
mcaplot	plot of category coordinates
mcaprojection	MCA dimension projection plot
estat coordinates	display of category coordinates
estat subinertia	matrix of inertias of the active variables (after JCA only)
estat summarize	estimation sample summary
screeplot	plot principal inertias (eigenvalues)
mds, mdslong, and mdsmat	
estat config	coordinates of the approximating configuration
estat correlations	correlations between dissimilarities and approximating distances
estat pairwise	pairwise dissimilarities, approximating distances, and raw residuals
estat quantiles	quantiles of the residuals per object
estat stress	Kruskal stress (loss) measure (only after classical MDS)
†estat summarize	estimation sample summary
mdsconfig	plot of approximating configuration
mdsshepard	Shepard diagram
screeplot	plot eigenvalues (only after classical MDS)
mfp	
fracplot	plot data and fit from most recently fit fractional polynomial model
fracpred	create variable containing prediction, deviance residuals, or SEs of fitted values
mi estimate and mi estimate using	
mi test	tests on coefficients
mi testtransform	tests on transformed coefficients
mi predict	linear predictions
mi predictnl	nonlinear predictions
nlogit	
estat alternatives	alternative summary statistics

Command	Description
pca and pcamat	
estat anti	anti-image correlation and covariance matrices
estat kmo	Kaiser–Meyer–Olkin measure of sampling adequacy
estat loadings	component-loading matrix in one of several normalizations
estat residuals	matrix of correlation or covariance residuals
estat rotatecompare	compare rotated and unrotated components
estat smc	squared multiple correlations between each variable and the rest
†estat summarize	estimation sample summary
loadingplot	plot component loadings
rotate	rotate component loadings
scoreplot	plot score variables
screeplot	plot eigenvalues
poisson	
estat gof	goodness-of-fit test
probit	
estat classification	report various summary statistics, including the classification table
estat gof	Pearson or Hosmer–Lemeshow goodness-of-fit test
lroc	compute area under ROC curve and graph the curve
lsens	graph sensitivity and specificity versus probability cutoff
procrustes	
estat compare	fit statistics for orthogonal, oblique, and unrestricted transformations
estat mvreg	display multivariate regression resembling unrestricted transformation
estat summarize	display summary statistics over the estimation sample
procoverlay	produce a Procrustes overlay graph
regress	
dfbeta	DFBETA influence statistics
estat hettest	tests for heteroskedasticity
estat imtest	information matrix test
estat ovtest	Ramsey regression specification-error test for omitted variables
estat szroeter	Szroeter's rank test for heteroskedasticity
estat vif	variance inflation factors for the independent variables
acprplot	augmented component-plus-residual plot
avplot	added-variable plot
avplots	all added-variables plots in one image
cprplot	component-plus-residual plot
lvr2plot	leverage-versus-squared-residual plot
rvfplot	residual-versus-fitted plot
rvpplot	residual-versus-predictor plot

† estat summarize is not available after mdsmat or pcamat.

Command	Description
regress postestimation time series	
estat archlm	test for ARCH effects in the residuals
estat bgodfrey	Breusch–Godfrey test for higher-order serial correlation
estat durbinalt	Durbin's alternative test for serial correlation
estat dwatson	Durbin–Watson d statistic to test for first-order serial correlation
rocfit	
rocplot	plot the fitted ROC curve and simultaneous confidence bands
rocreg	
estat nproc	nonparametric ROC curve estimation, keeping fit information from rocreg
rocregplot	plot marginal and covariate-specific ROC curves
sem	
estat eqgof	equation-level goodness of fit
estat eqtest	equation-level Wald tests
estat framework	display results in modeling framework
estat ggof	group-level goodness of fit
estat ginvariant	test of invariance of parameters across groups
estat gof	overall goodness of fit
estat mindices	modification indices
estat residuals	matrices of residuals
estat scoretests	score tests
estat stable	assess stability of nonrecursive systems
estat stdize:	tests standardized parameters
estat teffects	decomposition of effects
stcox	
estat concordance	compute the concordance probability
estat phtest	test proportional-hazards assumption based on Schoenfeld residuals
stcurve	plot the survivor, hazard, and cumulative hazard functions
stcrreg	
stcurve	plot the cumulative subhazard and cumulative incidence functions
streg	
stcurve	plot the survivor, hazard, and cumulative hazard functions

Command	Description
svar, var, and varbasic	
fcast compute	obtain dynamic forecasts
fcast graph	graph dynamic forecasts obtained from `fcast compute`
irf	create and analyze IRFs and FEVDs
vargranger	Granger causality tests
varlmar	LM test for autocorrelation in residuals
varnorm	test for normally distributed residuals
varsoc	lag-order selection criteria
varstable	check stability condition of estimates
varwle	Wald lag-exclusion statistics
ucm	
estat period	display cycle periods in time units
psdensity	estimate the spectral density
vec	
fcast compute	obtain dynamic forecasts
fcast graph	graph dynamic forecasts obtained from `fcast compute`
irf	create and analyze IRFs and FEVDs
veclmar	LM test for autocorrelation in residuals
vecnorm	test for normally distributed residuals
vecstable	check stability condition of estimates
xtabond, xtdpd, and xtdpdsys	
estat abond	test for autocorrelation
estat sargan	Sargan test of overidentifying restrictions
xtgee	
estat wcorrelation	estimated matrix of the within-group correlations
xtmelogit, xtmepoisson, and xtmixed	
estat group	summarize the composition of the nested groups
estat recovariance	display the estimated random-effects covariance matrix
xtreg	
xttest0	Breusch and Pagan LM test for random effects

Also see

[R] **contrast** — Contrasts and linear hypothesis tests after estimation

[R] **estat** — Postestimation statistics

[R] **estimates** — Save and manipulate estimation results

[R] **hausman** — Hausman specification test

[R] **lincom** — Linear combinations of estimators

[R] **linktest** — Specification link test for single-equation models

[R] **lrtest** — Likelihood-ratio test after estimation

[R] **margins** — Marginal means, predictive margins, and marginal effects

[R] **marginsplot** — Graph results from margins (profile plots, etc.)

[R] **nlcom** — Nonlinear combinations of estimators

[R] **predict** — Obtain predictions, residuals, etc., after estimation

[R] **predictnl** — Obtain nonlinear predictions, standard errors, etc., after estimation

[R] **pwcompare** — Pairwise comparisons

[R] **suest** — Seemingly unrelated estimation

[R] **test** — Test linear hypotheses after estimation

[R] **testnl** — Test nonlinear hypotheses after estimation

[U] **20 Estimation and postestimation commands**

Title

> **prefix commands** — Quick reference for prefix commands

Description

Prefix commands operate on other Stata commands. They modify the input, modify the output, and repeat execution of the other Stata command.

Command	Reference	Description
by	[D] **by**	run command on subsets of data
statsby	[D] **statsby**	same as by, but collect statistics from each run
rolling	[TS] **rolling**	run command on moving subsets and collect statistics
bootstrap	[R] **bootstrap**	run command on bootstrap samples
jackknife	[R] **jackknife**	run command on jackknife subsets of data
permute	[R] **permute**	run command on random permutations
simulate	[R] **simulate**	run command on manufactured data
svy	[SVY] **svy**	run command and adjust results for survey sampling
mi estimate	[MI] **mi estimate**	run command on multiply imputed data and adjust results for multiple imputation (MI)
nestreg	[R] **nestreg**	run command with accumulated blocks of regressors, and report nested model comparison tests
stepwise	[R] **stepwise**	run command with stepwise variable inclusion/exclusion
xi	[R] **xi**	run command after expanding factor variables and interactions; for most commands, using factor variables is preferred to using xi (see [U] **11.4.3 Factor variables**)
fracpoly	[R] **fracpoly**	run command with fractional polynomials of one regressor
mfp	[R] **mfp**	run command with multiple fractional polynomial regressors
capture	[P] **capture**	run command and capture its return code
noisily	[P] **quietly**	run command and show the output
quietly	[P] **quietly**	run command and suppress the output
version	[P] **version**	run command under specified version

The last group—capture, noisily, quietly, and version—have to do with programming Stata, and for historical reasons, capture, noisily, and quietly allow you to omit the colon.

Also see

[U] **11.1.10 Prefix commands**

Title

> **import and export data** — Quick reference for importing and exporting data

Description

This entry provides a quick reference for determining which method to use for reading non-Stata data into memory and for exporting Stata data from memory to other formats. See [U] **21 Inputting and importing data** for more details on reading non-Stata data into memory.

Remarks

Remarks are presented under the following headings:

> *Summary of the different import methods*
> > *import excel*
> > *insheet*
> > *odbc*
> > *infile (free format)—infile without a dictionary*
> > *infix (fixed format)*
> > *infile (fixed format)—infile with a dictionary*
> > *import sasxport*
> > *haver (Windows only)*
> > *xmluse*
> *Summary of the different export methods*
> > *export excel*
> > *outsheet*
> > *odbc*
> > *outfile*
> > *export sasxport*
> > *xmlsave*

Summary of the different import methods

import excel

- import excel reads worksheets from Microsoft Excel (.xls and .xlsx) files.
- Entire worksheets can be read, or custom cell ranges can be read.
- See [D] **import excel**.

insheet

- insheet reads text files created by a spreadsheet or a database program.
- The data must be tab-separated or comma-separated, but not both simultaneously. A custom delimiter may also be specified.
- An observation must be on only one line.
- The first line in the file can optionally contain the names of the variables.
- See [D] **insheet**.

odbc

○ ODBC, an acronym for Open DataBase Connectivity, is a standard for exchanging data between programs. Stata supports the ODBC standard for importing data via the odbc command and can read from any ODBC data source on your computer.

○ See [D] **odbc**.

infile (free format)—infile without a dictionary

○ The data can be space-separated, tab-separated, or comma-separated.

○ Strings with embedded spaces or commas must be enclosed in quotes (even if tab- or comma-separated).

○ An observation can be on more than one line, or there can even be multiple observations per line.

○ See [D] **infile (free format)**.

infix (fixed format)

○ The data must be in fixed-column format.

○ An observation can be on more than one line.

○ infix has simpler syntax than infile (fixed format).

○ See [D] **infix (fixed format)**.

infile (fixed format)—infile with a dictionary

○ The data may be in fixed-column format.

○ An observation can be on more than one line.

○ ASCII or EBCDIC data can be read.

○ infile (fixed format) has the most capabilities for reading data.

○ See [D] **infile (fixed format)**.

import sasxport

○ import sasxport reads SAS XPORT Transport format files.

○ import sasxport will also read value label information from a formats.xpf XPORT file, if available.

○ See [D] **import sasxport**.

haver (Windows only)

○ haver reads Haver Analytics (http://www.haver.com/) database files.

○ haver is available only for Windows and requires a corresponding DLL (DLXAPI32.DLL) available from Haver Analytics.

○ See [TS] **haver**.

xmluse

○ `xmluse` reads extensible markup language (XML) files—highly adaptable text-format files derived from the standard generalized markup language (SGML).

○ `xmluse` can read either an Excel-format XML or a Stata-format XML file into Stata.

○ See [D] **xmlsave**.

Summary of the different export methods

export excel

○ `export excel` creates Microsoft Excel worksheets in `.xls` and `.xlsx` files.

○ Entire worksheets can be exported, or custom cell ranges can be overwritten.

○ See [D] **import excel**.

outsheet

○ `outsheet` creates comma-separated (CSV) or tab-delimited files that many other programs can read.

○ A custom delimiter may also be specified.

○ The first line of the file can optionally contain the names of the variables.

○ See [D] **outsheet**.

odbc

○ ODBC, an acronym for Open DataBase Connectivity, is a standard for exchanging data between programs. Stata supports the ODBC standard for exporting data via the `odbc` command and can write to any ODBC data source on your computer.

○ See [D] **odbc**.

outfile

○ `outfile` creates text-format datasets.

○ The data can be written in space-separated or comma-separated format.

○ Alternatively, the data can be written in fixed-column format.

○ See [D] **outfile**.

export sasxport

○ `export sasxport` saves SAS XPORT Transport format files.

○ `export sasxport` can also write value label information to a `formats.xpf` XPORT file.

○ See [D] **import sasxport**.

xmlsave

    ○ xmlsave writes extensible markup language (XML) files—highly adaptable text-format files derived from the standard generalized markup language (SGML).

    ○ xmlsave can write either an Excel-format XML or a Stata-format XML file.

    ○ See [D] **xmlsave**.

Also see

[D] **edit** — Browse or edit data with Data Editor

[D] **import** — Overview of importing data into Stata

[D] **export** — Overview of exporting data from Stata

[D] **infile (fixed format)** — Read text data in fixed format with a dictionary

[D] **infile (free format)** — Read unformatted text data

[D] **infix (fixed format)** — Read text data in fixed format

[D] **input** — Enter data from keyboard

[D] **insheet** — Read text data created by a spreadsheet

[D] **import excel** — Import and export Excel files

[D] **import sasxport** — Import and export datasets in SAS XPORT format

[D] **odbc** — Load, write, or view data from ODBC sources

[D] **outfile** — Export dataset in text format

[D] **outsheet** — Write spreadsheet-style dataset

[D] **xmlsave** — Export or import dataset in XML format

[TS] **haver** — Load data from Haver Analytics database

[U] **21 Inputting and importing data**

Title

2SIV	two-step instrumental variables
2SLS	two-stage least squares
3SLS	three-stage least squares
ADF	asymptotic distribution free
AF	attributable fraction for the population
AFE	attributable fraction among the exposed
AFT	accelerated failure time
AIC	Akaike information criterion
AIDS	almost ideal demand system
ANCOVA	analysis of covariance
ANOVA	analysis of variance
APE	average partial effects
AR	autoregressive
AR(1)	first-order autoregressive
ARCH	autoregressive conditional heteroskedasticity
ARFIMA	autoregressive fractionally integrated moving average
ARIMA	autoregressive integrated moving average
ARMA	autoregressive moving average
ARMAX	autoregressive moving-average exogenous
ASE	asymptotic standard error
ASL	achieved significance level
AUC	area under the time-versus-concentration curve
BC	bias corrected
BCa	bias corrected and accelerated
BE	between effects
BFGS	Broyden–Fletcher–Goldfarb–Shanno
BHHH	Berndt–Hall–Hall–Hausman
BIC	Bayesian information criterion
BLUP	best linear unbiased prediction
BRR	balanced repeated replication
CA	correspondence analysis
CCI	conservative confidence interval
CD	coefficient of determination
CDC	Centers for Disease Control and Prevention
CDF	cumulative distribution function
CES	constant elasticity of substitution
CFA	confirmatory factor analysis
CFI	comparative fit index
CI	confidence interval

CIF	cumulative incidence function
CMLE	conditional maximum likelihood estimates
ct	count time
cusum	cumulative sum
c.v.	coefficient of variation
DA	data augmentation
DEFF	design effect
DEFT	design effect (standard deviation metric)
DF	dynamic factor
df / d.f.	degree(s) of freedom
DFAR	dynamic factors with vector autoregressive errors
DFP	Davidon–Fletcher–Powell
DPD	dynamic panel data
EBCDIC	extended binary coded decimal interchange code
EGARCH	exponential GARCH
EGLS	estimated generalized least squares
EIM	expected information matrix
EM	expectation maximization
EPS	Encapsulated PostScript
ESS	error sum of squares
FCS	fully conditional specification
FD	first-differenced estimator
FDA	Food and Drug Administration
FE	fixed effects
FEVD	forecast-error variance decomposition
FGLS	feasible generalized least squares
FGNLS	feasible generalized nonlinear least squares
FIVE estimator	full-information instrumental-variables efficient estimator
FIML	full information maximum likelihood
flong	full long
flongsep	full long and separate
FMI	fraction of missing information
FP	fractional polynomial
FPC	finite population correction
GARCH	generalized autoregressive conditional heteroskedasticity
GEE	generalized estimating equations
GEV	generalized extreme value
GHK	Geweke–Hajivassiliou–Keane
GLIM	generalized linear interactive modeling
GLLAMM	generalized linear latent and mixed models

GLM	generalized linear models
GLS	generalized least squares
GMM	generalized method of moments
GUI	graphical user interface
HAC	heteroskedasticity- and autocorrelation-consistent
HR	hazard ratio
HRF	human readable form
IC	information criteria
ICD-9	International Classification of Diseases, Ninth Revision
IIA	independence of irrelevant alternatives
i.i.d.	independent and identically distributed
IQR	interquartile range
IR	incidence rate
IRF	impulse–response function
IRLS	iterated, reweighted least squares
IRR	incidence-rate ratio
IV	instrumental variables
JCA	joint correspondence analysis
LAPACK	linear algebra package
LAV	least absolute value
LDA	linear discriminant analysis
LIML	limited-information maximum likelihood
LM	Lagrange multiplier
LOO	leave one out
LOWESS	locally weighted scatterplot smoothing
LR	likelihood ratio
LSB	least-significant byte
MA	moving average
MAD	median absolute deviation
MANCOVA	multivariate analysis of covariance
MANOVA	multivariate analysis of variance
MAR	missing at random
MCA	multiple correspondence analysis
MCAR	missing completely at random
MCE	Monte Carlo error
MCMC	Markov chain Monte Carlo
MDS	multidimensional scaling
ME	multiple equation

MEFF	misspecification effect
MEFT	misspecification effect (standard deviation metric)
MFP	multivariable fractional polynomial
MI	multiple imputation
mi	multiple imputation
midp	mid-p-value
MIMIC	multiple indicators and multiple causes
MINQUE	minimum norm quadratic unbiased estimation
MIVQUE	minimum variance quadratic unbiased estimation
ML	maximum likelihood
MLE	maximum likelihood estimate
MLMV	maximum likelihood with missing values
mlong	marginal long
MM	method of moments
MNAR	missing not at random
MNP	multinomial probit
MPL	modified profile likelihood
MS	mean square
MSB	most-significant byte
MSE	mean squared error
MSL	maximum simulated likelihood
MSS	model sum of squares
MUE	median unbiased estimates
MVN	multivariate normal
MVREG	multivariate regression
NARCH	nonlinear ARCH
NHANES	National Health and Nutrition Examination Survey
NLS	nonlinear least squares
NPARCH	nonlinear power ARCH
NR	Newton–Raphson
ODBC	Open DataBase Connectivity
OIM	observed information matrix
OIRF	orthogonalized impulse–response function
OLE	Object Linking and Embedding (Microsoft product)
OLS	ordinary least squares
OPG	outer product of the gradient
OR	odds ratio
PA	population averaged
PARCH	power ARCH
PCA	principal component analysis
PCSE	panel-corrected standard error
p.d.f.	probability density function

PF	prevented fraction for the population
PFE	prevented fraction among the exposed
PH	proportional hazards
pk	pharmacokinetic data
PMM	predictive mean matching
PNG	Portable Network Graphics
PSU	primary sampling unit
QDA	quadratic discriminant analysis
QML	quasimaximum likelihood
rc	return code
RCT	randomized controlled trial
RE	random effects
REML	restricted (or residual) maximum likelihood
RESET	regression specification-error test
RMSE	root mean squared error
RMSEA	root mean squared error of approximation
ROC	receiver operating characteristic
ROP	rank-ordered probit
ROT	rule of thumb
RR	relative risk
RRR	relative-risk ratio
RSS	residual sum of squares
RUM	random utility maximization
RVI	relative variance increase
SAARCH	simple asymmetric ARCH
SARIMA	seasonal ARIMA
s.d.	standard deviation
SE / s.e.	standard error
SEM	structural equation modeling
SF	static factor
SFAR	static factors with vector autoregressive errors
SIF	Stata internal form
SIR	standardized incidence ratio
SJ	Stata Journal
SMCL	Stata Markup and Control Language
SMR	standardized mortality/morbidity ratio
SMSA	standard metropolitan statistical area
SOR	standardized odds ratio
SQL	Structured Query Language
SRD	standardized rate difference
SRMR	standardized root mean squared residual

SRR	standardized risk ratio
SRS	simple random sample/sampling
SRSWR	SRS with replacement
SSC	Statistical Software Components
SSCP	sum of squares and cross products
SSD	summary statistics data
SSU	secondary sampling unit
st	survival time
STB	Stata Technical Bulletin
STS	structural time series
SUR	seemingly unrelated regression
SURE	seemingly unrelated regression estimation
SVAR	structural vector autoregressive model
SVD	singular value decomposition
TARCH	threshold ARCH
TDT	transmission/disequilibrium test
TIFF	tagged image file format
TLI	Tucker–Lewis index
TSS	total sum of squares
UCM	unobserved-components model
VAR	vector autoregressive model
VAR(1)	first-order vector autoregressive
VARMA	vector autoregressive moving average
VARMA(1,1)	first-order vector autoregressive moving average
VCE	variance–covariance estimate
VECM	vector error-correction model
VIF	variance inflation factor
WLC	worst linear combination
WLF	worst linear function
WLS	weighted least squares
XML	Extensible Markup Language
ZINB	zero-inflated negative binomial
ZIP	zero-inflated Poisson
ZTNB	zero-truncated negative binomial
ZTP	zero-truncated Poisson

Vignettes index

Aalen, O. O. (1947–), [ST] **sts**
Akaike, H. (1927–2009), [R] **estat**
Arellano, M. (1957–), [XT] **xtabond**

Bartlett, M. S. (1910–2002), [TS] **wntestb**
Berkson, J. (1899–1982), [R] **logit**
Bliss, C. I. (1899–1979), [R] **probit**
Bond, S. R. (1963–), [XT] **xtabond**
Bonferroni, C. E. (1892–1960), [R] **correlate**
Box, G. E. P. (1919–), [TS] **arima**
Breusch, T. S. (1949–), [R] **regress postestimation time series**
Brier, G. W. (1913–1998), [R] **brier**

Cholesky, A.-L. (1875–1918), [M-5] **cholesky( )**
Cleveland, W. S. (1943–), [R] **lowess**
Cochran, W. G. (1909–1980), [SVY] **survey**
Cochrane, D. (1917–1983), [TS] **prais**
Cohen, J. (1923–1998), [R] **kappa**
Cornfield, J. (1912–1979), [ST] **epitab**
Cox, D. R. (1924–), [ST] **stcox**
Cox, G. M. (1900–1978), [R] **anova**
Cronbach, L. J. (1916–2001), [R] **alpha**

Dickey, D. A. (1945–), [TS] **dfuller**
Durbin, J. (1923–), [R] **regress postestimation time series**

Efron, B. (1938–), [R] **bootstrap**
Engle, R. F. (1942–), [TS] **arch**

Fisher, R. A. (1890–1962), [R] **anova**
Fourier, J. B. J. (1768–1830), [R] **cumul**
Fuller, W. A. (1931–), [TS] **dfuller**

Gabriel, K. R. (1929–2003), [MV] **biplot**
Galton, F. (1822–1911), [R] **regress**
Gauss, J. C. F. (1777–1855), [R] **regress**
Gnanadesikan, R. (1932–), [R] **diagnostic plots**
Godfrey, L. G. (1946–), [R] **regress postestimation time series**
Gompertz, B. (1779–1865), [ST] **streg**
Gosset, W. S. (1876–1937), [R] **ttest**
Granger, C. W. J. (1934–2009), [TS] **vargranger**
Greenwood, M. (1880–1949), [ST] **sts**

Hadamard, J. S. (1865–1963), [D] **functions**
Haenszel, W. M. (1910–1998), [ST] **strate**
Halton, J. H. (1931–), [M-5] **halton( )**
Hammersley, J. M. (1920–2004), [M-5] **halton( )**
Hartley, H. O. (1912–1980), [MI] **mi impute**
Hausman, J. A. (1946–), [R] **hausman**
Heckman, J. J. (1944–), [R] **heckman**
Henderson, C. R. (1911–1989), [XT] **xtmixed**

Hermite, C. (1822–1901), [M-5] **issymmetric( )**
Hesse, L. O. (1811–1874), [M-5] **moptimize( )**
Hessenberg, K. A. (1904–1959), [M-5] **hessenbergd( )**
Hilbert, D. (1862–1943), [M-5] **Hilbert( )**
Hotelling, H. (1895–1973), [MV] **hotelling**
Householder, A. S. (1904–1993), [M-5] **qrd( )**
Huber, P. J. (1934–), [U] **20 Estimation and postestimation commands**

Jaccard, P. (1868–1944), [MV] *measure_option*
Jacobi, C. G. J. (1804–1851), [M-5] **deriv( )**
Jeffreys, H. (1891–1989), [R] **ci**
Jenkins, G. M. (1933–1982), [TS] **arima**
Johansen, S. (1939–), [TS] **vecrank**

Kaiser, H. F. (1927–1992), [MV] **rotate**
Kaplan, E. L. (1920–2006), [ST] **sts**
Kendall, M. G. (1907–1983), [R] **spearman**
Kish, L. (1910–2000), [SVY] **survey**
Kolmogorov, A. N. (1903–1987), [R] **ksmirnov**
Kronecker, L. (1823–1891), [M-2] **op_kronecker**
Kruskal, J. B. (1928–2010), [MV] **mds**
Kruskal, W. H. (1919–2005), [R] **kwallis**

Laplace, P.-S. (1749–1827), [R] **regress**
Legendre, A.-M. (1752–1833), [R] **regress**
Lexis, W. (1837–1914), [ST] **stsplit**
Lorenz, M. O. (1876–1959), [R] **inequality**

Mahalanobis, P. C. (1893–1972), [MV] **hotelling**
Mann, H. B. (1905–2000), [R] **ranksum**
Mantel, N. (1919–2002), [ST] **strate**
Marquardt, D. W. (1929–1997), [M-5] **moptimize( )**
McFadden, D. L. (1937–), [R] **asclogit**
McNemar, Q. (1900–1986), [ST] **epitab**
Meier, P. (1924–), [ST] **sts**
Moore, E. H. (1862–1932), [M-5] **pinv( )**
Murrill, W. A. (1867–1957), [MV] **discrim knn**

Nelder, J. A. (1924–2010), [R] **glm**
Nelson, W. B. (1936–), [ST] **sts**
Newey, W. K. (1954–), [TS] **newey**
Newton, I. (1643–1727), [M-5] **optimize( )**
Neyman, J. (1894–1981), [R] **ci**

Orcutt, G. H. (1917–), [TS] **prais**

Pearson, K. (1857–1936), [R] **correlate**
Penrose, R. (1931–), [M-5] **pinv( )**
Perron, P. (1959–), [TS] **pperron**
Phillips, P. C. B. (1948–), [TS] **pperron**
Playfair, W. (1759–1823), [G-2] **graph pie**
Poisson, S.-D. (1781–1840), [R] **poisson**
Prais, S. J. (1928–), [TS] **prais**

Raphson, J. (1648–1715), [M-5] **optimize( )**
Rubin, D. B. (1943–), [MI] **intro substantive**

Scheffé, H. (1907–1977), [R] **oneway**
Schur, I. (1875–1941), [M-5] **schurd( )**
Schwarz, G. E. (1933–2007), [R] **estat**
Shapiro, S. S. (1930–), [R] **swilk**
Shepard, R. N. (1929–), [MV] **mds postestimation**
Shewhart, W. A. (1891–1967), [R] **qc**
Šidák, Z. (1933–1999), [R] **correlate**
Simpson, T. (1710–1761), [M-5] **optimize( )**
Smirnov, N. V. (1900–1966), [R] **ksmirnov**
Sneath, P. H. A. (1923–), [MV] *measure_option*
Sokal, R. R. (1926–), [MV] *measure_option*
Spearman, C. E. (1863–1945), [R] **spearman**

Theil, H. (1924–2000), [R] **reg3**
Thiele, T. N. (1838–1910), [R] **summarize**
Tobin, J. (1918–2002), [R] **tobit**
Toeplitz, O. (1881–1940), [M-5] **Toeplitz( )**
Tukey, J. W. (1915–2000), [R] **jackknife**

Vandermonde, A.-T. (1735–1796),
    [M-5] **Vandermonde( )**

Wald, A. (1902–1950), [TS] **varwle**
Wallis, W. A. (1912–1998), [R] **kwallis**
Ward, J. H., Jr. (1926–), [MV] **cluster linkage**
Watson, G. S. (1921–1998), [R] **regress postestimation
    time series**
Wedderburn, R. W. M. (1947–1975), [R] **glm**
Weibull, E. H. W. (1887–1979), [ST] **streg**
West, K. D. (1953–), [TS] **newey**
White, H. L., Jr. (1950–), [U] **20 Estimation and
    postestimation commands**
Whitney, D. R. (1915–2007), [R] **ranksum**
Wilcoxon, F. (1892–1965), [R] **signrank**
Wilk, M. B. (1922–), [R] **diagnostic plots**
Wilks, S. S. (1906–1964), [MV] **manova**
Wilson, E. B. (1879–1964), [R] **ci**
Winsten, C. B. (1923–2005), [TS] **prais**
Woolf, B. (1902–1983), [ST] **epitab**

Zellner, A. (1927–2010), [R] **sureg**

Author index

A

Aalen, O. O., [ST] **stcrreg postestimation**, [ST] **sts**

Abayomi, K., [MI] **intro substantive**, [MI] **mi impute**

Abraham, B., [TS] **tssmooth**, [TS] **tssmooth dexponential**, [TS] **tssmooth exponential**, [TS] **tssmooth hwinters**, [TS] **tssmooth shwinters**

Abraira-García, L., [ST] **epitab**

Abramowitz, M., [D] **functions**, [R] **contrast**, [R] **orthog**, [XT] **xtmelogit**, [XT] **xtmepoisson**

Abrams, K. R., [R] **meta**

Abramson, J. H., [R] **kappa**, [ST] **epitab**

Abramson, Z. H., [R] **kappa**, [ST] **epitab**

Achen, C. H., [R] **scobit**

Achenback, T. M., [MV] **mvtest**

Acock, A. C., [R] **alpha**, [R] **anova**, [R] **correlate**, [R] **nestreg**, [R] **oneway**, [R] **prtest**, [R] **ranksum**, [R] **ttest**

Adkins, L. C., [R] **heckman**, [R] **regress**, [R] **regress postestimation**, [TS] **arch**

Afifi, A. A., [R] **anova**, [R] **stepwise**, [U] **20.24 References**

Agresti, A., [R] **ci**, [R] **expoisson**, [R] **tabulate twoway**, [ST] **epitab**

Ahn, S. K., [TS] **vec intro**

Ahrens, J. H., [D] **functions**

Aielli, G. P., [TS] **mgarch**, [TS] **mgarch dcc**

Aigner, D., [R] **frontier**, [XT] **xtfrontier**

Aiken, L. S., [R] **pcorr**

Aisbett, C. W., [ST] **stcox**, [ST] **streg**

Aitchison, J., [R] **ologit**, [R] **oprobit**

Aitken, A. C., [R] **reg3**

Aitkin, M. A., [MV] **mvtest correlations**

Aivazian, S. A., [R] **ksmirnov**

Akaike, H., [MV] **factor postestimation**, [R] **BIC note**, [R] **estat**, [R] **glm**, [SEM] **References**, [ST] **streg**, [TS] **varsoc**

Albert, A., [MI] **mi impute**, [MV] **discrim**, [MV] **discrim logistic**

Albert, P. S., [XT] **xtgee**

Aldenderfer, M. S., [MV] **cluster**

Aldrich, J. H., [R] **logit**, [R] **probit**

Alexandersson, A., [R] **regress**

Alf, E., Jr., [R] **rocfit**, [R] **rocreg**

Alfaro, R., [MI] **intro**

Alldredge, J. R., [R] **pk**, [R] **pkcross**

Allen, M. J., [R] **alpha**

Allison, M. J., [MV] **manova**

Allison, P. D., [MI] **intro substantive**, [MI] **mi impute**, [R] **rologit**, [R] **testnl**, [ST] **discrete**, [XT] **xtlogit**, [XT] **xtpoisson**, [XT] **xtreg**

Alonzo, T. A., [R] **rocreg**, [R] **rocreg postestimation**, [R] **rocregplot**

Altman, D. G., [R] **anova**, [R] **fracpoly**, [R] **kappa**, [R] **kwallis**, [R] **meta**, [R] **mfp**, [R] **nptrend**, [R] **oneway**

Alvarez, J., [XT] **xtabond**

Alwin, D. F., [SEM] **References**

Ambler, G., [R] **fracpoly**, [R] **mfp**, [R] **regress**

Amemiya, T., [R] **glogit**, [R] **intreg**, [R] **ivprobit**, [R] **nlogit**, [R] **tobit**, [TS] **varsoc**, [XT] **xthtaylor**, [XT] **xtivreg**

Amisano, G., [TS] **irf create**, [TS] **var intro**, [TS] **var svar**, [TS] **vargranger**, [TS] **varwle**

An, S., [TS] **arfima**

Anderberg, M. R., [MV] **cluster**, [MV] *measure_option*

Andersen, E. B., [R] **clogit**

Andersen, P. K., [R] **glm**, [ST] **stcox**, [ST] **stcrreg**

Anderson, B. D. O., [TS] **sspace**

Anderson, E., [M-1] **LAPACK**, [M-5] **lapack( )**, [MV] **clustermat**, [MV] **discrim estat**, [MV] **discrim lda**, [MV] **discrim lda postestimation**, [MV] **mvtest**, [MV] **mvtest normality**, [P] **matrix eigenvalues**

Anderson, J. A., [MI] **mi impute**, [R] **ologit**, [R] **slogit**

Anderson, M. L., [ST] **stcrreg**

Anderson, R. E., [R] **rologit**

Anderson, R. L., [R] **anova**

Anderson, S., [R] **pkequiv**

Anderson, T. W., [MI] **intro substantive**, [MV] **discrim**, [MV] **manova**, [MV] **pca**, [R] **ivregress postestimation**, [TS] **vec**, [TS] **vecrank**, [XT] **xtabond**, [XT] **xtdpd**, [XT] **xtdpdsys**, [XT] **xtivreg**

Andrews, D. F., [D] **egen**, [MV] **discrim lda postestimation**, [MV] **discrim qda**, [MV] **discrim qda postestimation**, [MV] **manova**, [R] **rreg**

Andrews, D. W. K., [R] **ivregress**

Andrews, M., [XT] **xtmelogit**, [XT] **xtmepoisson**, [XT] **xtmixed**, [XT] **xtreg**

Ängquist, L., [R] **bootstrap**, [R] **permute**

Angrist, J. D., [R] **ivregress**, [R] **ivregress postestimation**, [R] **qreg**, [R] **regress**, [U] **20.24 References**

Anscombe, F. J., [R] **binreg postestimation**, [R] **glm**, [R] **glm postestimation**

Ansley, C. F., [TS] **arima**

Arbuthnott, J., [R] **signrank**

Archer, K. J., [R] **logistic**, [R] **logistic postestimation**, [R] **logit**, [R] **logit postestimation**, [SVY] **estat**

Arellano, M., [R] **gmm**, [XT] **xtabond**, [XT] **xtdpd**, [XT] **xtdpd postestimation**, [XT] **xtdpdsys**, [XT] **xtdpdsys postestimation**, [XT] **xtivreg**, [XT] **xtreg**

Arminger, G., [R] **suest**

Armitage, P., [R] **ameans**, [R] **expoisson**, [R] **pkcross**, [R] **sdtest**

Armstrong, R. D., [R] **qreg**

Arnold, B. C., [MI] **intro substantive**, [MI] **mi impute chained**

Arnold, S. F., [MV] **manova**
Arora, S. S., [XT] **xtivreg**, [XT] **xtreg**
Arseven, E., [MV] **discrim lda**
Arthur, M., [R] **symmetry**
Aten, B., [XT] **xtunitroot**
Atkinson, A. C., [D] **functions**, [R] **boxcox**, [R] **nl**
Azen, S. P., [R] **anova**, [U] **20.24 References**
Aznar, A., [TS] **vecrank**

B

Babiker, A., [R] **sampsi**, [ST] **epitab**, [ST] **stpower**,
    [ST] **stpower cox**, [ST] **sts test**
Babin, B. J., [R] **rologit**
Babu, A. J. G., [D] **functions**
Bai, Z., [M-1] **LAPACK**, [M-5] **lapack( )**, [P] **matrix
    eigenvalues**
Baillie, R. T., [TS] **arfima**
Baker, R. J., [R] **glm**
Baker, R. M., [R] **ivregress postestimation**
Bakker, A., [R] **mean**
Balaam, L. N., [R] **pkcross**
Balakrishnan, N., [D] **functions**
Baldus, W. P., [ST] **stcrreg**
Balestra, P., [XT] **xtivreg**
Baltagi, B. H., [R] **hausman**, [XT] **xt**, [XT] **xtabond**,
    [XT] **xtdpd**, [XT] **xtdpdsys**, [XT] **xthtaylor**,
    [XT] **xtivreg**, [XT] **xtmixed**, [XT] **xtpoisson**,
    [XT] **xtprobit**, [XT] **xtreg**, [XT] **xtreg
    postestimation**, [XT] **xtregar**, [XT] **xtunitroot**
Bamber, D., [R] **rocfit**, [R] **rocregplot**, [R] **roctab**
Bancroft, T. A., [R] **stepwise**
Banerjee, A., [XT] **xtunitroot**
Barbin, É., [M-5] **cholesky( )**
Barnard, G. A., [R] **spearman**, [R] **ttest**
Barnard, J., [MI] **intro substantive**, [MI] **mi estimate**,
    [MI] **mi estimate using**, [MI] **mi predict**,
    [MI] **mi test**
Barnett, A. G., [R] **glm**
Barnow, B. S., [R] **treatreg**
Barrison, I. G., [R] **binreg**
Barthel, F. M.-S., [ST] **stcox PH-assumption tests**,
    [ST] **stpower**, [ST] **stpower cox**
Bartlett, M. S., [MV] **factor**, [MV] **factor
    postestimation**, [MV] **Glossary**, [R] **oneway**,
    [TS] **wntestb**
Bartus, T., [R] **margins**
Basford, K. E., [G-2] **graph matrix**, [XT] **xtmelogit**,
    [XT] **xtmepoisson**, [XT] **xtmixed**
Basilevsky, A. T., [MV] **factor**, [MV] **pca**
Basmann, R. L., [R] **ivregress**, [R] **ivregress
    postestimation**
Bassett, G., Jr., [R] **qreg**
Basu, A., [R] **glm**
Bates, D. M., [XT] **xtmelogit**, [XT] **xtmepoisson**,
    [XT] **xtmixed**, [XT] **xtmixed postestimation**
Battese, G. E., [XT] **xtfrontier**

Baum, C. F., [D] **cross**, [D] **fillin**, [D] **joinby**,
    [D] **reshape**, [D] **separate**, [D] **stack**,
    [D] **xpose**, [M-1] **intro**, [MV] **mvtest**,
    [MV] **mvtest normality**, [P] **intro**, [P] **levelsof**,
    [R] **gmm**, [R] **heckman**, [R] **heckprob**,
    [R] **ivregress**, [R] **ivregress postestimation**,
    [R] **margins**, [R] **net**, [R] **net search**, [R] **regress
    postestimation**, [R] **regress postestimation time
    series**, [R] **ssc**, [TS] **time series**, [TS] **arch**,
    [TS] **arima**, [TS] **dfgls**, [TS] **rolling**, [TS] **tsset**,
    [TS] **var**, [TS] **wntestq**, [U] **11.7 References**,
    [U] **16.5 References**, [U] **18.13 References**,
    [U] **20.24 References**, [XT] **xtgls**, [XT] **xtreg**,
    [XT] **xtunitroot**
Bauwens, L., [TS] **mgarch**
Baxter, M., [TS] **tsfilter**, [TS] **tsfilter bk**, [TS] **tsfilter
    cf**
Bayart, D., [R] **qc**
Beale, E. M. L., [R] **stepwise**, [R] **test**
Beall, G., [MV] **mvtest**, [MV] **mvtest covariances**
Beaton, A. E., [R] **rreg**
Beck, N., [XT] **xtgls**, [XT] **xtpcse**
Becker, R. A., [G-2] **graph matrix**
Becketti, S., [P] **pause**, [R] **fracpoly**, [R] **runtest**,
    [R] **spearman**, [TS] **corrgram**
Beerstecher, E., [MV] **manova**
Beggs, S., [R] **rologit**
Belanger, A. J., [R] **sktest**
Bellman, R. E., [MV] **Glossary**
Bellocco, R., [ST] **epitab**
Belsley, D. A., [R] **estat**, [R] **regress postestimation**,
    [U] **18.13 References**
Beltrami, E., [M-5] **svd( )**
Bendel, R. B., [R] **stepwise**
Benedetti, J. K., [R] **tetrachoric**
Beniger, J. R., [G-2] **graph bar**, [G-2] **graph pie**,
    [G-2] **graph twoway histogram**, [R] **cumul**
Bentham, G., [XT] **xtmepoisson**
Bentler, P. M., [MV] **rotate**, [MV] **rotatemat**,
    [MV] **Glossary**, [SEM] **References**
Bera, A. K., [R] **sktest**, [TS] **arch**, [TS] **varnorm**,
    [TS] **vecnorm**, [XT] **xtreg**, [XT] **xtreg
    postestimation**, [XT] **xtregar**
Beran, J., [TS] **arfima**, [TS] **arfima postestimation**
Beran, R. J., [R] **regress postestimation time series**
Berglund, P. A., [SVY] **survey**, [SVY] **subpopulation
    estimation**
Berk, K. N., [R] **stepwise**
Berk, R. A., [R] **rreg**
Berkes, I., [TS] **mgarch**
Berkson, J., [R] **logit**, [R] **probit**
Bern, P. H., [R] **nestreg**
Bernaards, C. A., [MV] **rotatemat**
Bernasco, W., [R] **tetrachoric**
Berndt, E. K., [M-5] **optimize( )**, [R] **glm**, [TS] **arch**,
    [TS] **arima**
Berndt, E. R., [R] **treatreg**, [R] **truncreg**
Bernstein, I. H., [R] **alpha**
Berry, G., [R] **ameans**, [R] **expoisson**, [R] **sdtest**

Best, D. J., [D] **functions**
Bewley, R., [R] **reg3**
Beyer, W. H., [R] **qc**
Beyersman, J., [ST] **stcrreg**
Bhargava, A., [XT] **xtregar**
Bianchi, G., [TS] **tsfilter**, [TS] **tsfilter bw**
Bibby, J. M., [MI] **mi impute mvn**, [MV] **discrim**,
    [MV] **discrim lda**, [MV] **factor**, [MV] **manova**,
    [MV] **matrix dissimilarity**, [MV] **mds**,
    [MV] **mds postestimation**, [MV] **mdslong**,
    [MV] **mdsmat**, [MV] **mvtest**, [MV] **mvtest
    means**, [MV] **mvtest normality**, [MV] **pca**,
    [MV] **procrustes**, [P] **matrix dissimilarity**
Bickeböller, H., [R] **symmetry**
Bickel, P. J., [D] **egen**, [R] **rreg**
Binder, D. A., [MI] **intro substantive**, [P] **_robust**,
    [SVY] **svy estimation**, [SVY] **variance
    estimation**, [U] **20.24 References**
Birdsall, T. G., [R] **logistic postestimation**
Bischof, C., [M-1] **LAPACK**, [M-5] **lapack( )**,
    [P] **matrix eigenvalues**
Black, F., [TS] **arch**
Black, W. C., [R] **rologit**
Blackburne, E. F., III, [XT] **xtabond**, [XT] **xtdpd**,
    [XT] **xtdpdsys**
Blackford, S., [M-1] **LAPACK**, [M-5] **lapack( )**,
    [P] **matrix eigenvalues**
Blackwell, J. L., III, [R] **areg**, [XT] **xtgls**, [XT] **xtpcse**,
    [XT] **xtreg**
Bland, M., [R] **ranksum**, [R] **sdtest**, [R] **signrank**,
    [R] **spearman**
Blashfield, R. K., [MV] **cluster**
Blasius, J., [MV] **ca**, [MV] **mca**
Blasnik, M., [D] **clonevar**, [D] **split**, [D] **statsby**
Bleda, M.-J., [R] **alpha**
Bliss, C. I., [R] **probit**
Bloch, D. A., [R] **brier**
Bloomfield, P., [R] **qreg**, [TS] **arfima**
Blundell, R., [R] **gmm**, [R] **ivprobit**, [XT] **xtdpd**,
    [XT] **xtdpdsys**
BMDP, [R] **symmetry**
Boggess, M., [ST] **stcrreg**, [ST] **stcrreg postestimation**
Boice, J. D., Jr., [R] **bitest**, [ST] **epitab**, [ST] **epitab**
Boland, P. J., [R] **ttest**
Bolduc, D., [R] **asmprobit**
Bollen, K. A., [MV] **factor postestimation**, [R] **regress
    postestimation**, [SEM] **References**
Bollerslev, T., [TS] **arch**, [TS] **arima**, [TS] **mgarch**,
    [TS] **mgarch ccc**, [TS] **mgarch dvech**
Bond, S., [R] **gmm**, [XT] **xtabond**, [XT] **xtdpd**,
    [XT] **xtdpd postestimation**, [XT] **xtdpdsys**,
    [XT] **xtdpdsys postestimation**, [XT] **xtivreg**
Bonferroni, C. E., [R] **correlate**
Borenstein, M., [R] **meta**
Borg, I., [MV] **mds**, [MV] **mds postestimation**,
    [MV] **mdslong**, [MV] **mdsmat**
Borgan, Ø., [ST] **stcrreg**
Bornhorst, F., [XT] **xtunitroot**

Borowczyk, J., [M-5] **cholesky( )**
Boshuizen, H. C., [MI] **intro substantive**, [MI] **mi
    impute**, [MI] **mi impute chained**, [MI] **mi
    impute monotone**
Boswell, T., [ST] **streg postestimation**
Boswijk, H. P., [TS] **vec**
Bottai, M., [ST] **epitab**, [XT] **xtreg**
Bound, J., [R] **ivregress postestimation**
Bover, O., [XT] **xtdpd**, [XT] **xtdpdsys**
Bowerman, B. L., [TS] **tssmooth**, [TS] **tssmooth
    dexponential**, [TS] **tssmooth exponential**,
    [TS] **tssmooth hwinters**, [TS] **tssmooth
    shwinters**
Bowker, A. H., [R] **symmetry**
Box, G. E. P., [MV] **manova**, [MV] **mvtest
    covariances**, [R] **anova**, [R] **boxcox**,
    [R] **lnskew0**, [TS] **arfima**, [TS] **arima**,
    [TS] **corrgram**, [TS] **cumsp**, [TS] **dfuller**,
    [TS] **pergram**, [TS] **pperron**, [TS] **psdensity**,
    [TS] **wntestq**, [TS] **xcorr**
Box, J. F., [R] **anova**
Box-Steffensmeier, J. M., [ST] **stcox**, [ST] **streg**
Boyd, N. F., [R] **kappa**
Boyle, J. M., [P] **matrix symeigen**
Boyle, P., [XT] **xtmepoisson**
Brackstone, G., [R] **swilk**
Bradley, R. A., [R] **signrank**
Brady, A. R., [R] **logistic**, [R] **spikeplot**
Brady, T., [D] **edit**
Brand, J. P. L., [MI] **intro substantive**, [MI] **mi impute
    chained**
Brant, R., [R] **ologit**
Bray, R. J., [MV] **clustermat**
Bray, T. A., [D] **functions**
Breitung, J., [XT] **xtunitroot**
Brent, R. P., [MV] **mdsmat**, [MV] **mvtest means**
Breslow, N. E., [R] **clogit**, [R] **dstdize**, [R] **symmetry**,
    [ST] **epitab**, [ST] **epitab**, [ST] **stcox**, [ST] **stcox
    PH-assumption tests**, [ST] **sts**, [ST] **sts test**,
    [XT] **xtmelogit**, [XT] **xtmepoisson**
Breusch, T. S., [R] **mvreg**, [R] **regress postestimation**,
    [R] **regress postestimation time series**,
    [R] **sureg**, [TS] **Glossary**, [XT] **xtreg
    postestimation**
Brier, G. W., [R] **brier**
Brillinger, D. R., [R] **jackknife**
Brockwell, P. J., [TS] **corrgram**, [TS] **sspace**
Brook, R. H., [R] **brier**
Brown, B. W., [ST] **sts graph**
Brown, C. C., [ST] **epitab**
Brown, D. R., [R] **anova**, [R] **contrast**, [R] **loneway**,
    [R] **oneway**, [R] **pwcompare**
Brown, J. D., [MV] **manova**
Brown, L. D., [R] **ci**
Brown, M. B., [R] **sdtest**, [R] **tetrachoric**
Brown, S. E., [R] **symmetry**
Brown, T. A., [SEM] **References**
Browne, M. W., [MV] **procrustes**, [SEM] **References**

Bru, B., [R] **poisson**
Bruno, G. S. F., [XT] **xtabond**, [XT] **xtdpd**,
    [XT] **xtdpdsys**, [XT] **xtreg**
Bryk, A. S., [XT] **xtmepoisson**, [XT] **xtmixed**
Brzinsky-Fay, C., [G-2] **graph twoway rbar**
Buchholz, A., [ST] **stcrreg**
Buchner, D. M., [R] **ladder**
Buis, M. L., [G-3] *by_option*, [R] **logistic**, [R] **logit**,
    [R] **margins**
Buja, A., [U] **20.24 References**
Bunch, D. S., [R] **asmprobit**
Buot, M.-L. G., [MV] **mvtest means**
Burke, W. J., [R] **tobit**
Burkhauser, R. V., [MI] **intro substantive**
Burnam, M. A., [R] **lincom**, [R] **mlogit**, [R] **mprobit**,
    [R] **mprobit postestimation**, [R] **predictnl**,
    [R] **slogit**
Burns, A. F., [TS] **tsfilter**, [TS] **tsfilter bk**, [TS] **tsfilter**
    bw, [TS] **tsfilter cf**, [TS] **tsfilter hp**, [TS] **ucm**
Burr, I. W., [R] **qc**
Buskens, V., [R] **tabstat**
Butterworth, S., [TS] **tsfilter**, [TS] **tsfilter bw**

C

Cai, T., [R] **rocreg**
Cai, T. T., [R] **ci**
Cailliez, F., [MV] **mdsmat**
Cain, G. G., [R] **treatreg**
Caines, P. E., [TS] **sspace**
Califf, R. M., [ST] **stcox postestimation**
Caliński, T., [MV] **cluster**, [MV] **cluster stop**
Cameron, A. C., [R] **asclogit**, [R] **asmprobit**,
    [R] **bootstrap**, [R] **gmm**, [R] **heckman**,
    [R] **intreg**, [R] **ivregress**, [R] **ivregress**
    postestimation, [R] **logit**, [R] **mprobit**,
    [R] **nbreg**, [R] **ologit**, [R] **oprobit**, [R] **poisson**,
    [R] **probit**, [R] **qreg**, [R] **regress**, [R] **regress**
    postestimation, [R] **simulate**, [R] **sureg**,
    [R] **tnbreg**, [R] **tobit**, [R] **tpoisson**, [R] **zinb**
    postestimation, [R] **zip postestimation**, [XT] **xt**,
    [XT] **xtmixed**, [XT] **xtnbreg**, [XT] **xtpoisson**
Campbell, D. T., [SEM] **References**
Campbell, M. J., [R] **ci**, [R] **kappa**, [R] **logistic**
    postestimation, [R] **poisson**, [R] **tabulate**
    twoway, [ST] **stpower**, [ST] **stpower cox**,
    [ST] **stpower logrank**
Canette, I., [XT] **xtmixed**
Cappellari, L., [D] **corr2data**, [D] **egen**, [R] **asmprobit**
Cardell, S., [R] **rologit**
Carle, A. C., [XT] **xtmixed**
Carlile, T., [R] **kappa**
Carlin, J. B., [MI] **intro substantive**, [MI] **intro**,
    [MI] **mi estimate**, [MI] **mi impute**, [MI] **mi**
    impute mvn, [MI] **mi impute regress**,
    [R] **ameans**, [ST] **epitab**
Carnes, B. A., [ST] **streg**
Carpenter, J. R., [MI] **intro substantive**, [MI] **intro**,
    [MI] **mi impute**, [XT] **xtmelogit**

Carroll, J. B., [MV] **rotatemat**
Carroll, R. J., [R] **boxcox**, [R] **rreg**, [R] **sdtest**,
    [XT] **xtmixed**
Carson, R. T., [R] **tnbreg**, [R] **tpoisson**
Carter, S., [R] **frontier**, [R] **lrtest**, [R] **nbreg**,
    [ST] **stcox**, [ST] **streg**, [XT] **xt**, [XT] **xtmelogit**,
    [XT] **xtmepoisson**, [XT] **xtmixed**
Casals, J., [TS] **sspace**
Casella, G., [XT] **xtmixed**
Castillo, E., [MI] **intro substantive**, [MI] **mi impute**
    chained
Cattell, R. B., [MV] **factor postestimation**,
    [MV] **pca postestimation**, [MV] **procrustes**,
    [MV] **screeplot**
Caudill, S. B., [R] **frontier**, [XT] **xtfrontier**
Caulcutt, R., [R] **qc**
Cefalu, M., [ST] **stcox postestimation**, [ST] **stcurve**,
    [ST] **sts graph**
Chabert, J.-L., [M-5] **cholesky( )**
Chadwick, J., [R] **poisson**
Chakraborti, S., [R] **ksmirnov**
Chamberlain, G., [R] **clogit**, [R] **gmm**
Chambers, J. M., [G-2] **graph box**, [G-2] **graph**
    matrix, [G-3] *by_option*, [R] **diagnostic plots**,
    [R] **grmeanby**, [R] **lowess**, [U] **1.4 References**
Chang, I. M., [R] **margins**
Chang, Y., [TS] **sspace**
Chang, Y.-J., [XT] **xtivreg**, [XT] **xtreg**
Chao, E. C., [XT] **xtmelogit**, [XT] **xtmelogit**
    postestimation, [XT] **xtmepoisson**,
    [XT] **xtmepoisson postestimation**
Charlett, A., [R] **fracpoly**
Chatfield, C., [TS] **arima**, [TS] **corrgram**,
    [TS] **pergram**, [TS] **tssmooth**, [TS] **tssmooth**
    dexponential, [TS] **tssmooth exponential**,
    [TS] **tssmooth hwinters**, [TS] **tssmooth ma**,
    [TS] **tssmooth shwinters**, [TS] **Glossary**
Chatfield, M., [R] **anova**
Chatterjee, S., [R] **poisson**, [R] **regress**, [R] **regress**
    postestimation, [TS] **prais**
Chen, X., [R] **logistic**, [R] **logistic postestimation**,
    [R] **logit**
Cheung, Y.-W., [TS] **dfgls**
Chiang, C. L., [ST] **ltable**
Chiburis, R., [R] **heckman**, [R] **heckprob**, [R] **oprobit**
Choi, B. C. K., [R] **rocfit**, [R] **rocreg postestimation**,
    [R] **rocregplot**, [R] **roctab**
Choi, I., [XT] **xtunitroot**
Choi, M.-D., [M-5] **Hilbert( )**
Choi, S. C., [MV] **discrim knn**
Cholesky, A.-L., [M-5] **cholesky( )**
Chou, R. Y., [TS] **arch**
Chow, G. C., [R] **contrast**
Chow, S.-C., [R] **pk**, [R] **pkcross**, [R] **pkequiv**,
    [R] **pkexamine**, [R] **pkshape**, [ST] **stpower**,
    [ST] **stpower exponential**
Christakis, N., [R] **rologit**
Christiano, L. J., [TS] **irf create**, [TS] **tsfilter**,
    [TS] **tsfilter cf**, [TS] **var svar**

Chu, C.-S. J., [XT] **xtunitroot**

Chu-Chun-Lin, S., [TS] **sspace**

Clark, V. A., [ST] **ltable**

Clarke, M. R. B., [MV] **factor**

Clarke, R. D., [R] **poisson**

Clarke-Pearson, D. L., [R] **roccomp**, [R] **rocreg**, [R] **roctab**

Clarkson, D. B., [R] **tabulate twoway**

Clayton, D. G., [D] **egen**, [R] **cloglog**, [R] **cumul**, [ST] **epitab**, [ST] **stptime**, [ST] **strate**, [ST] **stsplit**, [ST] **sttocc**, [XT] **xtmelogit**, [XT] **xtmepoisson**

Cleland, J., [XT] **xtmelogit**

Clerget-Darpoux, F., [R] **symmetry**

Cleveland, W. S., [G-1] **graph intro**, [G-2] **graph box**, [G-2] **graph dot**, [G-2] **graph matrix**, [G-2] **graph twoway lowess**, [G-3] *by_option*, [R] **diagnostic plots**, [R] **lowess**, [R] **lpoly**, [R] **sunflower**, [U] **1.4 References**

Cleves, M. A., [MI] **mi estimate**, [R] **binreg**, [R] **dstdize**, [R] **logistic**, [R] **logit**, [R] **roccomp**, [R] **rocfit**, [R] **rocreg**, [R] **rocreg postestimation**, [R] **rocregplot**, [R] **roctab**, [R] **sdtest**, [R] **symmetry**, [ST] **survival analysis**, [ST] **st**, [ST] **stcox**, [ST] **stcrreg**, [ST] **stcrreg postestimation**, [ST] **stcurve**, [ST] **stdescribe**, [ST] **stpower**, [ST] **stpower exponential**, [ST] **stpower logrank**, [ST] **streg**, [ST] **sts**, [ST] **stset**, [ST] **stsplit**, [ST] **stvary**

Cliff, N., [MV] **canon postestimation**

Clogg, C. C., [R] **suest**

Clopper, C. J., [R] **ci**

Cobb, G. W., [R] **anova**

Cochran, W. G., [P] **levelsof**, [R] **ameans**, [R] **anova**, [R] **correlate**, [R] **dstdize**, [R] **mean**, [R] **oneway**, [R] **poisson**, [R] **probit**, [R] **proportion**, [R] **ranksum**, [R] **ratio**, [R] **signrank**, [R] **total**, [SVY] **survey**, [SVY] **estat**, [SVY] **subpopulation estimation**, [SVY] **svyset**, [SVY] **variance estimation**

Cochrane, D., [TS] **prais**

Coelli, T. J., [R] **frontier**, [XT] **xtfrontier**

Coffey, C., [MI] **intro substantive**

Cohen, J., [R] **kappa**, [R] **pcorr**

Cohen, P., [R] **pcorr**

Coleman, J. S., [R] **poisson**

Collett, D., [R] **clogit**, [R] **logistic**, [R] **logistic postestimation**, [ST] **stci**, [ST] **stcox postestimation**, [ST] **stcrreg postestimation**, [ST] **stpower**, [ST] **stpower logrank**, [ST] **streg postestimation**, [ST] **sts test**, [ST] **stsplit**

Collins, E., [SVY] **survey**, [SVY] **svy estimation**

Comrey, A. L., [MV] **rotate**, [MV] **rotatemat**, [MV] **Glossary**

Comte, F., [TS] **mgarch**

Cone-Wesson, B., [R] **rocreg**, [R] **rocreg postestimation**, [R] **rocregplot**

Cong, R., [R] **tobit**, [R] **tobit postestimation**, [R] **treatreg**, [R] **truncreg**

Conover, W. J., [R] **centile**, [R] **ksmirnov**, [R] **kwallis**, [R] **nptrend**, [R] **sdtest**, [R] **spearman**, [R] **tabulate twoway**

Conroy, R. M., [R] **intreg**

Conway, M. R., [XT] **xtlogit**, [XT] **xtprobit**

Cook, A., [R] **ci**

Cook, I., [U] **1.4 References**

Cook, N. R., [R] **rocreg**

Cook, R. D., [P] **_predict**, [R] **boxcox**, [R] **regress postestimation**

Cooper, M. C., [MV] **cluster**, [MV] **cluster programming subroutines**, [MV] **cluster stop**

Cornfield, J., [ST] **epitab**

Corten, R., [MV] **mds**

Coster, D., [R] **contrast**

Coull, B. A., [R] **ci**

Coviello, V., [ST] **stcrreg**, [ST] **stcrreg postestimation**, [ST] **sttocc**

Cox, C., [SEM] **References**

Cox, C. S., [SVY] **survey**, [SVY] **svy estimation**

Cox, D. R., [R] **boxcox**, [R] **exlogistic**, [R] **expoisson**, [R] **lnskew0**, [ST] **ltable**, [ST] **stcox**, [ST] **stcox PH-assumption tests**, [ST] **stcrreg**, [ST] **stpower**, [ST] **stpower cox**, [ST] **streg**, [ST] **streg postestimation**, [ST] **sts**

Cox, G. M., [P] **levelsof**, [R] **anova**

Cox, M. A. A., [MV] **biplot**, [MV] **ca**, [MV] **mds**, [MV] **mds postestimation**, [MV] **mdsmat**, [MV] **procrustes**, [MV] **Glossary**

Cox, N. J., [D] **by**, [D] **clonevar**, [D] **contract**, [D] **describe**, [D] **destring**, [D] **drop**, [D] **ds**, [D] **duplicates**, [D] **egen**, [D] **fillin**, [D] **format**, [D] **functions**, [D] **missing values**, [D] **rename**, [D] **reshape**, [D] **sample**, [D] **separate**, [D] **split**, [D] **statsby**, [G-1] **graph intro**, [G-2] **graph bar**, [G-2] **graph box**, [G-2] **graph dot**, [G-2] **graph twoway dot**, [G-2] **graph twoway function**, [G-2] **graph twoway histogram**, [G-2] **graph twoway kdensity**, [G-2] **graph twoway lowess**, [G-2] **graph twoway lpoly**, [G-2] **graph twoway pcarrow**, [G-2] **graph twoway pcspike**, [G-2] **graph twoway scatter**, [G-3] *added_line_options*, [G-3] *aspect_option*, [G-3] *axis_label_options*, [G-3] *axis_scale_options*, [G-3] *by_option*, [G-4] *linestyle*, [MV] **mvtest**, [MV] **mvtest normality**, [P] **forvalues**, [P] **levelsof**, [P] **matrix define**, [P] **unab**, [R] **ci**, [R] **cumul**, [R] **diagnostic plots**, [R] **histogram**, [R] **inequality**, [R] **kappa**, [R] **kdensity**, [R] **ladder**, [R] **lowess**, [R] **lpoly**, [R] **net**, [R] **net search**, [R] **regress postestimation**, [R] **search**, [R] **serrbar**, [R] **sktest**, [R] **smooth**, [R] **spikeplot**, [R] **ssc**, [R] **stem**, [R] **summarize**, [R] **sunflower**, [R] **tabulate oneway**, [R] **tabulate twoway**, [TS] **tsline**, [TS] **tssmooth hwinters**, [TS] **tssmooth shwinters**, [U] **11.7 References**, [U] **12.10 References**, [U] **13.12 References**, [U] **17.10 Reference**,

Cox, N. J., *continued*
    [U] **23.5 Reference**, [U] **24.8 References**,
    [XT] **xtdescribe**
Cox, T. F., [MV] **biplot**, [MV] **ca**, [MV] **mds**,
    [MV] **mds postestimation**, [MV] **mdsmat**,
    [MV] **procrustes**, [MV] **Glossary**
Cozad, J. B., [MV] **discrim lda**
Cragg, J. G., [R] **ivregress postestimation**
Cramer, E. M., [MV] **procrustes**
Cramér, H., [R] **tabulate twoway**
Cramer, J. S., [R] **logit**
Crawford, C. B., [MV] **rotate**, [MV] **rotatemat**,
    [MV] **Glossary**
Critchley, F., [MV] **mdsmat**
Cronbach, L. J., [R] **alpha**
Croux, C., [R] **rreg**
Crowder, M. J., [ST] **stcrreg**, [ST] **streg**
Crowe, P. R., [G-2] **graph box**
Crowley, J., [ST] **stcox**, [ST] **stcrreg**, [ST] **stset**
Cudeck, R., [SEM] **References**
Cui, J., [R] **symmetry**, [ST] **stcox**, [ST] **streg**,
    [XT] **xtgee**
Cummings, P., [R] **binreg**, [R] **glm**, [ST] **epitab**
Curran, P. J., [SEM] **References**
Curtis, J. T., [MV] **clustermat**
Curts-García, J., [R] **smooth**
Cutler, S. J., [ST] **ltable**
Cuzick, J., [R] **kappa**, [R] **nptrend**
Czekanowski, J., [MV] *measure_option*

D

D'Agostino, R. B., [MV] **mvtest normality**, [R] **sktest**
D'Agostino, R. B., Jr., [R] **sktest**
Daniel, C., [R] **diagnostic plots**, [R] **oneway**
Daniel, R., [MI] **intro substantive**, [MI] **mi impute**,
    [MI] **mi impute chained**, [MI] **mi impute**
    monotone
Danuso, F., [R] **nl**
Das, S., [XT] **xtunitroot**
DasGupta, A., [R] **ci**
Davey Smith, G., [R] **meta**
David, H. A., [D] **egen**, [R] **spearman**, [R] **summarize**
David, J. S., [TS] **arima**
Davidon, W. C., [M-5] **optimize( )**
Davidson, R., [R] **boxcox**, [R] **cnsreg**, [R] **gmm**,
    [R] **intreg**, [R] **ivregress**, [R] **ivregress**
    postestimation, [R] **mlogit**, [R] **nl**, [R] **nlsur**,
    [R] **reg3**, [R] **regress**, [R] **regress postestimation**
    time series, [R] **tobit**, [R] **truncreg**, [TS] **arch**,
    [TS] **arima**, [TS] **prais**, [TS] **sspace**,
    [TS] **varlmar**, [TS] **Glossary**, [XT] **xtgls**,
    [XT] **xtpcse**
Davis, R. A., [TS] **corrgram**, [TS] **sspace**
Davison, A. C., [R] **bootstrap**
Day, N. E., [R] **clogit**, [R] **dstdize**, [R] **symmetry**,
    [ST] **epitab**
Day, W. H. E., [MV] **cluster**

De Hoyos, R. E., [XT] **xtreg**
de Irala-Estévez, J., [R] **logistic**
De Jong, P., [TS] **dfactor**, [TS] **sspace**, [TS] **sspace**
    postestimation, [TS] **ucm**
de Leeuw, J., [MV] **ca postestimation**
De Luca, G., [R] **biprobit**, [R] **heckprob**, [R] **probit**
De Stavola, B. L., [ST] **stcox**, [ST] **stset**
de Wolf, I., [R] **rologit**
Deaton, A., [R] **nlsur**, [U] **20.24 References**
Deb, P., [R] **nbreg**
DeGroot, M. H., [TS] **arima**
Dehon, C., [R] **correlate**
Deistler, M., [TS] **sspace**
del Mar Zamora, M., [R] **heckprob**
del Rio, A., [TS] **tsfilter hp**
DeLong, D. M., [R] **roccomp**, [R] **rocreg**, [R] **roctab**
DeLong, E. R., [R] **roccomp**, [R] **rocreg**, [R] **roctab**
DeMaris, A., [R] **regress postestimation**
Demidenko, E., [XT] **xtmixed**
Demmel, J., [M-1] **LAPACK**, [M-5] **lapack( )**,
    [P] **matrix eigenvalues**
Demnati, A., [SVY] **direct standardization**,
    [SVY] **poststratification**, [SVY] **variance**
    estimation
Dempster, A. P., [MI] **intro substantive**, [MI] **mi**
    impute mvn, [XT] **xtmixed**
Denis, D., [G-2] **graph twoway scatter**
Desu, M. M., [ST] **stpower**, [ST] **stpower exponential**
Deville, J.-C., [SVY] **direct standardization**,
    [SVY] **poststratification**, [SVY] **variance**
    estimation
Devroye, L., [D] **functions**
Dewey, M. E., [R] **correlate**
Dice, L. R., [MV] *measure_option*
Dickens, R., [TS] **prais**
Dickey, D. A., [TS] **dfgls**, [TS] **dfuller**, [TS] **pperron**,
    [TS] **Glossary**
Dickson, E. R., [ST] **stcrreg**
Diebold, F. X., [TS] **arch**
Dieter, U., [D] **functions**
Digby, P. G. N., [R] **tetrachoric**
Diggle, P. J., [TS] **arima**, [TS] **wntestq**, [XT] **xtmixed**
Dijksterhuis, G. B., [MV] **procrustes**
DiNardo, J., [XT] **xtrc**
Ding, Z., [TS] **arch**
Dinno, A., [MV] **factor**, [MV] **pca**
Dixon, W. J., [R] **ttest**
Dobbin, K., [ST] **stpower**
Dobson, A. J., [R] **glm**
Dodd, L. E., [R] **rocreg**
Dohoo, I., [R] **regress**, [ST] **epitab**
Doll, R., [R] **poisson**, [ST] **epitab**
Donald, S. G., [R] **ivregress postestimation**
Dongarra, J. J., [M-1] **LAPACK**, [M-5] **lapack( )**,
    [P] **matrix eigenvalues**, [P] **matrix symeigen**
Donner, A., [R] **loneway**
Donoho, D. L., [R] **lpoly**

Doornik, J. A., [MV] **mvtest**, [MV] **mvtest normality**, [TS] **arfima**, [TS] **vec**

Dore, C. J., [R] **fracpoly**

Dorfman, D. D., [R] **rocfit**, [R] **rocreg**

Draper, N., [R] **eivreg**, [R] **oneway**, [R] **regress**, [R] **stepwise**

Driver, H. E., [MV] *measure_option*

Drukker, D. M., [R] **asmprobit**, [R] **boxcox**, [R] **frontier**, [R] **lrtest**, [R] **nbreg**, [R] **tobit**, [R] **treatreg**, [ST] **stcox**, [ST] **streg**, [TS] **arfima postestimation**, [TS] **sspace**, [TS] **vec**, [XT] **xt**, [XT] **xtmelogit**, [XT] **xtmepoisson**, [XT] **xtmixed**, [XT] **xtregar**

Du Croz, J., [M-1] **LAPACK**, [M-5] **lapack( )**, [P] **matrix eigenvalues**

Duan, N., [R] **heckman**

Dubes, R. C., [MV] **cluster**

Duda, R. O., [MV] **cluster**, [MV] **cluster stop**

Duncan, A. J., [R] **qc**

Duncan, O. D., [SEM] **References**

Dunlop, D. D., [R] **sampsi**

Dunn, G., [MV] **discrim**, [MV] **discrim qda postestimation**, [MV] **mca**, [R] **kappa**

Dunnett, C. W., [R] **mprobit**, [R] **pwcompare**

Dunnington, G. W., [R] **regress**

Dupont, W. D., [R] **logistic**, [R] **mkspline**, [R] **sunflower**, [ST] **epitab**, [ST] **stcox**, [ST] **stir**, [ST] **sts**

Durbin, J., [R] **ivregress postestimation**, [R] **regress postestimation time series**, [TS] **prais**, [TS] **ucm**, [TS] **Glossary**

Duren, P., [R] **regress**

Durlauf, S. N., [TS] **vec intro**, [TS] **vec**, [TS] **vecrank**

Duval, R. D., [R] **bootstrap**, [R] **jackknife**, [R] **rocreg**, [R] **rocregplot**

Dwyer, J. H., [XT] **xtreg**

E

Eaves, R. C., [SEM] **References**

Ecob, R., [MI] **mi estimate**

Eddings, W. D., [MI] **mi impute**

Edelsbrunner, H., [MV] **cluster**

Ederer, F., [ST] **ltable**

Edgington, E. S., [R] **runtest**

Edwards, A. L., [R] **anova**

Edwards, A. W. F., [R] **tetrachoric**

Edwards, J. H., [R] **tetrachoric**

Efron, B., [R] **bootstrap**, [R] **qreg**

Efroymson, M. A., [R] **stepwise**

Egger, M., [R] **meta**

Eichenbaum, M., [TS] **irf create**, [TS] **var svar**

Eisenhart, C., [R] **correlate**, [R] **runtest**

Elliott, G., [TS] **dfgls**, [TS] **Glossary**

Ellis, C. D., [R] **poisson**

Eltinge, J. L., [R] **test**, [SVY] **survey**, [SVY] **estat**, [SVY] **svy postestimation**, [SVY] **svydescribe**, [SVY] **variance estimation**

Emerson, J. D., [R] **lv**, [R] **stem**

Enas, G. G., [MV] **discrim knn**

Ender, P., [R] **marginsplot**

Ender, P. B., [MV] **canon**

Enders, W., [TS] **arch**, [TS] **arima**, [TS] **corrgram**

Engel, A., [R] **marginsplot**, [SVY] **survey**, [SVY] **estat**, [SVY] **subpopulation estimation**, [SVY] **svy**, [SVY] **svy brr**, [SVY] **svy estimation**, [SVY] **svy jackknife**, [SVY] **svy postestimation**, [SVY] **svy: tabulate oneway**, [SVY] **svy: tabulate twoway**, [SVY] **svydescribe**

Engle, R. F., [R] **regress postestimation time series**, [TS] **arch**, [TS] **arima**, [TS] **dfactor**, [TS] **mgarch**, [TS] **mgarch dcc**, [TS] **mgarch dvech**, [TS] **mgarch vcc**, [TS] **vec intro**, [TS] **vec**, [TS] **vecrank**

Erdreich, L. S., [R] **roccomp**, [R] **rocfit**, [R] **roctab**

Esman, R. M., [D] **egen**

Eubank, R. L., [R] **lpoly**

Evans, C. L., [TS] **irf create**, [TS] **var svar**

Evans, M. A., [R] **pk**, [R] **pkcross**

Everitt, B. S., [MV] **cluster**, [MV] **cluster stop**, [MV] **discrim**, [MV] **discrim qda postestimation**, [MV] **mca**, [MV] **pca**, [MV] **screeplot**, [R] **gllamm**, [R] **glm**, [U] **1.4 References**

Ewens, W. J., [R] **symmetry**

Ezekiel, M., [R] **regress postestimation**

Ezzati-Rice, T. M., [MI] **intro substantive**

F

Falcaro, M., [MV] **cluster dendrogram**

Fan, J., [R] **lpoly**

Fan, Y.-A., [R] **tabulate twoway**

Fang, K.-T., [R] **asmprobit**

Farbmacher, H., [R] **tpoisson**

Fay, R. E., [SVY] **survey**, [SVY] **svy sdr**, [SVY] **variance estimation**

Feinleib, M., [XT] **xtreg**

Feiveson, A. H., [R] **nlcom**, [R] **ranksum**, [ST] **stpower**

Feldman, J. J., [SVY] **survey**, [SVY] **svy estimation**

Feldt, L. S., [R] **anova**

Feller, W., [TS] **wntestb**

Feltbower, R., [ST] **epitab**

Feng, S., [MI] **intro substantive**

Ferguson, G. A., [MV] **rotate**, [MV] **rotatemat**, [MV] **Glossary**

Ferri, H. A., [R] **kappa**

Fibrinogen Studies Collaboration, [ST] **stcox postestimation**

Fidell, L. S., [MV] **discrim**, [MV] **discrim lda**

Fieller, E. C., [R] **pkequiv**

Fienberg, S. E., [R] **kwallis**, [R] **tabulate twoway**

Filon, L. N. G., [R] **correlate**

Findley, D. F., [R] **estat**

Findley, T. W., [R] **ladder**

Fine, J. P., [ST] **stcrreg**

Finlay, K., [R] **ivprobit**, [R] **ivregress**, [R] **ivtobit**
Finney, D. J., [R] **probit**, [R] **tabulate twoway**
Fiocco, M., [ST] **stcrreg**, [ST] **stcrreg postestimation**
Fiorentini, G., [TS] **mgarch**
Fiorio, C. V., [R] **kdensity**
Fiser, D. H., [R] **logistic postestimation**
Fishell, E., [R] **kappa**
Fisher, L. D., [MV] **factor**, [MV] **pca**, [R] **anova**,
    [R] **dstdize**, [R] **oneway**, [ST] **epitab**
Fisher, M. R., [XT] **xtcloglog**, [XT] **xtgee**,
    [XT] **xtintreg**, [XT] **xtlogit**, [XT] **xtprobit**,
    [XT] **xttobit**
Fisher, N. I., [R] **regress postestimation time series**
Fisher, R. A., [MV] **clustermat**, [MV] **discrim**,
    [MV] **discrim estat**, [MV] **discrim lda**,
    [MV] **Glossary**, [P] **levelsof**, [R] **anova**,
    [R] **anova**, [R] **ranksum**, [R] **signrank**,
    [R] **tabulate twoway**, [ST] **streg**
Fiske, D. W., [SEM] **References**
Fitzgerald, T. J., [TS] **tsfilter**, [TS] **tsfilter cf**
Fitzmaurice, G. M., [XT] **xtmixed**
Fix, E., [MV] **discrim knn**
Flannery, B. P., [D] **functions**, [G-2] **graph twoway**
    contour, [P] **matrix symeigen**, [R] **dydx**,
    [R] **vwls**, [TS] **arch**, [TS] **arima**
Fleiss, J. L., [R] **dstdize**, [R] **kappa**, [R] **sampsi**,
    [ST] **epitab**
Fleming, T. R., [ST] **stcox**, [ST] **sts test**
Fletcher, K., [R] **rocreg**, [R] **rocreg postestimation**,
    [R] **rocregplot**
Fletcher, R., [M-5] **optimize( )**
Folsom, R. C., [R] **rocreg**, [R] **rocreg postestimation**,
    [R] **rocregplot**
Ford, J. M., [R] **frontier**, [XT] **xtfrontier**
Forsythe, A. B., [R] **sdtest**
Forthofer, R. N., [R] **dstdize**
Foster, A., [R] **regress**
Foulkes, M. A., [ST] **stpower**, [ST] **stpower cox**,
    [ST] **stpower exponential**, [U] **1.4 References**
Fourier, J. B. J., [R] **cumul**
Fox, J., [R] **kdensity**, [R] **lv**
Fox, W. C., [R] **logistic postestimation**
Francia, R. S., [R] **swilk**
Frank, M. W., [XT] **xtabond**, [XT] **xtdpd**,
    [XT] **xtdpdsys**
Frankel, M. R., [P] **_robust**, [SVY] **variance**
    estimation, [U] **20.24 References**
Franklin, C. H., [D] **cross**
Franzese, R. J., Jr., [XT] **xtpcse**
Franzini, L., [XT] **xtregar**
Frechette, G. R., [XT] **xtprobit**
Freedman, L. S., [ST] **stpower**, [ST] **stpower cox**,
    [ST] **stpower exponential**, [ST] **stpower**
    logrank, [U] **1.4 References**
Freeman, D. H., Jr., [SVY] **svy: tabulate twoway**
Freeman, E. H., [SEM] **References**
Freeman, J. L., [ST] **epitab**, [SVY] **svy: tabulate**
    twoway

Freese, J., [R] **asroprobit**, [R] **clogit**, [R] **cloglog**,
    [R] **logistic**, [R] **logit**, [R] **mlogit**, [R] **mprobit**,
    [R] **nbreg**, [R] **ologit**, [R] **oprobit**, [R] **poisson**,
    [R] **probit**, [R] **regress**, [R] **regress**
    postestimation, [R] **tnbreg**, [R] **tpoisson**,
    [R] **zinb**, [R] **zip**, [U] **20.24 References**
Friedman, J., [MV] **discrim knn**
Friedman, M., [TS] **arima**
Friendly, M., [G-2] **graph twoway scatter**
Frison, L., [R] **sampsi**
Frölich, M., [R] **qreg**
Frome, E. L., [R] **qreg**
Frydenberg, M., [R] **dstdize**, [R] **roccomp**, [R] **roctab**
Fu, V. K., [R] **ologit**
Fuller, W. A., [MV] **factor**, [P] **_robust**, [R] **regress**,
    [R] **spearman**, [SVY] **svy: tabulate twoway**,
    [SVY] **variance estimation**, [TS] **dfgls**,
    [TS] **dfuller**, [TS] **pperron**, [TS] **psdensity**,
    [TS] **tsfilter**, [TS] **tsfilter bk**, [TS] **ucm**,
    [TS] **Glossary**, [U] **20.24 References**
Funkhouser, H. G., [G-2] **graph pie**
Fyler, D. C., [ST] **epitab**
Fyles, A., [ST] **stcrreg**, [ST] **stcrreg postestimation**

G

Gabriel, K. R., [MV] **biplot**
Gail, M. H., [P] **_robust**, [R] **rocreg**, [R] **rocreg**
    postestimation, [ST] **stcrreg**, [ST] **stpower**,
    [ST] **stpower exponential**, [ST] **strate**,
    [U] **1.4 References**, [U] **20.24 References**
Galati, J. C., [MI] **intro substantive**, [MI] **intro**,
    [MI] **mi estimate**
Galecki, A. T., [XT] **xtmixed**
Gall, J.-R. L., [R] **logistic**, [R] **logistic postestimation**
Gallant, A. R., [R] **ivregress**, [R] **nl**
Galton, F., [R] **correlate**, [R] **cumul**, [R] **regress**,
    [R] **summarize**
Gan, F. F., [R] **diagnostic plots**
Gange, S. J., [XT] **xtcloglog**, [XT] **xtgee**,
    [XT] **xtintreg**, [XT] **xtlogit**, [XT] **xtprobit**,
    [XT] **xttobit**
Gani, J., [TS] **wntestb**
Garbow, B. S., [P] **matrix symeigen**
Gardiner, J. S., [TS] **tssmooth**, [TS] **tssmooth**
    dexponential, [TS] **tssmooth exponential**,
    [TS] **tssmooth hwinters**, [TS] **tssmooth**
    shwinters
Gardner, E. S., Jr., [TS] **tssmooth dexponential**,
    [TS] **tssmooth hwinters**
Garrett, J. M., [R] **fracpoly**, [R] **logistic**, [R] **logistic**
    postestimation, [R] **regress postestimation**,
    [ST] **stcox PH-assumption tests**
Garsd, A., [R] **exlogistic**
Gart, J. J., [ST] **epitab**
Gasser, T., [R] **lpoly**
Gastwirth, J. L., [R] **sdtest**
Gates, R., [R] **asmprobit**
Gauss, J. C. F., [R] **regress**

Gauvreau, K., [R] dstdize, [R] logistic, [R] sampsi,
    [ST] ltable, [ST] sts
Gehan, E. A., [ST] sts test
Geisser, S., [R] anova
Gel, Y. R., [R] sdtest
Gelbach, J., [R] ivprobit, [R] ivtobit
Gelfand, A. E., [MI] mi impute chained
Gelman, A., [MI] intro substantive, [MI] mi impute,
    [MI] mi impute mvn, [MI] mi impute regress
Gelman, R., [R] margins
Geman, D., [MI] mi impute chained
Geman, S., [MI] mi impute chained
Genest, C., [R] swilk
Gentle, J. E., [D] functions, [R] anova, [R] nl
Genton, M. G., [R] sktest
Genz, A., [R] asmprobit
George, S. L., [ST] stpower, [ST] stpower exponential
Gerkins, V. R., [R] symmetry
Gerow, K. G., [SVY] survey
Geskus, R. B., [ST] stcrreg, [ST] stcrreg
    postestimation
Geweke, J., [R] asmprobit, [TS] dfactor
Giannini, C., [TS] irf create, [TS] var intro, [TS] var
    svar, [TS] vargranger, [TS] varwle
Gibbons, J. D., [R] ksmirnov, [R] spearman
Gichangi, A., [ST] stcrreg
Giesen, D., [R] tetrachoric
Gifi, A., [MV] mds
Gijbels, I., [R] lpoly
Gilbert, G. K., [MV] measure_option
Giles, D. E. A., [TS] prais
Gill, R. D., [ST] stcrreg
Gillham, N. W., [R] regress
Gillispie, C. C., [R] regress
Gini, R., [R] vwls, [ST] epitab
Ginther, O. J., [XT] xtmixed
Girshick, M. A., [MV] pca
Glass, R. I., [ST] epitab
Gleason, J. R., [D] cf, [D] describe, [D] functions,
    [D] generate, [D] infile (fixed format),
    [D] label, [D] notes, [D] order, [R] anova,
    [R] bootstrap, [R] ci, [R] correlate,
    [R] loneway, [R] summarize, [R] ttest,
    [ST] epitab
Gleick, J., [M-5] optimize( )
Gleser, G., [R] alpha
Glidden, D. V., [R] logistic, [ST] stcox
Gloeckler, L., [ST] discrete
Glosten, L. R., [TS] arch
Gnanadesikan, R., [MV] manova, [R] cumul,
    [R] diagnostic plots
Godambe, V. P., [SVY] variance estimation
Godfrey, L. G., [R] regress postestimation time series
Goeden, G. B., [R] kdensity
Goerg, S. J., [R] ksmirnov
Golbe, D. L., [D] label language, [D] merge,
    [U] 22.1 References
Goldberger, A. S., [R] intreg, [R] tobit, [R] treatreg

Goldblatt, A., [ST] epitab
Golden, C. D., [SVY] survey, [SVY] svy estimation
Goldfarb, D., [M-5] optimize( )
Goldman, N., [XT] xtmelogit
Goldstein, H., [XT] xtmelogit, [XT] xtmepoisson,
    [XT] xtmixed
Goldstein, R., [D] egen, [R] brier, [R] correlate,
    [R] inequality, [R] nl, [R] ologit, [R] oprobit,
    [R] ranksum, [R] regress postestimation,
    [XT] xtreg
Golub, G. H., [M-5] svd( ), [R] orthog, [R] tetrachoric,
    [TS] arfima, [TS] arfima postestimation
Gómez, V., [TS] tsfilter, [TS] tsfilter hp
Gompertz, B., [ST] streg
Gönen, M., [ST] stcox postestimation
Gonzalez, J. F., Jr., [SVY] estat, [SVY] subpopulation
    estimation, [SVY] svy bootstrap, [SVY] svy
    estimation
Gonzalo, J., [TS] vec intro, [TS] vecrank
Good, P. I., [G-1] graph intro, [R] permute,
    [R] symmetry, [R] tabulate twoway
Goodall, C., [R] lowess, [R] rreg
Goodman, L. A., [R] tabulate twoway
Gooley, T. A., [ST] stcrreg
Gordon, A. D., [MV] biplot, [MV] cluster,
    [MV] cluster stop, [MV] measure_option
Gordon, M. G., [R] binreg
Gorga, M. P., [R] rocreg, [R] rocreg postestimation,
    [R] rocregplot
Gorman, J. W., [R] stepwise
Gorsuch, R. L., [MV] factor, [MV] rotate,
    [MV] rotatemat
Gosset [Student, pseud.], W. S., [R] ttest
Gosset, W. S., [R] ttest
Gould, W. W., [D] datasignature, [D] datetime,
    [D] destring, [D] egen, [D] format, [D] icd9,
    [D] infile (fixed format), [D] merge,
    [D] putmata, [D] reshape, [M-1] how,
    [M-1] interactive, [M-2] exp, [M-2] goto,
    [M-2] struct, [M-2] subscripts, [M-2] syntax,
    [M-4] io, [M-4] stata, [M-5] deriv( ),
    [M-5] eigensystem( ), [M-5] geigensystem( ),
    [M-5] inbase( ), [M-5] moptimize( ),
    [M-5] st_addvar( ), [M-5] st_global( ),
    [M-5] st_local( ), [M-5] st_view( ), [MI] mi
    estimate, [P] intro, [P] _datasignature,
    [P] matrix eigenvalues, [P] matrix mkmat,
    [P] postfile, [P] _robust, [R] bootstrap,
    [R] dydx, [R] frontier, [R] grmeanby,
    [R] jackknife, [R] kappa, [R] logistic,
    [R] margins, [R] maximize, [R] mkspline,
    [R] ml, [R] net search, [R] nlcom, [R] ologit,
    [R] oprobit, [R] predictnl, [R] qreg, [R] regress,
    [R] rreg, [R] simulate, [R] sktest, [R] smooth,
    [R] swilk, [R] testnl, [ST] survival analysis,
    [ST] stcox, [ST] stcrreg, [ST] stcrreg
    postestimation, [ST] stdescribe, [ST] stpower,
    [ST] stpower exponential, [ST] stpower
    logrank, [ST] streg, [ST] stset, [ST] stsplit,
    [ST] stvary, [SVY] survey, [SVY] ml for svy,

Gould, W. W., *continued*
  [U] **13.12 References**, [U] **18.13 References**,
  [U] **22.1 References**, [U] **26.27 Reference**,
  [XT] **xtfrontier**
Gourieroux, C., [R] **hausman**, [R] **suest**, [R] **test**,
  [TS] **arima**, [TS] **mgarch ccc**, [TS] **mgarch dcc**,
  [TS] **mgarch vcc**
Govindarajulu, Z., [D] **functions**
Gower, J. C., [MV] **biplot**, [MV] **ca**, [MV] **mca**,
  [MV] *measure_option*, [MV] **procrustes**
Gradshteyn, I. S., [TS] **arfima**
Graham, J. W., [MI] **intro substantive**, [MI] **mi impute**
Grambsch, P. M., [ST] **stcox**, [ST] **stcox PH-**
  assumption tests, [ST] **stcox postestimation**,
  [ST] **stcrreg**
Granger, C. W. J., [TS] **arch**, [TS] **arfima**,
  [TS] **vargranger**, [TS] **vec intro**, [TS] **vec**,
  [TS] **vecrank**
Graubard, B. I., [R] **margins**, [R] **ml**, [R] **test**,
  [SVY] **survey**, [SVY] **direct standardization**,
  [SVY] **estat**, [SVY] **svy**, [SVY] **svy estimation**,
  [SVY] **svy postestimation**, [SVY] **svy: tabulate**
  twoway, [SVY] **variance estimation**,
  [XT] **xtmixed**
Gray, R. J., [ST] **stcrreg**
Graybill, F. A., [R] **centile**
Green, B. F., [MV] **discrim lda**, [MV] **procrustes**
Green, D. M., [R] **logistic postestimation**
Green, P. E., [MV] **cluster**
Greenacre, M. J., [MV] **ca**, [MV] **mca**, [MV] **mca**
  postestimation
Greenbaum, A., [M-1] **LAPACK**, [M-5] **lapack( )**,
  [P] **matrix eigenvalues**
Greene, W. H., [P] **matrix accum**, [R] **asclogit**,
  [R] **asmprobit**, [R] **biprobit**, [R] **clogit**,
  [R] **cnsreg**, [R] **frontier**, [R] **gmm**,
  [R] **heckman**, [R] **heckprob**, [R] **hetprob**,
  [R] **ivregress**, [R] **logit**, [R] **lrtest**, [R] **margins**,
  [R] **mkspline**, [R] **mlogit**, [R] **nlogit**, [R] **nlsur**,
  [R] **pcorr**, [R] **probit**, [R] **reg3**, [R] **regress**,
  [R] **regress postestimation time series**,
  [R] **sureg**, [R] **testnl**, [R] **treatreg**, [R] **truncreg**,
  [R] **zinb**, [R] **zip**, [TS] **arch**, [TS] **arima**,
  [TS] **corrgram**, [TS] **var**, [XT] **xt**, [XT] **xtgls**,
  [XT] **xtpcse**, [XT] **xtpoisson**, [XT] **xtrc**,
  [XT] **xtreg**
Greenfield, S., [MV] **factor**, [MV] **factor**
  postestimation, [R] **alpha**, [R] **lincom**,
  [R] **mlogit**, [R] **mprobit**, [R] **mprobit**
  postestimation, [R] **predictnl**, [R] **slogit**
Greenhouse, J. B., [ST] **epitab**
Greenhouse, S. W., [R] **anova**, [ST] **epitab**
Greenland, S., [R] **ci**, [R] **glogit**, [R] **mkspline**,
  [R] **ologit**, [R] **poisson**, [ST] **epitab**
Greenwood, M., [ST] **ltable**, [ST] **sts**
Greenwood, P., [MI] **intro substantive**
Gregoire, A., [R] **kappa**
Griffith, J. L., [R] **brier**
Griffith, R., [R] **gmm**

Griffiths, W. E., [R] **cnsreg**, [R] **estat**, [R] **glogit**,
  [R] **ivregress**, [R] **ivregress postestimation**,
  [R] **logit**, [R] **probit**, [R] **regress**, [R] **regress**
  postestimation, [R] **test**, [TS] **arch**, [TS] **prais**,
  [XT] **xtgls**, [XT] **xtpcse**, [XT] **xtrc**, [XT] **xtreg**
Griliches, Z., [XT] **xtgls**, [XT] **xtnbreg**, [XT] **xtpcse**,
  [XT] **xtpoisson**, [XT] **xtrc**
Grimmett, G., [M-5] **halton( )**
Grizzle, J. E., [R] **vwls**
Groenen, P. J. F., [MV] **mds**, [MV] **mds**
  postestimation, [MV] **mdslong**, [MV] **mdsmat**
Grogger, J. T., [R] **tnbreg**, [R] **tpoisson**
Gronau, R., [R] **heckman**
Groothuis-Oudshoorn, C. G. M., [MI] **intro**
  substantive, [MI] **mi impute chained**
Gropper, D. M., [R] **frontier**, [XT] **xtfrontier**
Gross, A. J., [ST] **ltable**
Grunfeld, Y., [XT] **xtgls**, [XT] **xtpcse**, [XT] **xtrc**
Guan, W., [R] **bootstrap**
Guerry, A.-M., [G-2] **graph twoway histogram**
Guilkey, D. K., [XT] **xtprobit**
Guillemot, M., [M-5] **cholesky( )**
Guimarães, P., [XT] **xtnbreg**
Gutierrez, R. G., [MI] **mi estimate**, [R] **frontier**,
  [R] **lpoly**, [R] **lrtest**, [R] **nbreg**, [ST] **survival**
  analysis, [ST] **stcox**, [ST] **stcrreg**, [ST] **stcrreg**
  postestimation, [ST] **stdescribe**, [ST] **stpower**,
  [ST] **stpower exponential**, [ST] **stpower**
  logrank, [ST] **streg**, [ST] **streg postestimation**,
  [ST] **stset**, [ST] **stsplit**, [ST] **stvary**,
  [XT] **xt**, [XT] **xtmelogit**, [XT] **xtmepoisson**,
  [XT] **xtmixed**

H

Haaland, J.-A., [G-1] **graph intro**
Haan, P., [R] **asmprobit**, [R] **mlogit**, [R] **mprobit**
Haas, K., [M-5] **moptimize( )**
Hadamard, J. S., [D] **functions**
Hadi, A. S., [R] **poisson**, [R] **regress**, [R] **regress**
  postestimation, [TS] **prais**
Hadorn, D. C., [R] **brier**
Hadri, K., [XT] **xtunitroot**
Haenszel, W., [ST] **epitab**, [ST] **strate**, [ST] **sts test**
Hahn, G. J., [M-5] **moptimize( )**
Hahn, J., [R] **ivregress postestimation**
Hair, J. F., Jr., [R] **rologit**
Hajian-Tilaki, K. O., [R] **rocreg**
Hajivassiliou, V. A., [R] **asmprobit**
Hakkio, C. S., [D] **egen**
Hald, A., [R] **qreg**, [R] **regress**, [R] **signrank**,
  [R] **summarize**
Haldane, J. B. S., [ST] **epitab**
Hall, A. D., [R] **frontier**
Hall, A. R., [R] **gmm**, [R] **gmm postestimation**,
  [R] **ivregress**, [R] **ivregress postestimation**
Hall, B. H., [M-5] **optimize( )**, [R] **glm**, [TS] **arch**,
  [TS] **arima**, [XT] **xtnbreg**, [XT] **xtpoisson**
Hall, N. S., [R] **anova**

Hall, P., [R] bootstrap, [R] regress postestimation time series
Hall, R. E., [M-5] optimize( ), [R] glm, [TS] arch, [TS] arima
Hall, W. J., [MV] biplot, [R] roccomp, [R] rocfit, [R] roctab
Haller, A. O., [SEM] References
Halley, E., [ST] ltable
Hallock, K., [R] qreg
Halton, J. H., [M-5] halton( )
Halvorsen, K. T., [R] tabulate twoway
Hamann, U., [MV] *measure_option*
Hamer, R. M., [MV] mds, [MV] mdslong, [MV] mdsmat
Hamerle, A., [R] clogit
Hamilton, J. D., [P] matrix eigenvalues, [R] gmm, [TS] time series, [TS] arch, [TS] arfima, [TS] arima, [TS] corrgram, [TS] dfuller, [TS] fcast compute, [TS] irf, [TS] irf create, [TS] pergram, [TS] pperron, [TS] psdensity, [TS] sspace, [TS] sspace postestimation, [TS] tsfilter, [TS] ucm, [TS] var intro, [TS] var, [TS] var svar, [TS] vargranger, [TS] varnorm, [TS] varsoc, [TS] varstable, [TS] varwle, [TS] vec intro, [TS] vec, [TS] vecnorm, [TS] vecrank, [TS] vecstable, [TS] xcorr, [TS] Glossary
Hamilton, L. C., [D] xpose, [G-1] graph intro, [MV] factor, [MV] screeplot, [R] bootstrap, [R] diagnostic plots, [R] ladder, [R] lv, [R] mlogit, [R] regress, [R] regress postestimation, [R] rreg, [R] simulate, [R] summarize, [R] ttest
Hammarling, S., [M-1] LAPACK, [M-5] lapack( ), [P] matrix eigenvalues
Hammersley, J. M., [M-5] halton( )
Hampel, F. R., [D] egen, [R] rreg, [U] 20.24 References
Hancock, G. R., [SEM] References
Hand, D. J., [MV] biplot, [MV] ca, [MV] discrim, [MV] mca
Handscomb, D. C., [M-5] halton( )
Hankey, B., [ST] strate
Hanley, J. A., [R] roccomp, [R] rocfit, [R] rocreg, [R] rocreg postestimation, [R] rocregplot, [R] roctab
Hannachi, A., [MV] pca
Hannan, E. J., [TS] sspace
Hansen, H., [MV] mvtest, [MV] mvtest normality
Hansen, L. P., [R] gmm, [R] ivregress, [R] ivregress postestimation, [XT] xtabond, [XT] xtdpd, [XT] xtdpdsys
Hao, L., [R] qreg
Harabasz, J., [MV] cluster, [MV] cluster stop
Harbord, R. M., [R] roccomp, [R] roctab, [XT] xtmelogit

Hardin, J. W., [D] statsby, [G-1] graph intro, [R] binreg, [R] biprobit, [R] estat, [R] glm, [R] glm postestimation, [R] regress postestimation, [TS] newey, [TS] prais, [XT] xtgee, [XT] xtmelogit postestimation, [XT] xtmepoisson postestimation, [XT] xtpoisson
Haritou, A., [R] suest
Harkness, J., [R] ivprobit, [R] ivtobit
Harley, J. B., [ST] stpower cox
Harman, H. H., [MV] factor, [MV] factor postestimation, [MV] rotate, [MV] rotatemat
Harrell, F. E., Jr., [R] mkspline, [R] ologit, [ST] stcox postestimation
Harrington, D. P., [ST] stcox, [ST] sts test
Harris, E. K., [MV] discrim, [MV] discrim logistic
Harris, R. D. F., [XT] xtunitroot
Harris, R. J., [MV] canon postestimation
Harris, R. L., [R] qc
Harris, T., [R] qreg
Harrison, D. A., [D] list, [G-2] graph twoway histogram, [R] histogram, [R] tabulate oneway, [R] tabulate twoway
Harrison, J. A., [R] dstdize
Harrison, J. M., [ST] stcrreg
Hart, P. E., [MV] cluster, [MV] cluster stop
Hartigan, J. A., [G-2] graph matrix
Hartley, H. O., [MI] intro substantive, [MI] mi impute
Harvey, A. C., [R] hetprob, [TS] intro, [TS] arch, [TS] arima, [TS] prais, [TS] psdensity, [TS] sspace, [TS] sspace postestimation, [TS] tsfilter, [TS] tsfilter hp, [TS] tssmooth hwinters, [TS] ucm, [TS] var svar, [U] 1.4 References
Harville, D. A., [XT] xtmixed
Hassell, J., [ST] sts
Hastie, T. J., [MV] discrim knn, [R] grmeanby, [R] slogit
Hastorf, A. H., [ST] epitab
Hauck, W. W., [R] pkequiv, [XT] xtcloglog, [XT] xtlogit, [XT] xtprobit
Haughton, J., [R] inequality
Hauser, M. A., [TS] arfima
Hausman, J. A., [M-5] optimize( ), [R] glm, [R] hausman, [R] ivregress postestimation, [R] nlogit, [R] rologit, [R] suest, [SEM] References, [TS] arch, [TS] arima, [XT] xthtaylor, [XT] xtnbreg, [XT] xtpoisson, [XT] xtreg postestimation
Hayashi, F., [R] gmm, [R] ivregress, [R] ivregress postestimation
Hayes, R. J., [R] permute
Haynam, G. E., [D] functions
Hays, R. D., [R] lincom, [R] mlogit, [R] mprobit, [R] mprobit postestimation, [R] predictnl, [R] slogit
Heagerty, P. J., [MV] factor, [MV] pca, [R] anova, [R] dstdize, [R] oneway, [ST] epitab, [XT] xtmixed

Heckman, J., [R] biprobit, [R] heckman, [R] heckman postestimation, [R] heckprob

Hedges, L. V., [R] meta

Hedley, D., [ST] stcrreg, [ST] stcrreg postestimation

Heeringa, S. G., [SVY] survey, [SVY] subpopulation estimation

Heinecke, K., [P] matrix mkmat

Heinonen, O. P., [ST] epitab

Heiss, F., [R] nlogit

Heitjan, D. F., [MI] intro substantive, [MI] mi impute

Heller, G., [ST] stcox postestimation

Henderson, B. E., [R] symmetry

Henderson, C. R., [XT] xtmixed

Hendrickson, A. E., [MV] rotate, [MV] rotatemat, [MV] Glossary

Hendrickx, J., [R] mlogit, [R] xi

Henry-Amar, M., [ST] ltable

Hensher, D. A., [R] nlogit

Henze, N., [MV] mvtest, [MV] mvtest normality

Hermite, C., [M-5] issymmetric( )

Herrin, J., [U] 18.13 References

Herriot, J. G., [M-5] spline3( )

Hertz, S., [ST] stsplit

Herzberg, A. M., [MV] discrim lda postestimation, [MV] discrim qda, [MV] discrim qda postestimation, [MV] manova

Hess, K. R., [ST] stcox PH-assumption tests, [ST] sts graph

Hesse, L. O., [M-5] moptimize( )

Hessenberg, K. A., [M-5] hessenbergd( )

Heston, A., [XT] xtunitroot

Heyde, C. C., [U] 1.4 References

Hickam, D. H., [R] brier

Higbee, K. T., [D] clonevar

Higgins, J. E., [R] anova

Higgins, J. P. T., [R] meta

Higgins, M. L., [TS] arch

Hilbe, J. M., [D] functions, [MV] discrim lda, [MV] manova, [MV] measure_option, [R] cloglog, [R] estat, [R] glm, [R] glm postestimation, [R] logistic, [R] logit, [R] nbreg, [R] poisson, [R] probit, [R] sampsi, [R] simulate, [R] tnbreg, [R] tpoisson, [XT] xtgee, [XT] xtmelogit postestimation, [XT] xtmepoisson postestimation, [XT] xtpoisson

Hilbert, D., [M-5] Hilbert( )

Hildreth, C., [TS] prais

Hilferty, M. M., [MV] mvtest normality

Hilgard, E. R., [ST] epitab

Hill, A. B., [R] poisson, [ST] epitab

Hill, R. C., [R] cnsreg, [R] estat, [R] glogit, [R] heckman, [R] ivregress, [R] ivregress postestimation, [R] logit, [R] probit, [R] regress, [R] regress postestimation, [R] test, [TS] arch, [TS] prais, [XT] xtgls, [XT] xtpcse, [XT] xtrc, [XT] xtreg

Hill, R. P., [ST] stcrreg, [ST] stcrreg postestimation

Hill, W. G., [ST] epitab

Hills, M., [D] egen, [R] cloglog, [R] cumul, [ST] epitab, [ST] stcox, [ST] stptime, [ST] strate, [ST] stset, [ST] stsplit, [ST] sttocc

Hinkley, D. V., [R] bootstrap

Hipel, K. W., [TS] arima, [TS] ucm

Hirji, K. F., [R] exlogistic, [R] expoisson

Hlouskova, J., [XT] xtunitroot

Hoaglin, D. C., [R] diagnostic plots, [R] lv, [R] regress postestimation, [R] smooth, [R] stem

Hocevar, D., [SEM] References

Hochberg, Y., [R] oneway

Hocking, R. R., [MI] intro substantive, [R] stepwise, [XT] xtmixed

Hodges, J. L., [MV] discrim knn

Hodrick, R. J., [TS] tsfilter, [TS] tsfilter hp

Hoechle, D., [XT] xtgls, [XT] xtpcse, [XT] xtreg, [XT] xtregar

Hoel, P. G., [R] bitest, [R] ttest

Hoffmann, J. P., [R] glm

Hofman, A. F., [ST] stcrreg

Hogben, L., [ST] sts

Hole, A. R., [R] asmprobit, [R] clogit, [R] mlogit, [R] mprobit

Holloway, L., [R] brier

Holm, S., [R] test

Holmes, D. J., [XT] xtmixed

Holmes, S., [R] bootstrap

Holmgren, J., [ST] epitab

Holt, C. C., [TS] tssmooth, [TS] tssmooth dexponential, [TS] tssmooth exponential, [TS] tssmooth hwinters, [TS] tssmooth shwinters

Holt, D., [SVY] survey, [SVY] estat

Holtz-Eakin, D., [XT] xtabond, [XT] xtdpd, [XT] xtdpdsys

Honoré, B. E., [XT] xttobit

Hood, W. C., [R] ivregress

Hooker, P. F., [ST] streg

Horst, P., [MV] factor postestimation, [MV] rotate, [MV] rotatemat

Horton, N. J., [MI] intro substantive, [XT] xtmixed

Horváth, L., [TS] mgarch

Hosking, J. R. M., [TS] arfima

Hosmer, D. W., Jr., [R] clogit, [R] clogit postestimation, [R] glm, [R] glogit, [R] lincom, [R] logistic, [R] logistic postestimation, [R] logit, [R] logit postestimation, [R] lrtest, [R] mlogit, [R] predictnl, [R] stepwise, [ST] stcox, [ST] stpower, [ST] stpower cox, [ST] streg, [XT] xtgee

Hossain, K. M., [ST] epitab

Hotelling, H., [MV] canon, [MV] hotelling, [MV] manova, [MV] pca, [MV] roccomp, [R] rocfit, [R] roctab

Hougaard, P., [ST] streg

Householder, A. S., [M-5] qrd( ), [MV] mds, [MV] mdslong, [MV] mdsmat

Hoşten, S., [MV] mvtest means
Hsiao, C., [XT] xt, [XT] xtabond, [XT] xtdpd,
    [XT] xtdpdsys, [XT] xtivreg, [XT] xtregar
Hsieh, F. Y., [ST] stpower, [ST] stpower cox,
    [ST] stpower logrank
Hu, M., [ST] stcox, [ST] stset
Huang, C., [R] sunflower
Huang, D. S., [R] nlsur, [R] sureg
Hubálek, Z., [MV] measure_option
Huber, P. J., [D] egen, [P] _robust, [R] qreg, [R] rreg,
    [R] suest, [U] 20.24 References
Huberty, C. J., [MV] candisc, [MV] discrim,
    [MV] discrim estat, [MV] discrim lda,
    [MV] discrim lda postestimation, [MV] discrim
    qda
Hubrich, K., [TS] vec intro, [TS] vecrank
Hughes, J. B., [MV] manova
for Human Resource Research, C., [XT] xt
Hunter, D. R., [R] qreg
Huq, M. I., [ST] epitab
Huq, N. M., [XT] xtmelogit
Hurd, M., [R] intreg, [R] tobit
Hurley, J. R., [MV] procrustes
Hurst, H. E., [TS] arfima
Hutto, C., [R] exlogistic
Huynh, H., [R] anova

I

Iglewicz, B., [R] lv
Ikebe, Y., [P] matrix symeigen
Im, K. S., [XT] xtunitroot
Isaacs, D., [R] fracpoly
Ishiguro, M., [R] BIC note
ISSP, [MV] ca, [MV] mca, [MV] mca postestimation

J

Jaccard, P., [MV] measure_option
Jackman, R. W., [R] regress postestimation
Jackson, J. E., [MV] pca, [MV] pca postestimation
Jacobi, C. G. J., [M-5] deriv( )
Jacobs, K. B., [R] symmetry
Jacobs, M., [D] duplicates
Jacoby, W. G., [MV] biplot
Jaeger, A., [TS] tsfilter, [TS] tsfilter hp
Jaeger, D. A., [R] ivregress postestimation
Jagannathan, R., [TS] arch
Jain, A. K., [MV] cluster
James, B. R., [R] rocreg, [R] rocreg postestimation
James, G. S., [MV] mvtest, [MV] mvtest means
James, I., [M-2] op_kronecker, [M-5] deriv( ),
    [M-5] issymmetric( ), [M-5] pinv( )
James, K. L., [R] rocreg, [R] rocreg postestimation
Janes, H., [R] rocfit, [R] rocreg, [R] rocreg
    postestimation, [R] rocregplot
Jang, D. S., [SVY] variance estimation

Jann, B., [P] mark, [R] estimates store, [R] ksmirnov,
    [R] saved results, [R] tabulate twoway,
    [SVY] svy: tabulate twoway
Jarque, C. M., [R] sktest, [TS] varnorm, [TS] vecnorm
Jeantheau, T., [TS] mgarch
Jeanty, P. W., [D] reshape
Jeffreys, H., [R] ci, [R] spearman
Jenkins, B., [M-5] hash1( )
Jenkins, G. M., [TS] arfima, [TS] arima,
    [TS] corrgram, [TS] cumsp, [TS] dfuller,
    [TS] pergram, [TS] pperron, [TS] psdensity,
    [TS] xcorr
Jenkins, S. P., [D] corr2data, [D] egen, [D] rename,
    [MI] intro substantive, [R] asmprobit, [R] do,
    [R] inequality, [ST] discrete, [ST] stcox
Jennrich, R. I., [MV] mvtest, [MV] mvtest
    correlations, [MV] rotate, [MV] rotatemat,
    [MV] Glossary
Jensen, A. R., [MV] rotate
Jensen, D. R., [MV] mvtest, [MV] mvtest means
Jerez, M., [TS] sspace
Jewell, N. P., [ST] epitab
Jick, H., [ST] epitab
Joe, H., [R] tabulate twoway, [XT] xtmelogit,
    [XT] xtmepoisson
Johansen, S., [TS] irf create, [TS] varlmar, [TS] vec
    intro, [TS] vec, [TS] veclmar, [TS] vecnorm,
    [TS] vecrank, [TS] vecstable
Johnson, D. E., [MV] manova, [R] anova, [R] contrast,
    [R] pwcompare
Johnson, L. A., [TS] tssmooth, [TS] tssmooth
    dexponential, [TS] tssmooth exponential,
    [TS] tssmooth hwinters, [TS] tssmooth
    shwinters
Johnson, M. E., [R] sdtest
Johnson, M. M., [R] sdtest
Johnson, N. L., [D] functions, [R] ksmirnov,
    [R] nbreg, [R] poisson, [U] 1.4 References
Johnson, R. A., [MV] canon, [MV] discrim,
    [MV] discrim estat, [MV] discrim lda,
    [MV] discrim lda postestimation, [MV] mvtest,
    [MV] mvtest correlations, [MV] mvtest
    covariances, [MV] mvtest means
Johnson, W., [MI] intro substantive, [SVY] survey
Johnston, J., [XT] xtrc
Jolliffe, D., [R] inequality, [R] qreg, [R] regress
Jolliffe, I. T., [MV] biplot, [MV] pca, [R] brier
Jones, A., [R] heckman, [R] logit, [R] probit
Jones, B. S., [ST] stcox, [ST] streg
Jones, D. R., [R] meta
Jones, M. C., [R] kdensity, [R] lpoly
Jones, P. S., [M-5] Vandermonde( )
Jordan, C, [M-5] svd( )
Jöreskog, K. G., [MV] factor postestimation,
    [SEM] References
Jorgensen, R. A., [ST] stcrreg
Jorner, U., [G-1] graph intro
Joyeux, R., [TS] arfima

Judge, G. G., [R] **estat**, [R] **glogit**, [R] **ivregress**, [R] **ivregress postestimation**, [R] **logit**, [R] **probit**, [R] **regress postestimation**, [R] **test**, [TS] **arch**, [TS] **prais**, [XT] **xtgls**, [XT] **xtpcse**, [XT] **xtrc**, [XT] **xtreg**

Judkins, D. R., [SVY] **svy brr**, [SVY] **svyset**, [SVY] **variance estimation**

Judson, D. H., [R] **poisson**, [R] **tabulate twoway**, [R] **tpoisson**

Jung, B. C., [XT] **xtmixed**

Juul, S., [R] **dstdize**, [R] **roccomp**, [R] **roctab**

K

Kachitvichyanukul, V., [D] **functions**

Kadane, J. B., [XT] **xtmelogit**, [XT] **xtmepoisson**

Kahn, H. A., [R] **dstdize**, [ST] **epitab**, [ST] **ltable**, [ST] **stcox**

Kaiser, H. F., [MV] **factor postestimation**, [MV] **pca postestimation**, [MV] **rotate**, [MV] **rotatemat**, [MV] **Glossary**

Kaiser, J., [R] **ksmirnov**, [R] **permute**, [R] **signrank**

Kalbfleisch, J. D., [ST] **ltable**, [ST] **stcox**, [ST] **stcox PH-assumption tests**, [ST] **stcox postestimation**, [ST] **streg**, [ST] **sts**, [ST] **sts test**, [ST] **stset**, [XT] **xtcloglog**, [XT] **xtlogit**, [XT] **xtprobit**

Kalman, R. E., [TS] **arima**

Kalmijn, M., [R] **tetrachoric**

Kantor, D., [D] **cf**, [D] **functions**

Kaplan, E. L., [ST] **stcrreg**, [ST] **stcrreg postestimation**, [ST] **sts**

Katz, J. N., [XT] **xtgls**, [XT] **xtpcse**

Kaufman, L., [MV] **cluster**, [MV] **clustermat**, [MV] **matrix dissimilarity**, [MV] *measure_option*, [P] **matrix dissimilarity**

Keane, M. P., [R] **asmprobit**

Keeler, E. B., [R] **brier**

Keiding, N., [ST] **stcrreg**, [ST] **stsplit**

Kemp, A. W., [D] **functions**, [R] **nbreg**, [R] **poisson**

Kemp, C. D., [D] **functions**

Kempthorne, P. J., [R] **regress postestimation**

Kendall, D. G., [MV] **mds**

Kendall, M. G., [MV] *measure_option*, [R] **centile**, [R] **spearman**, [R] **tabulate twoway**, [R] **tobit**

Kennedy, W. J., Jr., [P] **_robust**, [R] **anova**, [R] **nl**, [R] **regress**, [R] **stepwise**, [SVY] **svy: tabulate twoway**

Kenny, D. A., [SEM] **References**

Kent, J. T., [MI] **mi impute mvn**, [MV] **discrim**, [MV] **discrim lda**, [MV] **factor**, [MV] **manova**, [MV] **matrix dissimilarity**, [MV] **mds**, [MV] **mds postestimation**, [MV] **mdslong**, [MV] **mdsmat**, [MV] **mvtest**, [MV] **mvtest means**, [MV] **mvtest normality**, [MV] **pca**, [MV] **procrustes**, [P] **matrix dissimilarity**, [P] **_robust**, [U] **20.24 References**

Kenward, M. G., [MI] **intro substantive**, [MI] **mi impute**

Kettenring, J. R., [R] **diagnostic plots**

Keynes, J. M., [R] **ameans**

Khan, M. R., [ST] **epitab**

Khandker, S. R., [R] **inequality**

Khanti-Akom, S., [XT] **xthtaylor**

Khare, M., [MI] **intro substantive**

Kiernan, M., [R] **kappa**

Kim, I.-M., [TS] **vec intro**, [TS] **vec**, [TS] **vecrank**

Kim, J. O., [MV] **factor**

Kimber, A. C., [ST] **streg**

Kimbrough, J. W., [MV] **discrim knn**

Kinderman, A. J., [D] **functions**

King, M. L., [TS] **prais**

King, R. G., [TS] **tsfilter**, [TS] **tsfilter bk**, [TS] **tsfilter cf**, [TS] **tsfilter hp**, [TS] **vecrank**

Kirkwood, B. R., [R] **dstdize**, [R] **summarize**

Kish, L., [P] **_robust**, [R] **loneway**, [SVY] **survey**, [SVY] **estat**, [SVY] **variance estimation**, [U] **20.24 References**

Kitagawa, G., [R] **BIC note**

Kiviet, J. F., [XT] **xtabond**

Klar, J., [R] **logistic postestimation**

Klecka, W. R., [MV] **discrim**, [MV] **discrim lda**

Kleiber, C., [R] **inequality**

Klein, J. P., [ST] **stci**, [ST] **stcox**, [ST] **stcox postestimation**, [ST] **stcrreg**, [ST] **stpower**, [ST] **stpower cox**, [ST] **streg**, [ST] **sts**, [ST] **sts graph**, [ST] **sts test**

Klein, L. R., [R] **reg3**, [R] **regress postestimation time series**

Klein, M., [R] **binreg**, [R] **clogit**, [R] **logistic**, [R] **lrtest**, [R] **mlogit**, [R] **ologit**, [XT] **xtgee**

Kleinbaum, D. G., [R] **binreg**, [R] **clogit**, [R] **logistic**, [R] **lrtest**, [R] **mlogit**, [R] **ologit**, [ST] **epitab**, [XT] **xtgee**

Kleiner, B., [G-2] **graph box**, [G-2] **graph matrix**, [G-3] *by_option*, [R] **diagnostic plots**, [R] **lowess**, [U] **1.4 References**

Kleinman, K. P., [MI] **intro substantive**

Klema, V. C., [P] **matrix symeigen**

Kline, R. B., [SEM] **References**

Kluegel, J. R., [SEM] **References**

Kmenta, J., [R] **eivreg**, [R] **ivregress**, [R] **regress**, [TS] **arch**, [TS] **prais**, [TS] **rolling**, [XT] **xtpcse**

Knook, D. L., [MI] **intro substantive**, [MI] **mi impute**, [MI] **mi impute chained**, [MI] **mi impute monotone**

Knuth, D., [D] **functions**

Koch, G. G., [R] **anova**, [R] **kappa**, [R] **vwls**, [SVY] **svy: tabulate twoway**

Koehler, A. B., [TS] **tssmooth**, [TS] **tssmooth dexponential**, [TS] **tssmooth exponential**, [TS] **tssmooth hwinters**, [TS] **tssmooth shwinters**

Koehler, K. J., [R] **diagnostic plots**

Koenker, R., [R] **qreg**, [R] **regress postestimation**

Kohler, U., [D] **input**, [G-2] **graph twoway rbar**, [MV] **biplot**, [R] **kdensity**, [R] **regress**, [R] **regress postestimation**

Kohn, R., [TS] **arima**

Kolenikov, S., [MV] **factor**, [SVY] **svy bootstrap**, [SVY] **variance estimation**

Kolev, G. I., [P] **scalar**, [U] **11.7 References**

Kolmogorov, A. N., [R] **ksmirnov**

Kontopantelis, E., [R] **meta**

Koopman, S. J., [TS] **ucm**

Koopmans, T. C., [R] **ivregress**

Korin, B. P., [MV] **mvtest**

Korn, E. L., [R] **margins**, [R] **ml**, [R] **test**, [SVY] **survey**, [SVY] **direct standardization**, [SVY] **estat**, [SVY] **svy**, [SVY] **svy estimation**, [SVY] **svy postestimation**, [SVY] **svy: tabulate twoway**, [SVY] **variance estimation**, [XT] **xtmixed**

Kotz, S., [D] **functions**, [R] **inequality**, [R] **ksmirnov**, [R] **nbreg**, [R] **nlogit**, [R] **poisson**, [U] **1.4 References**

Krakauer, H., [ST] **ltable**

Krall, J. M., [ST] **stpower cox**

Kramer, C. Y., [MV] **mvtest**, [MV] **mvtest means**

Krauss, N., [SVY] **estat**, [SVY] **subpopulation estimation**, [SVY] **svy bootstrap**, [SVY] **svy estimation**

Kreidberg, M. B., [ST] **epitab**

Kreuter, F., [R] **kdensity**, [R] **regress**, [R] **regress postestimation**, [SVY] **survey**

Krishnaiah, P. R., [MV] **mvtest**

Krishnamoorthy, K., [MV] **mvtest**, [MV] **mvtest means**

Kroeber, A. L., [MV] *measure_option*

Kronecker, L., [M-2] **op_kronecker**

Kroner, K. F., [TS] **arch**

Krus, D. J., [MV] **canon postestimation**

Krushelnytskyy, B., [R] **inequality**, [R] **qreg**

Kruskal, J. B., [MV] **mds**, [MV] **mds postestimation**, [MV] **mdslong**, [MV] **mdsmat**, [MV] **Glossary**

Kruskal, W. H., [R] **kwallis**, [R] **ranksum**, [R] **spearman**, [R] **tabulate twoway**

Kshirsagar, A. M., [MV] **discrim lda**, [MV] **pca**

Kuehl, R. O., [R] **anova**, [R] **contrast**, [R] **oneway**

Kuh, E., [R] **estat**, [R] **regress postestimation**, [R] **regress postestimation**, [U] **18.13 References**

Kulczynski, S., [MV] *measure_option*

Kumbhakar, S. C., [R] **frontier**, [XT] **xtfrontier**

Kung, D. S., [R] **qreg**

Künsch, H. R., [U] **20.24 References**

Kupper, L. L., [ST] **epitab**

Kutner, M. H., [R] **pkcross**, [R] **pkequiv**, [R] **pkshape**, [R] **regress postestimation**

Kwiatkowski, D., [XT] **xtunitroot**

L

Lachenbruch, P. A., [MV] **discrim**, [MV] **discrim estat**, [MV] **discrim lda**, [R] **diagnostic plots**

Lachin, J. M., [ST] **stpower**, [ST] **stpower cox**, [ST] **stpower exponential**, [U] **1.4 References**

Lacy, M. G., [R] **permute**

Lafontaine, F., [R] **boxcox**

Lahiri, K., [R] **tobit**, [XT] **xtgls**

Lai, K. S., [TS] **dfgls**

Lai, S., [R] **exlogistic**

Laird, N. M., [MI] **intro substantive**, [MI] **mi impute mvn**, [R] **expoisson**, [XT] **xtmelogit**, [XT] **xtmepoisson**, [XT] **xtmixed**

Lakatos, E., [ST] **stpower**, [ST] **stpower exponential**, [ST] **stpower logrank**

Lal, R., [D] **functions**

Lambert, D., [R] **zip**

Lambert, P. C., [R] **poisson**, [ST] **stcox**, [ST] **stcrreg**, [ST] **stptime**, [ST] **streg**

LaMotte, L. R., [XT] **xtmixed**

Lan, K. K. G., [ST] **stpower**, [ST] **stpower exponential**, [ST] **stpower logrank**

Lance, G. N., [MV] **cluster**

Landau, S., [MV] **cluster**, [MV] **cluster stop**

Landis, J. R., [R] **kappa**

Lane, M. A., [SVY] **survey**, [SVY] **svy estimation**

Lane, P. W., [R] **margins**

Lange, K., [R] **qreg**

Lange, S. M., [ST] **stcrreg**

Langford, I. H., [XT] **xtmepoisson**

Langholz, B., [ST] **sttocc**

Laplace, P. S., [R] **regress**

Larrimore, J., [MI] **intro substantive**

Larsen, W. A., [R] **regress postestimation**

Lash, T. L., [R] **ci**, [R] **glogit**, [R] **poisson**, [ST] **epitab**

Latouche, A., [ST] **stcrreg**

Laurent, S., [TS] **mgarch**

Lauritsen, J. M., [D] **labelbook**, [D] **list**

Lauritzen, S. L., [R] **summarize**

Lavori, P. W., [ST] **stpower**, [ST] **stpower cox**

Lawless, J. F., [ST] **ltable**, [ST] **stpower**

Lawley, D. N., [MV] **canon**, [MV] **factor**, [MV] **factor postestimation**, [MV] **manova**, [MV] **mvtest**, [MV] **mvtest correlations**, [MV] **pca**

Layard, R., [XT] **xtabond**, [XT] **xtdpd**, [XT] **xtdpdsys**, [XT] **xtivreg**

Ledermann, W., [M-5] **schurd( )**

Ledolter, J., [TS] **tssmooth**, [TS] **tssmooth dexponential**, [TS] **tssmooth exponential**, [TS] **tssmooth hwinters**, [TS] **tssmooth shwinters**

Lee, E. S., [R] **dstdize**

Lee, E. T., [R] **roccomp**, [R] **rocfit**, [R] **roctab**, [ST] **streg**

Lee, J. C., [MV] **mvtest**

Lee, J. W., [XT] **xtmixed**

Lee, K. J., [MI] **intro substantive**, [MI] **mi impute**

Lee, K. L., [ST] **stcox postestimation**

Lee, L. F., [XT] **xtreg**

Lee, P., [ST] **streg**

Lee, T.-C., [R] **estat**, [R] **glogit**, [R] **ivregress**, [R] **ivregress postestimation**, [R] **logit**, [R] **probit**, [R] **regress postestimation**, [R] **test**, [TS] **arch**, [TS] **prais**, [XT] **xtgls**, [XT] **xtpcse**,

Lee, T.-C., *continued*
  [XT] **xtrc**, [XT] **xtreg**
Lee, W. C., [R] **roctab**
Leese, M., [MV] **cluster**, [MV] **cluster stop**
Legendre, A.-M., [R] **regress**
Lehmann, E. L., [R] **oneway**
Leisenring, W., [ST] **stcrreg**
Lemeshow, S., [R] **clogit**, [R] **clogit postestimation**,
  [R] **glm**, [R] **glogit**, [R] **lincom**, [R] **logistic**,
  [R] **logistic postestimation**, [R] **logit**,
  [R] **logit postestimation**, [R] **lrtest**,
  [R] **mlogit**, [R] **predictnl**, [R] **stepwise**,
  [ST] **stcox**, [ST] **stpower**, [ST] **stpower**
  cox, [ST] **streg**, [SVY] **survey**, [SVY] **estat**,
  [SVY] **poststratification**, [XT] **xtgee**
Leonard, M., [XT] **xtgee**
Leone, F. C., [D] **functions**
Lepkowski, J. M., [MI] **intro substantive**, [MI] **mi**
  impute, [MI] **mi impute chained**, [MI] **mi**
  impute logit, [MI] **mi impute mlogit**, [MI] **mi**
  impute monotone, [MI] **mi impute ologit**,
  [MI] **mi impute poisson**, [MI] **mi impute**
  truncreg
Leroy, A. M., [R] **qreg**, [R] **regress postestimation**,
  [R] **rreg**
Lesaffre, E., [MV] **discrim logistic**
LeSage, G., [ST] **stcrreg**
Leser, C. E. V., [TS] **tsfilter**, [TS] **tsfilter hp**
Levene, H., [R] **sdtest**
Levin, A., [XT] **xtunitroot**
Levin, B., [R] **dstdize**, [R] **kappa**, [R] **sampsi**,
  [ST] **epitab**
Levin, W., [ST] **stcrreg**, [ST] **stcrreg postestimation**
Levinsohn, J., [R] **frontier**
Levy, D., [R] **sunflower**
Levy, M., [MI] **intro substantive**, [MI] **mi impute**
Levy, P. S., [SVY] **survey**, [SVY] **poststratification**
Lewis, D., [MI] **mi estimate**
Lewis, H. G., [R] **heckman**
Lewis, I. G., [R] **binreg**
Lewis, J. D., [R] **fracpoly**
Lexis, W. H., [ST] **stsplit**
Leyland, A. H., [XT] **xtmelogit**, [XT] **xtmepoisson**
Li, G., [R] **rreg**
Li, K.-H., [MI] **intro substantive**, [MI] **mi estimate**,
  [MI] **mi impute mvn**, [MI] **mi test**
Li, N., [MI] **intro substantive**
Li, Q., [XT] **xtivreg**, [XT] **xtreg postestimation**,
  [XT] **xtregar**
Li, W., [R] **pkcross**, [R] **pkequiv**, [R] **pkshape**
Liang, K.-Y., [XT] **xtcloglog**, [XT] **xtgee**, [XT] **xtlogit**,
  [XT] **xtmelogit**, [XT] **xtmepoisson**,
  [XT] **xtmixed**, [XT] **xtnbreg**, [XT] **xtpoisson**,
  [XT] **xtprobit**
Lieberman, O., [TS] **mgarch**
Likert, R. A., [R] **alpha**
Lilien, D. M., [TS] **arch**
Lilienfeld, D. E., [ST] **epitab**

Lim, G. C., [R] **cnsreg**, [R] **regress**, [R] **regress**
  postestimation, [TS] **arch**
Lin, C.-F., [XT] **xtunitroot**
Lin, D. Y., [P] **_robust**, [ST] **stcox**, [ST] **stcrreg**,
  [SVY] **svy estimation**, [U] **20.24 References**
Lin, X., [XT] **xtmelogit**, [XT] **xtmepoisson**
Lincoff, G. H., [MV] **discrim knn**
Linde-Zwirble, W., [D] **functions**
Lindelow, M., [SVY] **svy estimation**, [SVY] **svyset**
Lindley, D. V., [R] **ci**
Lindor, K. D., [ST] **stcrreg**
Lindsey, C., [R] **boxcox**, [R] **lowess**, [R] **regress**
  postestimation, [R] **stepwise**
Lindstrom, M. J., [XT] **xtcloglog**, [XT] **xtgee**,
  [XT] **xtintreg**, [XT] **xtlogit**, [XT] **xtprobit**,
  [XT] **xttobit**
Ling, S., [TS] **mgarch**
Lingoes, J. C., [MV] **mds**, [MV] **mdslong**,
  [MV] **mdsmat**
Linhart, J. M., [D] **format**, [M-5] **mindouble( )**,
  [R] **lpoly**, [ST] **sts**, [U] **13.12 References**
Lipset, S. M., [R] **histogram**
Lipsitz, S. R., [MI] **intro substantive**
Littell, R. C., [XT] **xtmelogit**
Little, R. J. A., [MI] **intro substantive**, [MI] **mi impute**
  mvn, [MI] **mi impute pmm**
Liu, J.-P., [R] **pk**, [R] **pkcross**, [R] **pkequiv**,
  [R] **pkexamine**, [R] **pkshape**, [ST] **stpower**
Liu, Q., [XT] **xtcloglog**, [XT] **xtintreg**, [XT] **xtlogit**,
  [XT] **xtmelogit**, [XT] **xtmepoisson**,
  [XT] **xtpoisson**, [XT] **xtprobit**, [XT] **xttobit**
Liu, T.-P., [SVY] **svy bootstrap**, [SVY] **variance**
  estimation
Ljung, G. M., [TS] **wntestq**
Locke, C. S., [R] **pkequiv**
Loftsgaarden, D. O., [MV] **discrim knn**
Lokshin, M., [R] **heckman**, [R] **heckprob**, [R] **oprobit**
Long, J. S., [D] **codebook**, [D] **label**, [D] **notes**,
  [R] **asroprobit**, [R] **clogit**, [R] **cloglog**,
  [R] **intreg**, [R] **logistic**, [R] **logit**, [R] **mlogit**,
  [R] **mprobit**, [R] **nbreg**, [R] **ologit**, [R] **oprobit**,
  [R] **poisson**, [R] **probit**, [R] **regress**, [R] **regress**
  postestimation, [R] **testnl**, [R] **tnbreg**,
  [R] **tobit**, [R] **tpoisson**, [R] **zinb**, [R] **zip**,
  [U] **12.10 References**, [U] **16.5 References**,
  [U] **20.24 References**
Longley, J. D., [R] **kappa**
Longton, G., [R] **rocfit**, [R] **rocreg**, [R] **rocreg**
  postestimation, [R] **rocregplot**
López-Feldman, A., [R] **inequality**
López-Vizcaíno, M. E., [ST] **epitab**
Lorenz, M. O., [R] **inequality**
Louis, T. A., [R] **tabulate twoway**
Lovell, C. A. K., [R] **frontier**, [XT] **xtfrontier**
Lovie, A. D., [R] **spearman**
Lovie, P., [R] **spearman**
Lu, J. Y., [TS] **prais**
Lucas, H. L., [R] **pkcross**
Luce, R. D., [R] **rologit**

Luckman, B., [MV] **screeplot**
Ludwig, J., [ST] **stcrreg**
Lumley, T. S., [MV] **factor**, [MV] **pca**, [R] **anova**,
    [R] **dstdize**, [R] **oneway**, [ST] **epitab**
Luniak, M., [MV] **biplot**
Lunn, M., [ST] **stcrreg**
Lunt, M., [R] **ologit**, [R] **slogit**
Lurie, M. B., [MV] **manova**
Lütkepohl, H., [M-5] **Dmatrix( )**, [M-5] **Kmatrix( )**,
    [M-5] **Lmatrix( )**, [R] **estat**, [R] **glogit**,
    [R] **ivregress**, [R] **ivregress postestimation**,
    [R] **logit**, [R] **probit**, [R] **regress postestimation**,
    [R] **test**, [TS] **time series**, [TS] **arch**,
    [TS] **dfactor**, [TS] **fcast compute**, [TS] **irf**,
    [TS] **irf create**, [TS] **mgarch dvech**, [TS] **prais**,
    [TS] **sspace**, [TS] **sspace postestimation**,
    [TS] **var intro**, [TS] **var**, [TS] **var svar**,
    [TS] **varbasic**, [TS] **vargranger**, [TS] **varnorm**,
    [TS] **varsoc**, [TS] **varstable**, [TS] **varwle**,
    [TS] **vec intro**, [TS] **vecnorm**, [TS] **vecrank**,
    [TS] **vecstable**, [XT] **xtgls**, [XT] **xtpcse**,
    [XT] **xtrc**, [XT] **xtreg**

M

Ma, G., [R] **roccomp**, [R] **rocfit**, [R] **roctab**
Machin, D., [R] **ci**, [R] **kappa**, [R] **tabulate twoway**,
    [ST] **stpower**, [ST] **stpower cox**, [ST] **stpower
    logrank**
Mack, T. M., [R] **symmetry**
MacKinnon, J. G., [P] **_robust**, [R] **boxcox**, [R] **cnsreg**,
    [R] **gmm**, [R] **intreg**, [R] **ivregress**, [R] **ivregress
    postestimation**, [R] **mlogit**, [R] **nl**, [R] **nlsur**,
    [R] **reg3**, [R] **regress**, [R] **regress postestimation
    time series**, [R] **tobit**, [R] **truncreg**, [TS] **arch**,
    [TS] **arima**, [TS] **dfuller**, [TS] **pperron**,
    [TS] **prais**, [TS] **sspace**, [TS] **varlmar**,
    [TS] **Glossary**, [U] **20.24 References**, [XT] **xtgls**,
    [XT] **xtpcse**
MacLaren, M. D., [D] **functions**
MacMahon, B., [ST] **epitab**
MacRae, K. D., [R] **binreg**
MaCurdy, T. E., [XT] **xthtaylor**
Madans, J. H., [SVY] **survey**, [SVY] **svy estimation**
Madansky, A., [R] **runtest**
Maddala, G. S., [R] **nlogit**, [R] **tobit**, [R] **treatreg**,
    [TS] **vec intro**, [TS] **vec**, [TS] **vecrank**,
    [XT] **xtgls**, [XT] **xtunitroot**
Magnus, J. R., [TS] **var svar**
Magnusson, L. M., [R] **ivprobit**, [R] **ivregress**,
    [R] **ivtobit**
Mahalanobis, P. C., [MV] **discrim lda**, [MV] **hotelling**,
    [MV] **Glossary**
Mair, C. S., [XT] **xtmepoisson**
Mallows, C. L., [R] **regress postestimation**
Manchul, L., [ST] **stcrreg**, [ST] **stcrreg postestimation**
Mandelbrot, B., [TS] **arch**
Mander, A., [R] **anova**, [R] **symmetry**, [ST] **stsplit**
Mangel, M., [TS] **varwle**
Manly, B. F. J., [MV] **discrim qda postestimation**

Mann, H. B., [R] **kwallis**, [R] **ranksum**
Manning, W. G., [R] **heckman**
Manski, C. F., [R] **gmm**
Mantel, H., [SVY] **svy bootstrap**, [SVY] **variance
    estimation**
Mantel, N., [R] **stepwise**, [ST] **epitab**, [ST] **strate**,
    [ST] **sts test**
Maravall, A., [TS] **tsfilter hp**
Marcellino, M., [XT] **xtunitroot**
Marchenko, Y. V., [MI] **intro substantive**, [MI] **mi
    estimate**, [MI] **mi impute**, [R] **anova**,
    [R] **loneway**, [R] **oneway**, [R] **sktest**,
    [ST] **survival analysis**, [ST] **stcox**, [ST] **stcrreg**,
    [ST] **stcrreg postestimation**, [ST] **stdescribe**,
    [ST] **stpower**, [ST] **stpower exponential**,
    [ST] **stpower logrank**, [ST] **streg**, [ST] **stset**,
    [ST] **stsplit**, [ST] **stvary**, [XT] **xtmelogit**,
    [XT] **xtmepoisson**, [XT] **xtmixed**
Marden, J. I., [R] **rologit**
Mardia, K. V., [MI] **mi impute mvn**, [MV] **discrim**,
    [MV] **discrim lda**, [MV] **factor**, [MV] **manova**,
    [MV] **matrix dissimilarity**, [MV] **mds**,
    [MV] **mds postestimation**, [MV] **mdslong**,
    [MV] **mdsmat**, [MV] **mvtest**, [MV] **mvtest
    means**, [MV] **mvtest normality**, [MV] **pca**,
    [MV] **procrustes**, [P] **matrix dissimilarity**
Mark, D. B., [ST] **stcox postestimation**
Markowski, C. A., [R] **sdtest**
Markowski, E. P., [R] **sdtest**
Marks, H. M., [ST] **sts**
Marquardt, D. W., [M-5] **moptimize( )**,
    [M-5] **optimize( )**
Marr, J. W., [ST] **stsplit**
Marsaglia, G., [D] **functions**, [D] **functions**
Marschak, J., [R] **ivregress**
Marsh, H. W., [SEM] **References**
Martin, W., [R] **regress**, [ST] **epitab**
Martínez, M. A., [R] **logistic**
Marubini, E., [ST] **stcrreg**, [ST] **stpower**, [ST] **stpower
    logrank**, [ST] **sts test**
Mascher, K., [R] **rocreg**, [R] **rocreg postestimation**,
    [R] **rocregplot**
Massey, F. J., Jr., [R] **ttest**
Massey, J. T., [R] **marginsplot**, [SVY] **survey**,
    [SVY] **estat**, [SVY] **subpopulation estimation**,
    [SVY] **svy**, [SVY] **svy brr**, [SVY] **svy
    estimation**, [SVY] **svy jackknife**, [SVY] **svy
    postestimation**, [SVY] **svy: tabulate oneway**,
    [SVY] **svy: tabulate twoway**, [SVY] **svydescribe**
Master, I. M., [R] **exlogistic**
Mastrucci, M. T., [R] **exlogistic**
Matthews, J. N. S., [R] **ameans**, [R] **expoisson**,
    [R] **sdtest**
Mátyás, L., [R] **gmm**
Maurer, K., [R] **marginsplot**, [SVY] **survey**,
    [SVY] **estat**, [SVY] **subpopulation estimation**,
    [SVY] **svy**, [SVY] **svy brr**, [SVY] **svy
    estimation**, [SVY] **svy jackknife**, [SVY] **svy
    postestimation**, [SVY] **svy: tabulate oneway**,

Maurer, K., *continued*
  [SVY] **svy: tabulate twoway**, [SVY] **svydescribe**
Maxwell, A. E., [MV] **factor**, [MV] **factor postestimation**, [R] **symmetry**
May, S., [ST] **stcox**, [ST] **stpower**, [ST] **stpower cox**, [ST] **streg**
Mazýa, V., [D] **functions**
McAleer, M., [TS] **mgarch**
McCarthy, P. J., [SVY] **survey**, [SVY] **svy bootstrap**, [SVY] **svy brr**, [SVY] **variance estimation**
McCleary, S. J., [R] **regress postestimation**
McClish, D. K., [R] **rocreg**
McCullagh, P., [R] **binreg**, [R] **binreg postestimation**, [R] **glm**, [R] **glm postestimation**, [R] **ologit**, [R] **rologit**, [XT] *vce_options*, [XT] **xtgee**, [XT] **xtmelogit postestimation**, [XT] **xtmepoisson postestimation**, [XT] **xtpoisson**
McCulloch, C. E., [R] **logistic**, [ST] **stcox**, [XT] **xtmelogit**, [XT] **xtmepoisson**, [XT] **xtmixed**
McCullough, B. D., [TS] **corrgram**
McDonald, A., [XT] **xtmepoisson**
McDonald, J. A., [R] **sunflower**
McDonald, J. F., [R] **tobit**, [R] **tobit postestimation**
McDowell, A., [R] **marginsplot**, [SVY] **survey**, [SVY] **estat**, [SVY] **subpopulation estimation**, [SVY] **svy**, [SVY] **svy brr**, [SVY] **svy estimation**, [SVY] **svy jackknife**, [SVY] **svy postestimation**, [SVY] **svy: tabulate oneway**, [SVY] **svy: tabulate twoway**, [SVY] **svydescribe**
McDowell, A. W., [R] **sureg**, [TS] **arima**
McFadden, D. L., [R] **asclogit**, [R] **asmprobit**, [R] **clogit**, [R] **hausman**, [R] **maximize**, [R] **nlogit**, [R] **suest**
McGilchrist, C. A., [ST] **stcox**, [ST] **streg**
McGill, R., [R] **sunflower**
McGinnis, R. E., [R] **symmetry**
McGuire, T. J., [R] **dstdize**
McKelvey, R. D., [R] **ologit**
McKenney, A., [M-1] **LAPACK**, [M-5] **lapack( )**, [P] **matrix eigenvalues**
McLachlan, G. J., [MV] **discrim**, [MV] **discrim estat**, [MV] **discrim knn**, [MV] **discrim lda**, [XT] **xtmelogit**, [XT] **xtmepoisson**, [XT] **xtmixed**
McLeod, A. I., [TS] **arima**, [TS] **ucm**
McNeil, B. J., [R] **roccomp**, [R] **rocfit**, [R] **rocreg**, [R] **rocreg postestimation**, [R] **rocregplot**, [R] **roctab**
McNeil, D., [R] **poisson**, [ST] **stcrreg**
McNemar, Q., [ST] **epitab**
Mead, R., [M-5] **optimize( )**
Meeusen, W., [R] **frontier**, [XT] **xtfrontier**
Mehta, C. R., [R] **exlogistic**, [R] **exlogistic postestimation**, [R] **expoisson**, [R] **tabulate twoway**
Meier, P., [ST] **stcrreg**, [ST] **stcrreg postestimation**, [ST] **sts**

Meiselman, D., [TS] **arima**
Melly, B., [R] **qreg**
Mendenhall, W., III, [SVY] **survey**
Meng, X.-L., [MI] **intro substantive**, [MI] **mi estimate**, [MI] **mi impute**, [MI] **mi test**
Mensing, R. W., [R] **anova postestimation**
Merryman, S., [XT] **xtunitroot**
Metz, C. E., [R] **logistic postestimation**
Meulders, M., [MI] **intro substantive**, [MI] **mi impute**
Meyer, B. D., [ST] **discrete**
Miao, W., [R] **sdtest**
Michel-Pajus, A., [M-5] **cholesky( )**
Michels, K. M., [R] **anova**, [R] **contrast**, [R] **loneway**, [R] **oneway**, [R] **pwcompare**
Michener, C. D., [MV] *measure_option*
Mickey, M. R., [MV] **discrim estat**
Midthune, D., [SVY] **estat**, [SVY] **svy estimation**
Mielke, P., [R] **brier**
Miettinen, O. S., [ST] **epitab**
Milan, L., [MV] **ca**, [MV] **factor**, [MV] **mca**, [MV] **pca**
Miller, A. B., [R] **kappa**
Miller, H. W., [SVY] **survey**, [SVY] **svy estimation**
Miller, J. I., [TS] **sspace**
Miller, R. G., Jr., [D] **functions**, [R] **diagnostic plots**, [R] **oneway**, [R] **pwcompare**
Milligan, G. W., [MV] **cluster**, [MV] **cluster programming subroutines**, [MV] **cluster stop**
Milliken, G. A., [MV] **manova**, [R] **anova**, [R] **contrast**, [R] **margins**, [R] **pwcompare**, [XT] **xtmelogit**
Milosevic, M., [ST] **stcrreg**, [ST] **stcrreg postestimation**
Miranda, A., [R] **gllamm**, [R] **heckprob**, [R] **ivprobit**, [R] **ivtobit**, [R] **logistic**, [R] **logit**, [R] **nbreg**, [R] **ologit**, [R] **oprobit**, [R] **poisson**, [R] **probit**
Mitchell, C., [R] **exlogistic**
Mitchell, M. N., [D] **data management**, [D] **by**, [D] **egen**, [D] **reshape**, [G-1] **graph intro**, [R] **logistic**, [R] **logistic postestimation**, [R] **logit**, [R] **marginsplot**, [U] **11.7 References**, [U] **13.12 References**, [U] **22.1 References**
Mitchell, W. C., [TS] **tsfilter**, [TS] **tsfilter bk**, [TS] **tsfilter bw**, [TS] **tsfilter cf**, [TS] **tsfilter hp**, [TS] **ucm**
Moeschberger, M. L., [ST] **stci**, [ST] **stcox**, [ST] **stcox postestimation**, [ST] **stcrreg**, [ST] **stpower**, [ST] **stpower cox**, [ST] **streg**, [ST] **sts**, [ST] **sts graph**, [ST] **sts test**
Moffitt, R. A., [R] **tobit**, [R] **tobit postestimation**
Molenberghs, G., [XT] **xtmixed**
Moler, C. B., [P] **matrix symeigen**
Monahan, J. F., [D] **functions**
Monfort, A., [R] **hausman**, [R] **suest**, [R] **test**, [TS] **arima**, [TS] **mgarch ccc**, [TS] **mgarch dcc**, [TS] **mgarch vcc**
Monshouwer, K., [MV] **mvtest**
Monson, R. R., [R] **bitest**, [ST] **epitab**, [ST] **epitab**

Montgomery, D. C., [TS] **tssmooth**, [TS] **tssmooth dexponential**, [TS] **tssmooth exponential**, [TS] **tssmooth hwinters**, [TS] **tssmooth shwinters**

Montoya, D., [R] **rocreg**, [R] **rocreg postestimation**, [R] **rocregplot**

Mood, A. M., [R] **centile**

Moon, H. R., [XT] **xtunitroot**

Mooney, C. Z., [R] **bootstrap**, [R] **jackknife**, [R] **rocreg**, [R] **rocregplot**

Moore, E. H., [M-5] **pinv( )**

Moore, J. B., [TS] **sspace**

Moore, R. J., [D] **functions**

Moran, J. L., [R] **dstdize**

Morgenstern, H., [ST] **epitab**, [ST] **epitab**

Mori, M., [ST] **stcrreg**

Morris, C., [R] **bootstrap**

Morris, J. N., [ST] **stsplit**

Morris, N. F., [R] **binreg**

Morrison, D. F., [MV] **clustermat**, [MV] **discrim lda**, [MV] **discrim logistic**, [MV] **discrim logistic postestimation**, [MV] **manova**

Morrow, A., [ST] **epitab**

Mortimore, P., [MI] **mi estimate**

Mosier, C. I., [MV] **procrustes**

Moskowitz, M., [R] **kappa**

Mosteller, F., [R] **jackknife**, [R] **regress**, [R] **regress postestimation**, [R] **rreg**

Moulton, L. H., [R] **permute**

Muellbauer, J., [R] **nlsur**

Mueller, C. W., [MV] **factor**

Mueller, R. O., [MV] **discrim lda**, [SEM] **References**

Muirhead, R. J., [MV] **pca**

Mulaik, S. A., [MV] **factor**, [MV] **rotate**

Mullahy, J., [R] **gmm**, [R] **zinb**, [R] **zip**

Müller, H.-G., [R] **lpoly**, [ST] **sts graph**

Mundlak, Y., [XT] **xtivreg**, [XT] **xtregar**

Munnell, A., [XT] **xtmixed**

Muro, J., [R] **heckprob**

Murphy, A. H., [R] **brier**

Murphy, J. L., [XT] **xtprobit**

Murphy, R. S., [SVY] **survey**, [SVY] **svy estimation**

Murray, R. M., [XT] **xtmelogit**

Murray-Lyon, I. M., [R] **binreg**

Murrill, W. A., [MV] **discrim knn**

Murtaugh, P. A., [ST] **stcrreg**

Mussolino, M. E., [SVY] **survey**, [SVY] **svy estimation**

Muthén, B., [SEM] **References**

Muñoz, J., [R] **exlogistic**

Myland, J. C., [D] **functions**

N

Nachtsheim, C. J., [R] **pkcross**, [R] **pkequiv**, [R] **pkshape**, [R] **regress postestimation**

Nadarajah, S., [R] **nlogit**

Nadaraya, E. A., [R] **lpoly**

Nagel, R., [MV] **discrim lda**

Nagler, J., [R] **scobit**

Naiman, D. Q., [R] **qreg**

Nannicini, T., [R] **treatreg**

Nardi, G., [ST] **epitab**

Narendranathan, W., [XT] **xtregar**

Narula, S. C., [R] **qreg**

Nash, J. C., [G-2] **graph box**

Nash, J. D., [D] **infile (fixed format)**, [D] **merge**

Naylor, J. C., [XT] **xtcloglog**, [XT] **xtintreg**, [XT] **xtlogit**, [XT] **xtmelogit**, [XT] **xtmepoisson**, [XT] **xtpoisson**, [XT] **xtprobit**, [XT] **xttobit**

Nee, J. C. M., [R] **kappa**

Neely, S. T., [R] **rocreg**, [R] **rocreg postestimation**, [R] **rocregplot**

Neff, R. K., [ST] **epitab**

Neimann, H., [MV] **mdsmat**

Nel, D. G., [MV] **mvtest**, [MV] **mvtest means**

Nelder, J. A., [M-5] **optimize( )**, [R] **binreg**, [R] **binreg postestimation**, [R] **glm**, [R] **glm postestimation**, [R] **margins**, [R] **ologit**, [XT] *vce_options*, [XT] **xtgee**, [XT] **xtmelogit postestimation**, [XT] **xtmepoisson postestimation**, [XT] **xtpoisson**

Nelson, C. R., [R] **ivregress postestimation**

Nelson, D. B., [TS] **arch**, [TS] **arima**, [TS] **mgarch**

Nelson, E. C., [MV] **factor**, [MV] **factor postestimation**, [R] **alpha**, [R] **lincom**, [R] **mlogit**, [R] **mprobit**, [R] **mprobit postestimation**, [R] **predictnl**, [R] **slogit**

Nelson, F. D., [R] **logit**, [R] **probit**

Nelson, W., [ST] **stcrreg postestimation**, [ST] **sts**

Nelson, W. C., [MV] **mvtest correlations**

Neter, J., [R] **pkcross**, [R] **pkequiv**, [R] **pkshape**, [R] **regress postestimation**

Neudecker, H., [TS] **var svar**

Neuhaus, J. M., [XT] **xtcloglog**, [XT] **xtintreg**, [XT] **xtlogit**, [XT] **xtmelogit**, [XT] **xtmepoisson**, [XT] **xtmixed**, [XT] **xtprobit**

Nevels, K., [MV] **procrustes**

Newbold, P., [TS] **arima**, [TS] **vec intro**

Newey, W. K., [R] **glm**, [R] **gmm**, [R] **ivprobit**, [R] **ivregress**, [R] **ivtobit**, [TS] **newey**, [TS] **pperron**, [XT] **xtabond**, [XT] **xtdpd**, [XT] **xtdpdsys**, [XT] **xtunitroot**

Newman, S. C., [R] **poisson**, [ST] **epitab**, [ST] **stcox**, [ST] **sts**

Newson, R., [D] **contract**, [D] **generate**, [D] **statsby**, [R] **centile**, [R] **glm**, [R] **inequality**, [R] **kwallis**, [R] **mkspline**, [R] **ranksum**, [R] **signrank**, [R] **spearman**, [R] **tabulate twoway**, [ST] **stcox postestimation**

Newton, H. J., [R] **kdensity**, [TS] **arima**, [TS] **corrgram**, [TS] **cumsp**, [TS] **dfuller**, [TS] **pergram**, [TS] **wntestb**, [TS] **xcorr**, [XT] **xtgee**

Newton, I., [M-5] **optimize( )**

Newton, M. A., [XT] **xtcloglog**, [XT] **xtgee**, [XT] **xtintreg**, [XT] **xtlogit**, [XT] **xtprobit**, [XT] **xttobit**

Neyman, J., [R] **ci**

Ng, E. S. W., [XT] **xtmelogit**

Ng, S., [TS] **dfgls**

Nicewander, W. A., [R] **correlate**

Nichols, A., [R] **ivregress**, [R] **reg3**, [R] **treatreg**, [XT] **xtmixed**, [XT] **xtrc**, [XT] **xtreg**

Nickell, S. J., [R] **gmm**, [XT] **xtabond**, [XT] **xtdpd**, [XT] **xtdpdsys**, [XT] **xtivreg**, [XT] **xtunitroot**

Nielsen, B., [TS] **varsoc**, [TS] **vec intro**

Nolan, D., [R] **diagnostic plots**

Nordlund, D. J., [MV] **discrim lda**

Norton, S. J., [R] **rocreg**, [R] **rocreg postestimation**, [R] **rocregplot**

Nunnally, J. C., [R] **alpha**

O

O'Connell, P. G. J., [XT] **xtunitroot**

O'Connell, R. T., [TS] **tssmooth**, [TS] **tssmooth dexponential**, [TS] **tssmooth exponential**, [TS] **tssmooth hwinters**, [TS] **tssmooth shwinters**

O'Donnell, C. J., [XT] **xtfrontier**

O'Donnell, O., [SVY] **svy estimation**, [SVY] **svyset**

O'Fallon, W. M., [R] **logit**

Oakes, D., [ST] **ltable**, [ST] **stcox**, [ST] **stcox PH-assumption tests**, [ST] **stpower**, [ST] **streg**, [ST] **sts**

Obstfeld, M., [XT] **xtunitroot**

Ochiai, A., [MV] *measure_option*

Odum, E. P., [MV] **clustermat**

Oehlert, G. W., [R] **nlcom**, [R] **rocreg postestimation**, [R] **rocregplot**

Oh, K.-Y., [XT] **xtunitroot**

Oldham, K. B., [D] **functions**

Olivier, D., [R] **expoisson**

Olkin, I., [MV] **hotelling**, [R] **kwallis**, [TS] **wntestb**

Olsen, M. K., [MI] **intro substantive**

Olshansky, S. J., [ST] **streg**

Olson, J. M., [R] **symmetry**

Ooms, M., [TS] **arfima**

Orcutt, G. H., [TS] **prais**

Ord, J. K., [R] **centile**, [R] **mean**, [R] **proportion**, [R] **qreg**, [R] **ratio**, [R] **summarize**, [R] **total**

Orsini, N., [R] **mkspline**, [ST] **epitab**, [XT] **xtreg**

Osbat, C., [XT] **xtunitroot**

Osterwald-Lenum, M., [TS] **vecrank**

Ostle, B., [R] **anova postestimation**

Ott, R. L., [SVY] **survey**

Over, M., [R] **regress**, [XT] **xtivreg**

P

Pacheco, J. M., [R] **dstdize**

Pagan, A. R., [R] **frontier**, [R] **mvreg**, [R] **regress postestimation**, [R] **sureg**, [TS] **Glossary**, [XT] **xtreg postestimation**

Pagano, M., [R] **dstdize**, [R] **logistic**, [R] **margins**, [R] **sampsi**, [R] **tabulate twoway**, [ST] **ltable**, [ST] **sts**

Paik, M. C., [R] **dstdize**, [R] **kappa**, [R] **sampsi**, [ST] **epitab**

Palma, W., [TS] **arfima**, [TS] **arfima postestimation**

Palta, M., [XT] **xtcloglog**, [XT] **xtgee**, [XT] **xtintreg**, [XT] **xtlogit**, [XT] **xtprobit**, [XT] **xttobit**

Pampel, F. C., [R] **logistic**, [R] **logit**, [R] **probit**

Panis, C., [R] **mkspline**

Pantazis, N., [XT] **xtmixed**

Park, H. J., [P] **_robust**, [R] **regress**, [SVY] **svy: tabulate twoway**

Park, J. Y., [R] **boxcox**, [R] **margins**, [R] **nlcom**, [R] **predictnl**, [R] **rocreg postestimation**, [R] **rocregplot**, [R] **testnl**, [TS] **sspace**, [TS] **vec intro**, [TS] **vec**, [TS] **vecrank**

Parks, W. P., [R] **exlogistic**

Parmar, M. K. B., [ST] **stcox**, [ST] **stpower**, [ST] **stpower cox**, [ST] **streg**

Parner, E. T., [R] **glm**, [ST] **stcox**

Parzen, E., [R] **estat**, [R] **kdensity**

Pasquini, J., [R] **vwls**, [ST] **epitab**

Patel, N. R., [R] **exlogistic**, [R] **exlogistic postestimation**, [R] **expoisson**, [R] **tabulate twoway**

Paterson, L., [XT] **xtmelogit**

Patterson, H. D., [R] **pkcross**

Patterson, K., [XT] **xtunitroot**

Paul, C., [R] **logistic**

Paulsen, J., [TS] **varsoc**, [TS] **vec intro**

Pearce, M. S., [R] **logistic**, [ST] **epitab**

Pearson, E. S., [R] **ci**, [R] **ttest**

Pearson, K., [G-2] **graph twoway histogram**, [MV] **mds**, [MV] *measure_option*, [MV] **pca**, [R] **correlate**, [R] **tabulate twoway**

Peen, C., [MV] **procrustes**

Pendergast, J. F., [XT] **xtcloglog**, [XT] **xtgee**, [XT] **xtintreg**, [XT] **xtlogit**, [XT] **xtprobit**, [XT] **xttobit**

Penrose, R., [M-5] **pinv( )**

Pepe, M. S., [R] **roc**, [R] **roccomp**, [R] **rocfit**, [R] **rocreg**, [R] **rocreg postestimation**, [R] **rocregplot**, [R] **roctab**, [ST] **stcrreg**

Peracchi, F., [R] **regress**, [R] **regress postestimation**

Pérez-Hernández, M. A., [R] **kdensity**

Pérez-Hoyos, S., [R] **lrtest**

Pérez-Santiago, M. I., [ST] **epitab**

Perkins, A. M., [R] **ranksum**

Perrin, E., [MV] **factor**, [MV] **factor postestimation**, [R] **alpha**, [R] **lincom**, [R] **mlogit**, [R] **mprobit**, [R] **mprobit postestimation**, [R] **predictnl**, [R] **slogit**

Perron, P., [TS] **dfgls**, [TS] **pperron**, [TS] **Glossary**

Persson, R., [G-1] **graph intro**

Pesaran, M. H., [XT] **xtunitroot**

Pesarin, F., [R] **tabulate twoway**

Peterson, B., [R] **ologit**

Peterson, W. W., [R] logistic postestimation
Petitclerc, M., [R] kappa
Petkova, E., [R] suest
Peto, J., [ST] sts test
Peto, R., [ST] stcox, [ST] streg, [ST] sts test
Petrin, A., [R] frontier
Pfeffer, R. I., [R] symmetry
Pfeffermann, D., [XT] xtmixed
Phillips, P. C. B., [R] boxcox, [R] margins, [R] nlcom,
    [R] predictnl, [R] regress postestimation
    time series, [R] rocreg postestimation,
    [R] rocregplot, [R] testnl, [TS] pperron,
    [TS] vargranger, [TS] vec intro, [TS] vec,
    [TS] vecrank, [TS] Glossary, [XT] xtunitroot
Piantadosi, S., [P] _robust, [U] 20.24 References
Pickles, A., [MV] cluster dendrogram, [R] gllamm,
    [R] glm, [XT] xtgee, [XT] xtmelogit,
    [XT] xtmepoisson, [XT] xtreg
Pierce, D. A., [TS] wntestq, [XT] xtcloglog,
    [XT] xtintreg, [XT] xtlogit, [XT] xtmelogit,
    [XT] xtmepoisson, [XT] xtpoisson,
    [XT] xtprobit, [XT] xttobit
Pierson, R. A., [XT] xtmixed
Pike, M. C., [R] symmetry, [ST] ltable, [ST] streg
Pillai, K. C. S., [MV] canon, [MV] manova
Pindyck, R. S., [R] biprobit, [R] heckprob
Pinheiro, J. C., [XT] xtmelogit, [XT] xtmelogit
    postestimation, [XT] xtmepoisson,
    [XT] xtmepoisson postestimation,
    [XT] xtmixed, [XT] xtmixed postestimation
Pintilie, M., [ST] stcrreg, [ST] stcrreg postestimation
Pisati, M., [TS] time series
Pischke, J.-S., [R] ivregress, [R] ivregress
    postestimation, [R] qreg, [R] regress,
    [U] 20.24 References
Pitarakis, J.-Y., [TS] vecrank
Pitblado, J. S., [M-5] deriv( ), [M-5] moptimize( ),
    [P] intro, [P] _robust, [R] frontier, [R] lpoly,
    [R] maximize, [R] ml, [ST] sts, [SVY] survey,
    [SVY] ml for svy, [XT] xtfrontier
Plackett, R. L., [R] ameans, [R] regress, [R] rologit,
    [R] summarize, [R] ttest
Playfair, W., [G-2] graph bar, [G-2] graph pie
Plosser, C. I., [TS] vecrank
Plummer, W. D., Jr., [R] sunflower, [ST] epitab
Pocock, S. J., [R] sampsi
Poi, B. P., [M-5] deriv( ), [M-5] moptimize( ),
    [P] intro, [P] _robust, [R] bootstrap, [R] bstat,
    [R] frontier, [R] ivregress, [R] ivregress
    postestimation, [R] maximize, [R] ml, [R] nl,
    [R] nlsur, [R] reg3, [SVY] survey, [SVY] ml
    for svy, [XT] xtfrontier, [XT] xtrc
Poirier, D. J., [R] biprobit
Poisson, S. D., [R] poisson
Pollock, D. S. G., [TS] tsfilter, [TS] tsfilter bk,
    [TS] tsfilter bw, [TS] tsfilter cf, [TS] tsfilter hp
Pollock, P. H., III, [R] histogram
Ponce de Leon, A., [R] roccomp, [R] roctab
Porter, T. M., [R] correlate

Portes, A., [SEM] References
Posten, H. O., [D] functions
Powell, M. J. D., [M-5] optimize( )
Powers, D. A., [R] logit, [R] probit
Prais, S. J., [TS] prais
Preece, D. A., [R] ttest
Pregibon, D., [R] glm, [R] linktest, [R] logistic,
    [R] logistic postestimation, [R] logit, [R] logit
    postestimation
Prentice, R. L., [ST] discrete, [ST] ltable, [ST] stcox,
    [ST] stcox PH-assumption tests, [ST] stcox
    postestimation, [ST] streg, [ST] sts, [ST] sts
    test, [ST] stset, [XT] xtgee
Prescott, E. C., [TS] tsfilter, [TS] tsfilter hp
Press, W. H., [D] functions, [G-2] graph twoway
    contour, [P] matrix symeigen, [R] dydx,
    [R] vwls, [TS] arch, [TS] arima
Priestley, M. B., [TS] psdensity, [TS] tsfilter, [TS] ucm
Prosser, R., [XT] xtmixed
Pryor, D. B., [ST] stcox postestimation
Punj, G. N., [R] rologit
Putter, H., [ST] stcrreg, [ST] stcrreg postestimation

Q

Qaqish, B., [XT] xtgee
Quesenberry, C. P., [MV] discrim knn

R

Rabe-Hesketh, S., [MV] pca, [MV] screeplot,
    [R] gllamm, [R] glm, [R] heckprob,
    [R] ivprobit, [R] ivtobit, [R] logistic,
    [R] logit, [R] nbreg, [R] ologit, [R] oprobit,
    [R] poisson, [R] probit, [XT] xtcloglog,
    [XT] xtgee, [XT] xtintreg, [XT] xtlogit,
    [XT] xtmelogit, [XT] xtmelogit postestimation,
    [XT] xtmepoisson, [XT] xtmepoisson
    postestimation, [XT] xtmixed, [XT] xtpoisson,
    [XT] xtprobit, [XT] xtreg, [XT] xttobit
Raciborski, R., [MV] cluster, [R] poisson, [R] tpoisson
Radmacher, R. D., [ST] stpower
Raftery, A. E., [R] BIC note, [R] estat, [R] glm,
    [SEM] References
Raghunathan, T. E., [MI] intro substantive, [MI] mi
    estimate, [MI] mi impute, [MI] mi impute
    chained, [MI] mi impute logit, [MI] mi impute
    mlogit, [MI] mi impute monotone, [MI] mi
    impute ologit, [MI] mi impute poisson, [MI] mi
    impute truncreg, [MI] mi test
Ramalheira, C., [R] ameans, [ST] ltable
Ramsey, J. B., [R] regress postestimation
Rao, C. R., [MV] factor, [MV] hotelling,
    [MV] manova, [XT] xtmixed
Rao, D. S. P., [XT] xtfrontier
Rao, J. N. K., [SVY] direct standardization,
    [SVY] poststratification, [SVY] svy bootstrap,
    [SVY] svy: tabulate twoway, [SVY] variance
    estimation

Rao, T. R., [MV] *measure_option*

Raphson, J., [M-5] **optimize( )**

Rasbash, J., [XT] **xtmelogit**, [XT] **xtmixed**

Ratcliffe, S. J., [XT] **xtgee**

Ratkowsky, D. A., [R] **nl**, [R] **pk**, [R] **pkcross**

Raudenbush, S. W., [XT] **xtmepoisson**, [XT] **xtmixed**

Ravn, M. O., [TS] **tsfilter**, [TS] **tsfilter hp**

Raykov, T., [SEM] **References**

Rebelo, S. T., [TS] **tsfilter**, [TS] **tsfilter hp**

Redelmeier, D. A., [R] **brier**

Reeves, D., [R] **meta**

Reichenheim, M. E., [R] **kappa**, [R] **roccomp**, [R] **roctab**

Reid, C., [M-5] **Hilbert( )**, [R] **ci**

Reid, N., [ST] **stcox**

Reilly, M., [R] **logistic**, [ST] **epitab**

Reinfurt, K. H., [MV] **mvtest correlations**

Reinsch, C., [P] **matrix symeigen**

Reinsch, C. H., [M-5] **spline3( )**

Reinsel, G. C., [TS] **arfima**, [TS] **arima**, [TS] **corrgram**, [TS] **cumsp**, [TS] **dfuller**, [TS] **pergram**, [TS] **pperron**, [TS] **psdensity**, [TS] **vec intro**, [TS] **xcorr**

Reiter, J. P., [MI] **intro substantive**, [MI] **intro**, [MI] **mi estimate**, [MI] **mi estimate using**, [MI] **mi test**

Relles, D. A., [R] **rreg**

Rencher, A. C., [MV] **biplot**, [MV] **ca**, [MV] **candisc**, [MV] **canon**, [MV] **canon postestimation**, [MV] **cluster**, [MV] **discrim**, [MV] **discrim estat**, [MV] **discrim knn**, [MV] **discrim lda**, [MV] **discrim lda postestimation**, [MV] **discrim logistic**, [MV] **discrim qda**, [MV] **discrim qda postestimation**, [MV] **factor**, [MV] **manova**, [MV] **mca**, [MV] **mvtest**, [MV] **mvtest correlations**, [MV] **mvtest covariances**, [MV] **mvtest means**, [MV] **mvtest normality**, [MV] **pca**, [MV] **screeplot**, [R] **anova postestimation**

Research Triangle Institute, [SVY] **svy: tabulate twoway**

Revankar, N. S., [R] **frontier**, [XT] **xtfrontier**

Richards, D. S. P., [MV] **mvtest means**

Richardson, W., [R] **ttest**

Richter, J. R., [ST] **stpower**

Riffenburgh, R. H., [R] **ksmirnov**, [R] **kwallis**

Riley, A. R., [D] **filefilter**, [D] **list**, [R] **net search**

Rising, B., [D] **functions**, [U] **12.10 References**

Rivers, D., [R] **ivprobit**

Robbins, N. B., [G-2] **graph dot**

Roberson, P. K., [R] **logistic postestimation**

Robins, J. M., [ST] **epitab**

Robins, R. P., [TS] **arch**

Robinson, A., [M-5] **Toeplitz( )**

Robyn, D. L., [G-2] **graph bar**, [G-2] **graph pie**, [G-2] **graph twoway histogram**, [R] **cumul**

Rodgers, J. L., [R] **correlate**

Rodríguez, G., [R] **nbreg**, [R] **poisson**, [XT] **xtmelogit**

Rogers, D. J., [MV] *measure_option*

Rogers, W. H., [D] **egen**, [P] **_robust**, [R] **brier**, [R] **glm**, [R] **heckman**, [R] **lincom**, [R] **mlogit**, [R] **mprobit**, [R] **mprobit postestimation**, [R] **nbreg**, [R] **poisson**, [R] **predictnl**, [R] **qreg**, [R] **regress**, [R] **rocreg**, [R] **rreg**, [R] **sktest**, [R] **slogit**, [R] **suest**, [ST] **stcox PH-assumption tests**, [ST] **stcox postestimation**, [U] **20.24 References**

Rogoff, K., [XT] **xtunitroot**

Rohlf, F. J., [MV] **cluster**, [MV] *measure_option*

Rombouts, J. V. K., [TS] **mgarch**

Romney, A. K., [MV] **ca**

Ronchetti, E. M., [D] **egen**

Ronning, G., [R] **clogit**

Roodman, D., [D] **collapse**, [XT] **xtdpd**, [XT] **xtdpdsys**

Room, T., [TS] **arima**

Rosati, R. A., [ST] **stcox postestimation**

Rose, D. W., [MV] **discrim knn**

Rose, J. M., [R] **nlogit**

Rosen, H. S., [XT] **xtabond**, [XT] **xtdpd**, [XT] **xtdpdsys**

Rosenthal, R., [R] **contrast**

Rosner, B., [R] **sampsi**

Rosnow, R. L., [R] **contrast**

Ross, G. J. S., [R] **nl**

Rossi, P. E., [R] **sureg**

Rossi, S. S., [ST] **stcrreg**

Rothenberg, T. J., [TS] **dfgls**, [TS] **sspace**, [TS] **var svar**, [TS] **vec**, [TS] **Glossary**

Rothkopf, E. Z., [MV] **mdslong**

Rothman, K. J., [R] **ci**, [R] **dstdize**, [R] **glogit**, [R] **poisson**, [ST] **epitab**

Rothstein, H. R., [R] **meta**

Rothwell, S. T., [SVY] **survey**, [SVY] **svy estimation**

Rousseeuw, P. J., [D] **egen**, [MV] **cluster**, [MV] **clustermat**, [MV] **matrix dissimilarity**, [MV] *measure_option*, [P] **matrix dissimilarity**, [R] **qreg**, [R] **regress postestimation**, [R] **rreg**

Rovine, M. J., [R] **correlate**

Roy, S. N., [MV] **canon**, [MV] **manova**

Royall, R. M., [P] **_robust**, [U] **20.24 References**

Royston, P., [D] **list**, [D] **sort**, [G-2] **graph twoway lowess**, [G-2] **graph twoway scatter**, [MI] **intro substantive**, [MI] **intro**, [MI] **mi estimate**, [MI] **mi estimate using**, [MI] **mi export**, [MI] **mi export ice**, [MI] **mi import**, [MI] **mi import ice**, [MI] **mi impute**, [MI] **mi impute chained**, [MI] **mi impute intreg**, [MI] **mi impute monotone**, [MI] **mi impute nbreg**, [MI] **mi predict**, [R] **bootstrap**, [R] **centile**, [R] **cusum**, [R] **diagnostic plots**, [R] **dotplot**, [R] **dydx**, [R] **estat**, [R] **fracpoly**, [R] **fracpoly postestimation**, [R] **glm**, [R] **glm**, [R] **kdensity**, [R] **lnskew0**, [R] **lowess**, [R] **mfp**, [R] **ml**, [R] **nl**, [R] **regress**, [R] **sampsi**, [R] **sktest**, [R] **smooth**, [R] **swilk**, [ST] **epitab**, [ST] **stcox**, [ST] **stcox PH-assumption tests**, [ST] **stpower**, [ST] **stpower cox**, [ST] **streg**

Rubin, D. B., [MI] **intro substantive**, [MI] **mi estimate**, [MI] **mi estimate using**, [MI] **mi impute**, [MI] **mi impute chained**, [MI] **mi impute logit**, [MI] **mi impute monotone**, [MI] **mi impute mvn**, [MI] **mi impute pmm**, [MI] **mi impute regress**, [MI] **mi predict**, [MI] **mi test**, [R] **contrast**, [XT] **xtmixed**

Rubin, H., [R] **ivregress postestimation**

Rubinfeld, D. L., [R] **biprobit**, [R] **heckprob**

Rubinstein, L. V., [ST] **stpower**, [ST] **stpower exponential**, [U] **1.4 References**

Rudebusch, G. D., [R] **ivregress postestimation**

Runkle, D. E., [TS] **arch**

Ruppert, D., [R] **boxcox**, [R] **rreg**, [XT] **xtmixed**

Rush, M., [D] **egen**

Russell, P. F., [MV] *measure_option*

Rutherford, E., [R] **poisson**

Rutherford, M. J., [R] **poisson**, [ST] **stptime**

Ruud, P. A., [R] **gmm**, [R] **rologit**, [R] **suest**

Ryan, P., [D] **egen**, [D] **pctile**, [U] **11.7 References**

Ryan, T. P., [R] **qc**

Ryzhik, I. M., [TS] **arfima**

S

Saikkonen, P., [TS] **vec intro**, [TS] **vecrank**

Sakamoto, Y., [R] **BIC note**

Salgado-Ugarte, I. H., [R] **kdensity**, [R] **lowess**, [R] **smooth**

Salim, A., [R] **logistic**, [ST] **epitab**

Salvador, M., [TS] **vecrank**

Samaniego, F. J., [TS] **varwle**

Sammon, J. W., Jr., [MV] **mds**, [MV] **mdslong**, [MV] **mdsmat**, [MV] **Glossary**

Sammons, P., [MI] **mi estimate**

Sampson, A. R., [MV] **hotelling**

Sanders, F., [R] **brier**

Santner, T. J., [ST] **stpower**, [ST] **stpower exponential**, [U] **1.4 References**

Santos Silva, J. M. C., [R] **gmm**

Sarabia, J. M., [MI] **intro substantive**, [MI] **mi impute chained**

Sarafidis, V., [XT] **xtreg**

Sargan, J. D., [R] **ivregress postestimation**, [TS] **prais**

Sargent, T. J., [TS] **dfactor**

Särndal, C.-E., [SVY] **variance estimation**

Sasieni, P., [D] **list**, [D] **memory**, [R] **dotplot**, [R] **lowess**, [R] **nptrend**, [R] **smooth**, [ST] **stcox**

Satterthwaite, F. E., [R] **ttest**, [SVY] **variance estimation**

Sauerbrei, W., [R] **bootstrap**, [R] **estat**, [R] **fracpoly**, [R] **mfp**

Savage, I. R., [ST] **sts test**

Savin, N. E., [R] **regress postestimation time series**

Saw, S. L. C., [R] **qc**

Sawa, T., [R] **estat**

Saxl, I., [R] **correlate**

Schaalje, G. B., [R] **anova postestimation**

Schabenberger, O., [XT] **xtmelogit**

Schafer, J. L., [MI] **intro substantive**, [MI] **mi estimate**, [MI] **mi impute**, [MI] **mi impute monotone**, [MI] **mi impute mvn**, [MI] **mi impute truncreg**

Schaffer, C. M., [MV] **cluster**

Schaffer, M. E., [R] **ivregress**, [R] **ivregress postestimation**

Schank, T., [XT] **xtmelogit**, [XT] **xtmepoisson**, [XT] **xtmixed**, [XT] **xtreg**

Scheaffer, R. L., [SVY] **survey**

Scheffé, H., [R] **anova**, [R] **oneway**

Schenker, N., [MI] **intro substantive**, [MI] **mi impute**, [MI] **mi impute pmm**, [MI] **mi impute regress**

Schlesselman, J. J., [R] **boxcox**, [ST] **epitab**

Schlossmacher, E. J., [R] **qreg**

Schmeiser, B. W., [D] **functions**

Schmidt, C. H., [R] **brier**

Schmidt, E., [M-5] **svd( )**

Schmidt, P., [R] **frontier**, [R] **regress postestimation**, [XT] **xtfrontier**, [XT] **xtunitroot**

Schmidt, T. J., [D] **egen**, [TS] **tsfilter**

Schneider, H., [R] **sdtest**

Schneider, W., [TS] **sspace**

Schnell, D., [P] **_robust**, [R] **regress**, [SVY] **svy: tabulate twoway**

Schoenfeld, D. A., [ST] **stcox**, [ST] **stcox postestimation**, [ST] **stpower**, [ST] **stpower cox**, [ST] **stpower exponential**, [ST] **stpower logrank**, [ST] **streg**, [U] **1.4 References**

Schonlau, M., [R] **glm**, [R] **logistic**, [R] **logit**, [R] **poisson**, [R] **regress**

Schuirmann, D. J., [R] **pkequiv**

Schumacher, M., [ST] **stcrreg**

Schumm, L. P., [D] **sort**

Schur, I., [M-5] **schurd( )**

Schwarz, G., [MV] **factor postestimation**, [R] **BIC note**, [R] **estat**, [SEM] **References**

Schwert, G. W., [TS] **dfgls**

Scott, A. J., [SVY] **estat**, [SVY] **svy: tabulate twoway**

Scott, C., [SVY] **estat**, [SVY] **subpopulation estimation**, [SVY] **svy bootstrap**, [SVY] **svy estimation**

Scott, D. W., [R] **kdensity**

Scott, G. B., [R] **exlogistic**

Scotto, M. G., [R] **diagnostic plots**, [ST] **streg**

Searle, S. R., [R] **contrast**, [R] **margins**, [R] **pwcompare**, [R] **pwmean**, [XT] **xtmelogit**, [XT] **xtmepoisson**, [XT] **xtmixed**

Sears, R. R., [ST] **epitab**

Seber, G. A. F., [MV] **biplot**, [MV] **manova**, [MV] **mvtest**, [MV] **mvtest means**, [MV] **mvtest normality**

Seed, P. T., [R] **ci**, [R] **correlate**, [R] **logistic postestimation**, [R] **roccomp**, [R] **roctab**, [R] **sampsi**, [R] **sdtest**, [R] **spearman**

Seidler, J., [R] **correlate**

Self, S. G., [XT] **xtmelogit**, [XT] **xtmepoisson**, [XT] **xtmixed**

Selvin, S., [R] **poisson**, [ST] **ltable**, [ST] **stcox**

Sempos, C. T., [R] dstdize, [ST] epitab, [ST] ltable, [ST] stcox
Semykina, A., [R] inequality, [R] qreg
Seneta, E., [R] correlate, [U] 1.4 References
Senn, S. J., [R] glm, [R] ttest
Sentana, E., [TS] mgarch
Serachitopol, D. M., [ST] sts graph
Serfling, R. J., [TS] irf create
Shah, B. V., [SVY] direct standardization, [SVY] poststratification, [SVY] variance estimation
Shanno, D. F., [M-5] optimize( )
Shao, J., [ST] stpower, [ST] stpower exponential, [SVY] survey, [SVY] svy jackknife, [SVY] variance estimation
Shapiro, S., [ST] epitab
Shapiro, S. S., [R] swilk
Shaposhnikova, T., [D] functions
Shavelson, R. J., [R] alpha
Shea, J., [R] ivregress postestimation
Sheather, S. J., [R] boxcox, [R] lowess, [R] lpoly, [R] regress postestimation, [R] stepwise
Sheldon, T. A., [R] meta
Shepard, R. N., [MV] mds postestimation
Shewhart, W. A., [R] qc
Shiboski, S. C., [R] logistic, [ST] stcox
Shiller, R. J., [R] tobit
Shimizu, M., [R] kdensity, [R] lowess
Shin, Y., [XT] xtunitroot
Shrout, P. E., [R] kappa
Shults, J., [XT] xtgee
Shumway, R. H., [TS] arima
Sibson, R., [MV] cluster
Šidák, Z., [R] correlate, [R] oneway
Silvennoinen, A., [TS] mgarch, [TS] mgarch ccc
Silverman, B. W., [R] kdensity
Silvey, S. D., [R] ologit, [R] oprobit
Simon, R., [ST] stpower
Simonoff, J. S., [R] kdensity, [R] tnbreg, [R] tpoisson
Simor, I. S., [R] kappa
Simpson, T., [M-5] optimize( )
Sims, C. A., [TS] dfactor, [TS] irf create, [TS] var svar, [TS] vec intro, [TS] vec, [TS] vecrank
Singleton, K. J., [R] gmm
Singleton, R., Jr., [SEM] References
Sininger, Y., [R] rocreg, [R] rocreg postestimation, [R] rocregplot
Skinner, C. J., [SVY] survey, [SVY] estat, [SVY] svy estimation, [SVY] variance estimation, [XT] xtmixed
Skovlund, E., [ST] stpower, [ST] stpower cox
Skrondal, A., [R] gllamm, [R] glm, [U] 1.4 References, [XT] xtcloglog, [XT] xtgee, [XT] xtintreg, [XT] xtlogit, [XT] xtmelogit, [XT] xtmelogit postestimation, [XT] xtmepoisson, [XT] xtmepoisson postestimation, [XT] xtmixed, [XT] xtpoisson, [XT] xtprobit, [XT] xttobit

Slaymaker, E., [P] file
Slone, D., [ST] epitab
Smans, M., [XT] xtmepoisson
Smeeton, N. C., [R] ranksum, [R] signrank
Smirnov, N. V., [R] ksmirnov
Smith, A. F. M., [MI] mi impute chained, [XT] xtcloglog, [XT] xtintreg, [XT] xtlogit, [XT] xtmelogit, [XT] xtmepoisson, [XT] xtpoisson, [XT] xtprobit, [XT] xttobit
Smith, B. T., [P] matrix symeigen
Smith, C. A. B., [MV] discrim estat, [MV] discrim qda
Smith, H., [MV] manova, [R] eivreg, [R] oneway, [R] regress, [R] stepwise
Smith, J. M., [R] fracpoly
Smith, R. J., [R] ivprobit
Smith, R. L., [ST] streg
Smith, T. M. F., [SVY] survey
Smith-Vikos, T., [MV] discrim knn
Smullyan, R., [MV] mds
Smythe, B., [ST] sts
Sneath, P. H. A., [MV] measure_option
Snedecor, G. W., [R] ameans, [R] anova, [R] correlate, [R] oneway, [R] ranksum, [R] signrank
Snell, E. J., [R] exlogistic, [R] expoisson, [ST] stcox, [ST] stcox PH-assumption tests, [ST] streg postestimation
Snowden, C. B., [SVY] svy bootstrap, [SVY] variance estimation
Sobel, M. E., [SEM] References
Sokal, R. R., [MV] measure_option
Solenberger, P., [MI] intro substantive, [MI] mi impute, [MI] mi impute chained, [MI] mi impute logit, [MI] mi impute mlogit, [MI] mi impute monotone, [MI] mi impute ologit, [MI] mi impute poisson, [MI] mi impute truncreg
Song, F., [R] meta
Song, S. H., [XT] xtmixed
Soon, T. W., [R] qc
Sörbom, D., [MV] factor postestimation, [SEM] References
Sorensen, D., [M-1] LAPACK, [M-5] lapack( ), [P] matrix eigenvalues
Sørensen, T., [MV] measure_option
Sorrentino, R., [TS] tsfilter, [TS] tsfilter bw
Sosa-Escudero, W., [XT] xtreg, [XT] xtreg postestimation, [XT] xtregar
Sotoca, S., [TS] sspace
Sowell, F., [TS] arfima
Spanier, J., [D] functions
Späth, H., [MV] cluster
Spearman, C., [MV] factor, [R] spearman
Speed, F. M., [R] margins
Speed, T., [R] diagnostic plots
Spence, I., [G-2] graph pie
Sperling, R., [TS] arch, [TS] arima, [TS] dfgls, [TS] wntestq

Spiegelhalter, D. J., [R] **brier**

Spieldman, R. S., [R] **symmetry**

Spitzer, J. J., [R] **boxcox**

Sprent, P., [R] **ranksum**, [R] **signrank**

Sribney, W. M., [P] **matrix mkmat**, [R] **orthog**, [R] **ranksum**, [R] **signrank**, [R] **stepwise**, [R] **test**, [SVY] **estat**, [SVY] **svy postestimation**, [SVY] **svy: tabulate twoway**, [SVY] **svydescribe**

Staelin, R., [R] **rologit**

Stahel, W. A., [D] **egen**

Stahl, D., [MV] **cluster**, [MV] **cluster stop**

Staiger, D., [R] **ivregress postestimation**

Starmer, C. F., [R] **vwls**

Starnes, C. E., [SEM] **References**

Startz, R., [R] **ivregress postestimation**

Stegun, I. A., [D] **functions**, [R] **contrast**, [R] **orthog**, [XT] **xtmelogit**, [XT] **xtmepoisson**

Steichen, T. J., [D] **duplicates**, [R] **kappa**, [R] **kdensity**, [R] **sunflower**

Steiger, W., [R] **qreg**

Stein, C., [R] **bootstrap**

Stephenson, D. B., [MV] **pca**, [R] **brier**

Stepniewska, K. A., [R] **nptrend**

Stern, H. S., [MI] **intro substantive**, [MI] **mi impute mvn**, [MI] **mi impute regress**

Sterne, J. A. C., [MI] **intro**, [R] **dstdize**, [R] **meta**, [R] **summarize**, [ST] **stcox**

Stevens, E. H., [MV] **mvtest**

Stevenson, R. E., [R] **frontier**

Stewart, G. W., [M-5] **svd( )**, [P] **matrix svd**

Stewart, J., [ST] **ltable**

Stewart, M. B., [R] **intreg**, [R] **oprobit**, [R] **tobit**, [XT] **xtprobit**

Stigler, S. M., [R] **ameans**, [R] **ci**, [R] **correlate**, [R] **kwallis**, [R] **qreg**, [R] **regress**

Stillman, S., [R] **ivregress**, [R] **ivregress postestimation**

Stine, R., [R] **bootstrap**

Stock, J. H., [R] **ivregress**, [R] **ivregress postestimation**, [TS] **time series**, [TS] **arch**, [TS] **dfactor**, [TS] **dfgls**, [TS] **irf create**, [TS] **rolling**, [TS] **sspace**, [TS] **var intro**, [TS] **var**, [TS] **var svar**, [TS] **vec intro**, [TS] **vec**, [TS] **vecrank**, [TS] **Glossary**, [XT] **xthtaylor**, [XT] **xtreg**

Stoll, B. J., [ST] **epitab**

Stoll, L., [MI] **mi estimate**

Stolley, P. D., [ST] **epitab**

Storer, B. E., [ST] **stcrreg**

Stork, D. G., [MV] **cluster**, [MV] **cluster stop**

Stoto, M. A., [R] **lv**

Stover, L., [R] **rocreg**, [R] **rocreg postestimation**, [R] **rocregplot**

Strain, D. O., [XT] **xtmixed**

Street, J. O., [R] **rreg**

Stroup, W. W., [XT] **xtmelogit**

Stryhn, H., [R] **regress**, [ST] **epitab**

Stuart, A., [R] **centile**, [R] **mean**, [R] **proportion**, [R] **qreg**, [R] **ratio**, [R] **summarize**, [R] **symmetry**, [R] **tobit**, [R] **total**, [SVY] **survey**

Student, *see* Gosset, W. S.

Stuetzle, W., [R] **sunflower**

Støvring, H., [M-2] **pointers**

Suárez, C., [R] **heckprob**

Sullivan, G., [P] **_robust**, [R] **regress**, [SVY] **svy: tabulate twoway**

Summers, G. F., [SEM] **References**

Summers, R., [XT] **xtunitroot**

Sun, W., [MI] **intro substantive**

Sutton, A. J., [R] **meta**

Svennerholm, A. M., [ST] **epitab**

Swagel, P., [U] **21.6 Reference**

Swamy, P. A. V. B., [XT] **xtivreg**, [XT] **xtrc**, [XT] **xtreg**

Swed, F. S., [R] **runtest**

Sweeting, T. J., [ST] **streg**

Swensson, B., [SVY] **variance estimation**

Swets, J. A., [R] **logistic postestimation**

Sylvester, J. J., [M-5] **svd( )**

Szroeter, J., [R] **regress postestimation**

T

Tabachnick, B. G., [MV] **discrim**, [MV] **discrim lda**

Taka, M. T., [R] **pkcross**

Tamhane, A. C., [R] **oneway**, [R] **sampsi**

Tan, S. B., [ST] **stpower**, [ST] **stpower logrank**

Tan, S. H., [ST] **stpower**, [ST] **stpower logrank**

Tan, W. Y., [P] **_robust**, [U] **20.24 References**

Tanimoto, T. T., [MV] *measure_option*

Taniuchi, T., [R] **kdensity**

Tanner, M. A., [MI] **intro substantive**, [MI] **mi impute mvn**

Tanner, W. P., Jr., [R] **logistic postestimation**

Tanur, J. M., [R] **kwallis**

Tapia, R. A., [R] **kdensity**

Tarlov, A. R., [MV] **factor**, [MV] **factor postestimation**, [R] **alpha**, [R] **lincom**, [R] **mlogit**, [R] **mprobit**, [R] **mprobit postestimation**, [R] **predictnl**, [R] **slogit**

Tarone, R. E., [ST] **epitab**, [ST] **sts test**, [U] **1.4 References**

Taub, A. J., [XT] **xtreg**

Taylor, C., [R] **gllamm**, [R] **glm**, [XT] **xtgee**, [XT] **xtreg**

Taylor, J. M. G., [MI] **intro substantive**, [MI] **mi impute**, [MI] **mi impute pmm**, [MI] **mi impute regress**

Taylor, W. E., [XT] **xthtaylor**

ten Berge, J. M. F., [MV] **procrustes**

ter Bogt, T., [MV] **mvtest**

Teräsvirta, T., [TS] **mgarch**, [TS] **mgarch ccc**

Teukolsky, S. A., [D] **functions**, [G-2] **graph twoway contour**, [P] **matrix symeigen**, [R] **dydx**, [R] **vwls**, [TS] **arch**, [TS] **arima**

Thall, P. F., [XT] **xtmepoisson**
Theil, H., [R] **ivregress**, [R] **reg3**, [TS] **prais**
Therneau, T. M., [ST] **stcox**, [ST] **stcox PH-assumption**
 tests, [ST] **stcox postestimation**, [ST] **stcrreg**
Thiele, T. N., [R] **summarize**
Thomas, D. C., [ST] **sttocc**
Thomas, D. G., [ST] **epitab**
Thomas, D. R., [SVY] **svy: tabulate twoway**
Thompson, B., [MV] **canon postestimation**
Thompson, J. C., [R] **diagnostic plots**
Thompson, J. R., [R] **kdensity**, [R] **poisson**,
 [ST] **stptime**
Thompson, M. L., [R] **rocreg**
Thompson, S. K., [SVY] **survey**
Thompson, W. A., Jr., [XT] **xtmixed**
Thomson, G. H., [MV] **factor postestimation**,
 [MV] **Glossary**
Thorndike, F., [R] **poisson**
Thurstone, L. L., [MV] **rotate**, [R] **rologit**
Tibshirani, R. J., [MV] **discrim knn**, [R] **bootstrap**,
 [R] **qreg**
Tidmarsh, C. E., [R] **fracpoly**
Tierney, L., [XT] **xtmelogit**, [XT] **xtmepoisson**
Tilford, J. M., [R] **logistic postestimation**
Tilling, K., [ST] **stcox**
Timm, N. H., [MV] **manova**
Tippett, L. H. C., [ST] **streg**
Tobías, A., [R] **alpha**, [R] **logistic postestimation**,
 [R] **lrtest**, [R] **poisson**, [R] **roccomp**, [R] **roctab**,
 [R] **sdtest**, [ST] **streg**
Tobin, J., [R] **tobit**
Toeplitz, O., [M-5] **Toeplitz( )**
Toman, R. J., [R] **stepwise**
Tong, H., [R] **estat**
Toplis, P. J., [R] **binreg**
Torgerson, W. S., [MV] **mds**, [MV] **mdslong**,
 [MV] **mdsmat**
Tosetto, A., [R] **logistic**, [R] **logit**
Touloupoulou, T., [XT] **xtmelogit**
Touloumi, G., [XT] **xtmixed**
Train, G. F., [SVY] **survey**, [SVY] **svy sdr**,
 [SVY] **variance estimation**
Train, K. E., [R] **asmprobit**
Trapido, E., [R] **exlogistic**
Trefethen, L. N., [M-5] **svd( )**
Treiman, D. J., [R] **eivreg**, [R] **mlogit**
Trewn, J., [MV] **mds**
Trichopoulos, D., [ST] **epitab**
Trimbur, T. M., [TS] **psdensity**, [TS] **tsfilter**,
 [TS] **tsfilter hp**, [TS] **ucm**
Trivedi, P. K., [R] **asclogit**, [R] **asmprobit**,
 [R] **bootstrap**, [R] **gmm**, [R] **heckman**,
 [R] **intreg**, [R] **ivregress**, [R] **ivregress**
 postestimation, [R] **logit**, [R] **mprobit**,
 [R] **nbreg**, [R] **ologit**, [R] **oprobit**, [R] **poisson**,
 [R] **probit**, [R] **qreg**, [R] **regress**, [R] **regress**
 postestimation, [R] **simulate**, [R] **sureg**,
 [R] **tnbreg**, [R] **tobit**, [R] **tpoisson**,

Trivedi, P. K., *continued*
 [R] **zinb postestimation**, [R] **zip postestimation**,
 [XT] **xt**, [XT] **xtmixed**, [XT] **xtnbreg**,
 [XT] **xtpoisson**
Tsay, R. S., [TS] **varsoc**, [TS] **vec intro**
Tse, Y. K., [TS] **mgarch**, [TS] **mgarch vcc**
Tsiatis, A. A., [R] **exlogistic**, [ST] **stcrreg**
Tsui, A. K. C., [TS] **mgarch**, [TS] **mgarch vcc**
Tu, D., [SVY] **survey**, [SVY] **svy jackknife**,
 [SVY] **variance estimation**
Tufte, E. R., [G-2] **graph bar**, [G-2] **graph pie**,
 [R] **stem**
Tukey, J. W., [D] **egen**, [G-2] **graph box**, [G-2] **graph**
 matrix, [P] **if**, [R] **jackknife**, [R] **ladder**,
 [R] **linktest**, [R] **lv**, [R] **regress**, [R] **regress**
 postestimation, [R] **rreg**, [R] **smooth**,
 [R] **spikeplot**, [R] **stem**, [SVY] **svy jackknife**
Tukey, P. A., [G-2] **graph box**, [G-2] **graph matrix**,
 [G-3] *by_option*, [R] **diagnostic plots**,
 [R] **lowess**, [U] **1.4 References**
Twisk, J. W. R., [XT] **xtgee**, [XT] **xtlogit**, [XT] **xtreg**
Tyler, D. E., [MV] **pca**
Tyler, J. H., [R] **regress**
Tzavalis, E., [XT] **xtunitroot**

U

Uebersax, J. S., [R] **tetrachoric**
Uhlendorff, A., [R] **asmprobit**, [R] **mlogit**, [R] **mprobit**
Uhlig, H., [TS] **tsfilter**, [TS] **tsfilter hp**
University Group Diabetes Program, [R] **glogit**,
 [ST] **epitab**
Upton, G., [U] **1.4 References**
Upward, R., [XT] **xtmelogit**, [XT] **xtmepoisson**,
 [XT] **xtmixed**, [XT] **xtreg**
Ureta, M., [XT] **xtreg**
Uthoff, V. A., [ST] **stpower cox**
Utts, J. M., [R] **ci**

V

Vach, W., [ST] **stcrreg**
Væth, M., [ST] **stpower**, [ST] **stpower cox**
Vail, S. C., [XT] **xtmepoisson**
Valliant, R., [SVY] **survey**
Valman, H. B., [R] **fracpoly**
Valsecchi, M. G., [ST] **stcrreg**, [ST] **stpower**,
 [ST] **stpower logrank**, [ST] **sts test**
van Belle, G., [MV] **factor**, [MV] **pca**, [R] **anova**,
 [R] **dstdize**, [R] **oneway**, [ST] **epitab**
van Buuren, S., [MI] **intro substantive**, [MI] **mi**
 impute, [MI] **mi impute chained**, [MI] **mi**
 impute logit, [MI] **mi impute mlogit**, [MI] **mi**
 impute monotone, [MI] **mi impute ologit**,
 [MI] **mi impute poisson**
Van de Ven, W. P. M. M., [R] **biprobit**, [R] **heckprob**
van den Broeck, J., [R] **frontier**, [XT] **xtfrontier**
van der Ende, J., [MV] **mvtest**
Van der Heijden, P. G. M., [MV] **ca postestimation**

Van der Merwe, C. A., [MV] **mvtest**, [MV] **mvtest means**

van Doorslaer, E., [SVY] **svy estimation**, [SVY] **svyset**

van Dorsselaer, S., [MV] **mvtest**

Van Hoewyk, J., [MI] **intro substantive**, [MI] **mi impute**, [MI] **mi impute chained**, [MI] **mi impute logit**, [MI] **mi impute mlogit**, [MI] **mi impute monotone**, [MI] **mi impute ologit**, [MI] **mi impute poisson**, [MI] **mi impute truncreg**

Van Kerm, P., [MV] **ca**, [P] **postfile**, [R] **inequality**, [R] **kdensity**

Van Loan, C. F., [R] **orthog**, [R] **tetrachoric**, [TS] **arfima**, [TS] **arfima postestimation**

Van Mechelen, I., [MI] **intro substantive**, [MI] **mi impute**

Van Pragg, B. M. S., [R] **biprobit**, [R] **heckprob**

Vandermonde, A.-T., [M-5] **Vandermonde( )**

Varadharajan-Krishnakumar, J., [XT] **xtivreg**

Velleman, P. F., [R] **regress postestimation**, [R] **smooth**

Verardi, V., [R] **correlate**, [R] **rreg**

Verbeke, G., [MI] **intro substantive**, [MI] **mi impute**, [XT] **xtmixed**

Verdurmen, J., [MV] **mvtest**

Vetterling, W. T., [D] **functions**, [G-2] **graph twoway contour**, [P] **matrix symeigen**, [R] **dydx**, [R] **vwls**, [TS] **arch**, [TS] **arima**

Vidmar, S., [R] **ameans**, [ST] **epitab**

Vittinghoff, E., [R] **logistic**, [ST] **stcox**

Vohr, B. R., [R] **rocreg**, [R] **rocreg postestimation**, [R] **rocregplot**

Vollebergh, W. A. M., [MV] **mvtest**

von Bortkewitsch, L., [R] **poisson**

von Eye, A., [R] **correlate**

Von Storch, H., [R] **brier**

Vondráček, J., [R] **correlate**

Vuong, Q. H., [R] **ivprobit**, [R] **zinb**, [R] **zip**

W

Wacholder, S., [R] **binreg**

Wagner, H. M., [R] **qreg**

Wagner, M., [XT] **xtunitroot**

Wagner, T., [MV] **mvtest**

Wagstaff, A., [SVY] **svy estimation**, [SVY] **svyset**

Wainer, H., [G-2] **graph pie**

Wald, A., [TS] **varwle**

Walker, A. J., [D] **functions**, [M-5] **runiform( )**

Walker, A. M., [ST] **epitab**

Walker, S., [ST] **sts test**

Wallgren, A., [G-1] **graph intro**

Wallgren, B., [G-1] **graph intro**

Wallis, W. A., [R] **kwallis**

Walters, S. J., [R] **ci**, [R] **kappa**, [R] **tabulate twoway**

Wand, M. P., [R] **kdensity**, [XT] **xtmixed**

Wang, D., [D] **duplicates**, [R] **ci**, [R] **dstdize**, [R] **prtest**

Wang, H., [ST] **stpower**, [ST] **stpower exponential**

Wang, J.-L., [ST] **sts graph**

Wang, J. W., [ST] **streg**

Wang, Y., [R] **asmprobit**

Wang, Z., [R] **logistic postestimation**, [R] **lrtest**, [R] **stepwise**, [ST] **epitab**

Ward, J. H., Jr., [MV] **cluster**, [MV] **cluster linkage**

Ware, J. H., [ST] **sts test**, [XT] **xtmelogit**, [XT] **xtmepoisson**, [XT] **xtmixed**

Ware, J. E., Jr., [MV] **factor**, [MV] **factor postestimation**, [R] **alpha**, [R] **lincom**, [R] **mlogit**, [R] **mprobit**, [R] **mprobit postestimation**, [R] **predictnl**, [R] **slogit**

Warren, K., [ST] **epitab**

Waterson, E. J., [R] **binreg**

Watson, G. S., [R] **lpoly**, [R] **regress postestimation time series**, [TS] **prais**, [TS] **Glossary**

Watson, M. W., [R] **ivregress**, [TS] **time series**, [TS] **arch**, [TS] **dfactor**, [TS] **dfgls**, [TS] **irf create**, [TS] **rolling**, [TS] **sspace**, [TS] **var intro**, [TS] **var**, [TS] **var svar**, [TS] **vec intro**, [TS] **vec**, [TS] **vecrank**, [XT] **xtreg**

Weber, S., [R] **correlate**

Webster, A. D., [R] **fracpoly**

Wedderburn, R. W. M., [R] **glm**, [XT] **xtgee**

Weeks, D. G., [SEM] **References**

Weesie, J., [D] **generate**, [D] **joinby**, [D] **label**, [D] **label language**, [D] **labelbook**, [D] **list**, [D] **merge**, [D] **mvencode**, [D] **order**, [D] **recode**, [D] **rename**, [D] **reshape**, [D] **sample**, [MV] **ca postestimation**, [MV] **pca**, [P] **matrix define**, [R] **alpha**, [R] **constraint**, [R] **hausman**, [R] **ladder**, [R] **logistic postestimation**, [R] **reg3**, [R] **regress**, [R] **regress postestimation**, [R] **rologit**, [R] **simulate**, [R] **suest**, [R] **sureg**, [R] **tabstat**, [R] **tabulate twoway**, [R] **test**, [R] **tetrachoric**, [ST] **stsplit**, [U] **20.24 References**

Wei, L. J., [P] **_robust**, [ST] **stcox**, [ST] **stcrreg**, [SVY] **svy estimation**, [U] **20.24 References**

Wei, W. W. S., [TS] **psdensity**, [TS] **tsfilter**, [TS] **ucm**, [TS] **Glossary**

Weibull, W., [ST] **streg**

Weisberg, H. F., [R] **summarize**

Weisberg, S., [R] **boxcox**, [R] **regress**, [R] **regress postestimation**

Weiss, J., [MV] **mdsmat**

Weiss, M., [D] **functions**, [G-3] **by_option**, [R] **estimates table**, [U] **13.12 References**

Weisstein, E. W., [R] **rocreg postestimation**

Welch, B. L., [R] **ttest**

Welch, K. B., [XT] **xtmixed**

Weller, S. C., [MV] **ca**

Wellington, J. F., [R] **qreg**

Wells, K. B., [R] **lincom**, [R] **mlogit**, [R] **mprobit**, [R] **mprobit postestimation**, [R] **predictnl**, [R] **slogit**

Welsch, R. E., [R] **estat**, [R] **regress postestimation**, [U] **18.13 References**

Welsh, D., [M-5] **halton( )**

Wernow, J. B., [D] **destring**

West, B. T., [SVY] **survey**, [SVY] **subpopulation estimation**, [XT] **xtmixed**

West, K. D., [R] **glm**, [R] **gmm**, [R] **ivregress**, [TS] **newey**, [TS] **pperron**, [XT] **xtunitroot**

West, S., [ST] **epitab**

West, S. G., [R] **pcorr**

Westfall, R. S., [M-5] **optimize( )**

Westlake, W. J., [R] **pkequiv**

Weyl, H. K. H., [M-5] **svd( )**

Wheaton, B., [SEM] **References**

White, H., [P] **_robust**, [R] **regress**, [R] **regress postestimation**, [R] **rocreg**, [R] **suest**, [TS] **newey**, [TS] **prais**, [U] **20.23 A list of postestimation commands**, [U] **20.24 References**, [XT] **xtivreg**

White, I. R., [MI] **intro substantive**, [MI] **intro**, [MI] **mi estimate**, [MI] **mi estimate using**, [MI] **mi impute**, [MI] **mi impute chained**, [MI] **mi impute monotone**, [MI] **mi predict**, [R] **simulate**, [ST] **sts test**

White, K. J., [R] **boxcox**, [R] **regress postestimation time series**

White, P. O., [MV] **rotate**, [MV] **rotatemat**, [MV] **Glossary**

Whitehead, A., [XT] **xtunitroot**

Whitehouse, E., [R] **inequality**

Whiting, P., [R] **roccomp**, [R] **roctab**, [XT] **xtmelogit**

Whitney, D. R., [R] **kwallis**, [R] **ranksum**

Whittaker, J., [D] **functions**, [MV] **ca**, [MV] **factor**, [MV] **mca**, [MV] **pca**

Wichern, D. W., [MV] **canon**, [MV] **discrim**, [MV] **discrim estat**, [MV] **discrim lda**, [MV] **discrim lda postestimation**, [MV] **mvtest**, [MV] **mvtest correlations**, [MV] **mvtest covariances**, [MV] **mvtest means**

Wichura, M. J., [D] **functions**, [D] **functions**

Widen, J. E., [R] **rocreg**, [R] **rocreg postestimation**, [R] **rocregplot**

Wieand, S., [R] **rocreg**, [R] **rocreg postestimation**

Wiesner, R. H., [ST] **stcrreg**

Wiggins, V., [G-3] *axis_choice_options*

Wiggins, V. L., [R] **regress postestimation**, [R] **regress postestimation time series**, [TS] **arch**, [TS] **arima**, [TS] **sspace**

Wilcox, D. W., [R] **ivregress postestimation**

Wilcox, R. R., [D] **egen**

Wilcoxon, F., [R] **kwallis**, [R] **ranksum**, [R] **signrank**, [ST] **sts test**

Wilde, J., [R] **gmm**

Wilk, M. B., [R] **cumul**, [R] **diagnostic plots**, [R] **swilk**

Wilkinson, J. H., [P] **matrix symeigen**

Wilkinson, L., [ST] **sts**

Wilks, D. S., [R] **brier**

Wilks, S. S., [MV] **canon**, [MV] **hotelling**, [MV] **manova**

Williams, B., [SVY] **survey**

Williams, B. K., [MV] **discrim lda**

Williams, R., [R] **glm**, [R] **ologit**, [R] **oprobit**, [R] **pcorr**, [R] **stepwise**

Williams, T. O., Jr., [SEM] **References**

Williams, W. T., [MV] **cluster**

Wilson, E. B., [MV] **mvtest normality**, [R] **ci**

Wilson, M., [MV] **rotate**

Wilson, S. R., [R] **bootstrap**

Windmeijer, F., [R] **gmm**, [XT] **xtabond**, [XT] **xtdpd**, [XT] **xtdpdsys**

Winer, B. J., [R] **anova**, [R] **contrast**, [R] **loneway**, [R] **oneway**, [R] **pwcompare**

Winsten, C. B., [TS] **prais**

Winter, N. J. G., [G-2] **graph twoway scatter**, [P] **levelsof**, [SVY] **survey**

Winters, P. R., [TS] **tssmooth**, [TS] **tssmooth dexponential**, [TS] **tssmooth exponential**, [TS] **tssmooth hwinters**, [TS] **tssmooth shwinters**

Wish, M., [MV] **mds**, [MV] **mdslong**, [MV] **mdsmat**

Wittes, J., [ST] **stpower**

Wolfe, F., [R] **correlate**, [R] **spearman**

Wolfe, R., [R] **ologit**, [R] **oprobit**, [R] **tabulate twoway**

Wolfinger, R. D., [XT] **xtmelogit**

Wolfowitz, J., [TS] **varwle**

Wolfram, S., [ST] **streg**, [XT] **xtmelogit postestimation**

Wolfson, C., [R] **kappa**

Wolk, A., [ST] **epitab**

Wolpin, K. I., [R] **asmprobit**

Wolter, K. M., [SVY] **survey**, [SVY] **svy brr**, [SVY] **variance estimation**

Wong, W. H., [MI] **intro substantive**, [MI] **mi impute mvn**

Wood, A. M., [MI] **intro substantive**, [MI] **mi estimate**, [MI] **mi estimate using**, [MI] **mi impute**, [MI] **mi impute chained**, [MI] **mi predict**

Wood, F. S., [R] **diagnostic plots**

Woodard, D. E., [MV] **manova**, [R] **contrast**

Wooldridge, J. M., [R] **gmm**, [R] **intreg**, [R] **ivprobit**, [R] **ivregress**, [R] **ivregress postestimation**, [R] **ivtobit**, [R] **margins**, [R] **regress**, [R] **regress postestimation**, [R] **regress postestimation time series**, [R] **tobit**, [SEM] **References**, [TS] **arch**, [TS] **mgarch**, [TS] **mgarch dvech**, [TS] **prais**, [XT] **xt**, [XT] **xtivreg**, [XT] **xtpoisson**, [XT] **xtreg**

Woolf, B., [ST] **epitab**

Working, H., [R] **roccomp**, [R] **rocfit**, [R] **roctab**

Wretman, J., [SVY] **variance estimation**

Wright, J. H., [R] **ivregress**, [R] **ivregress postestimation**, [XT] **xthtaylor**

Wright, J. T., [R] **binreg**

Wright, P. G., [R] **ivregress**

Wu, C. F. J., [R] **qreg**, [SVY] **svy bootstrap**, [SVY] **variance estimation**

Wu, D.-M., [R] **ivregress postestimation**

Wu, P. X., [XT] **xtregar**

Wu, S., [XT] **xtunitroot**

X

Xie, Y., [R] **logit**, [R] **probit**
Xu, J., [R] **cloglog**, [R] **logistic**, [R] **logit**, [R] **mlogit**,
      [R] **ologit**, [R] **oprobit**, [R] **probit**

Y

Yang, K., [MV] **mds**
Yar, M., [TS] **tssmooth**, [TS] **tssmooth dexponential**,
      [TS] **tssmooth exponential**, [TS] **tssmooth**
      hwinters, [TS] **tssmooth shwinters**
Yates, F., [P] **levelsof**
Yates, J. F., [R] **brier**
Yee, T. W., [R] **slogit**
Yellott, J. I., Jr., [R] **rologit**
Yen, S., [ST] **epitab**
Yen, W. M., [R] **alpha**
Yeo, D., [SVY] **svy bootstrap**, [SVY] **variance**
      estimation
Yogo, M., [R] **ivregress**, [R] **ivregress postestimation**,
      [XT] **xthtaylor**
Young, F. W., [MV] **mds**, [MV] **mdslong**,
      [MV] **mdsmat**
Young, G., [MV] **mds**, [MV] **mdslong**, [MV] **mdsmat**
Ypma, T. J., [M-5] **optimize( )**
Yu, J., [MV] **mvtest**, [MV] **mvtest means**
Yue, K., [SVY] **svy bootstrap**, [SVY] **variance**
      estimation
Yule, G. U., [MV] *measure_option*
Yung, W., [SVY] **svy bootstrap**, [SVY] **variance**
      estimation

Z

Zabell, S., [R] **kwallis**
Zakoian, J. M., [TS] **arch**
Zappasodi, P., [MV] **manova**
Zavoina, W., [R] **ologit**
Zeger, S. L., [XT] **xtcloglog**, [XT] **xtgee**, [XT] **xtlogit**,
      [XT] **xtmixed**, [XT] **xtnbreg**, [XT] **xtpoisson**,
      [XT] **xtprobit**
Zelen, M., [R] **ttest**
Zellner, A., [R] **frontier**, [R] **nlsur**, [R] **reg3**, [R] **sureg**,
      [TS] **prais**, [XT] **xtfrontier**
Zelterman, D., [R] **tabulate twoway**
Zhao, L. P., [XT] **xtgee**
Zheng, X., [R] **gllamm**
Zimmerman, F., [R] **regress**
Zirkler, B., [MV] **mvtest**, [MV] **mvtest normality**
Zubin, J., [MV] *measure_option*
Zubkoff, M., [MV] **factor**, [MV] **factor postestimation**,
      [R] **alpha**, [R] **lincom**, [R] **mlogit**, [R] **mprobit**,
      [R] **mprobit postestimation**, [R] **predictnl**,
      [R] **slogit**
Zucchini, W., [R] **rocreg**
Zwiers, F. W., [R] **brier**

Subject index

Symbols

! (not), *see* logical operators
!= (not equal), *see* relational operators
\& (and), *see* logical operators
* abbreviation character, *see* abbreviations
*, clear subcommand, [D] **clear**
* comment indicator, [P] **comments**
? abbreviation character, *see* abbreviations
– abbreviation character, *see* abbreviations
-> operator, [M-2] **struct**
., class, [P] **class**
/* */ comment delimiter, [M-2] **comments**,
    [P] **comments**
// comment indicator, [M-2] **comments**, [P] **comments**
/// comment indicator, [P] **comments**
; delimiter, [P] **#delimit**
< (less than), *see* relational operators
<= (less than or equal), *see* relational operators
== (equality), *see* relational operators
> (greater than), *see* relational operators
>= (greater than or equal), *see* relational operators
~ (not), *see* logical operators
\char'176 abbreviation character, *see* abbreviations
~= (not equal), *see* relational operators
\orbar (or), *see* logical operators
100% sample, [SVY] **Glossary**

A

.a, .b, . . . , .z, *see* missing values
Aalen–Nelson cumulative hazard, *see* Nelson–Aalen
    cumulative hazard
abbrev() function, [D] **functions**, [M-5] **abbrev( )**
abbreviations, [U] **11.1.1 varlist**, [U] **11.4 varlists**,
    [U] **11.2 Abbreviation rules**
    unabbreviating command names, [P] **unabcmd**
    unabbreviating variable list, [P] **syntax**, [P] **unab**
abond, estat subcommand, [XT] **xtabond**
    postestimation, [XT] **xtdpd postestimation**,
    [XT] **xtdpdsys postestimation**
aborting command execution, [U] **9 The Break key**,
    [U] **10 Keyboard use**
about command, [R] **about**
abs() function, [D] **functions**, [M-5] **abs( )**
absolute value dissimilarity measure,
    [MV] *measure_option*
absolute value function, *see* abs() function
absorption in regression, [R] **areg**
ac command, [TS] **corrgram**
accelerated failure-time model, [ST] **streg**,
    [ST] **Glossary**
Access, Microsoft, reading data from, [D] **odbc**,
    [U] **21.4 Transfer programs**

accrual period, [ST] **stpower exponential**, [ST] **stpower**
    logrank, [ST] **Glossary**
accum, matrix subcommand, [P] **matrix accum**
acos() function, [D] **functions**, [M-5] **sin( )**
acosh() function, [D] **functions**, [M-5] **sin( )**
acprplot command, [R] **regress postestimation**
actuarial tables, *see* life tables
add,
    irf subcommand, [TS] **irf add**
    mi subcommand, [MI] **mi add**
    return subcommand, [P] **return**
added lines, $y=x$, [G-2] **graph twoway function**
added-variable plots, [G-2] **graph other**, [R] **regress**
    postestimation
addedlinestyle, [G-4] *addedlinestyle*
addgroup, ssd subcommand, [SEM] **ssd**
adding
    fits, *see* fits, adding
    lines, *see* lines, adding
    text, *see* text, adding
addition, [M-2] **op_arith**, [M-2] **op_colon**
    across observations, [D] **egen**
    across variables, [D] **egen**
    operator, *see* arithmetic operators
addplot() option, [G-3] *addplot_option*
ADF, *see* asymptotic distribution free
adf, *see* sem option method()
adjoint matrix, [M-2] **op_transpose**, [M-5] **conj( )**
adjugate matrix, [M-2] **op_transpose**, [M-5] **conj( )**
adjusted
    Kaplan–Meier survivor function, [ST] **sts**
    margins, [R] **margins**, [R] **marginsplot**
    means, [R] **contrast**, [R] **margins**, [R] **marginsplot**
    partial residual plot, [R] **regress postestimation**
administrative censoring, [ST] **Glossary**
ado,
    clear subcommand, [D] **clear**
    view subcommand, [R] **view**
.ado filename suffix, [U] **11.6 Filenaming conventions**
ado_d, view subcommand, [R] **view**
ado, [R] **net**
    describe command, [R] **net**
    dir command, [R] **net**
ado-files, [M-1] **ado**, [P] **sysdir**, [P] **version**,
    [U] **3.5 The Stata Journal**, [U] **17 Ado-files**,
    [U] **18.11 Ado-files**
    adding comments to, [P] **comments**
    debugging, [P] **trace**
    downloading, *see* files, downloading
    editing, [R] **doedit**
    installing, [R] **net**, [R] **sj**, [R] **ssc**, [U] **17.6 How do**
    I install an addition?
    location of, [R] **which**, [U] **17.5 Where does Stata**
    look for ado-files?
    long lines, [P] **#delimit**, [U] **18.11.2 Comments and**
    long lines in ado-files

ado-files, *continued*
    official, [R] **update**, [U] **28 Using the Internet to keep up to date**
    searching for, [R] **search**, [R] **ssc**
    updating user-written, [R] **adoupdate**
adopath, [P] **sysdir**, [U] **17.5 Where does Stata look for ado-files?**
    + command, [P] **sysdir**
    ++ command, [P] **sysdir**
    – command, [P] **sysdir**
ado-path, [M-5] **adosubdir( )**
adosize, [U] **18.11 Ado-files**
adosize, set subcommand, [R] **set**, [P] **sysdir**
adosubdir macro extended function, [P] **macro**
adosubdir( ) function, [M-5] **adosubdir( )**
ado uninstall command, [R] **net**
adoupdate command, [R] **adoupdate**
AFT, *see* accelerated failure-time model
agglomerative hierarchical clustering methods, [MV] **Glossary**
aggregate
    functions, [D] **egen**
    statistics, dataset of, [D] **collapse**
agreement, interrater, [R] **kappa**
AIC, *see* Akaike information criterion
Akaike information criterion, [R] **BIC note**, [R] **estat**, [R] **estimates stats**, [R] **glm**, [SEM] **estat gof**, [SEM] **example 4**, [SEM] **methods and formulas**, [ST] **streg**, [TS] **varsoc**
algebraic expressions, functions, and operators, [P] **matrix define**, [U] **13 Functions and expressions**
alignment of text, [G-3] *textbox_options*
alignmentstyle, [G-4] *alignmentstyle*
_all, [U] **11.1.1 varlist**
all,
    clear subcommand, [D] **clear**
    update subcommand, [R] **update**
all( ) function, [M-5] **all( )**
all macro extended function, [P] **macro**
allmissing option, *see* sem option allmissing
allof( ) function, [M-5] **all( )**
alpha coefficient, Cronbach's, [R] **alpha**
alpha command, [R] **alpha**
alphabetizing
    observations, [D] **gsort**, [D] **sort**
    variable names, [D] **order**
    variables, [D] **sort**
alphanumeric variables, *see* string variables, parsing, *see* string variables
alternative-specific
    conditional logit (McFadden's choice) model, [R] **asclogit**
    multinomial probit regression, [R] **asmprobit**
    rank-ordered probit regression, [R] **asroprobit**
alternatives, estat subcommand, [R] **asclogit postestimation**, [R] **asmprobit postestimation**, [R] **asroprobit postestimation**

ameans command, [R] **ameans**
analysis of covariance, *see* ANCOVA
analysis of variance, *see* ANOVA
analysis step, [MI] **intro substantive**, [MI] **mi estimate**, *also see* estimation
analysis time, [ST] **Glossary**
analysis-of-variance test of normality, [R] **swilk**
analytic weights, [U] **11.1.6 weight**, [U] **20.22.2 Analytic weights**
anchoring, [SEM] **Glossary**, *see* constraints, normalization
ANCOVA, [R] **anova**
and operator, [U] **13.2.4 Logical operators**
Anderberg coefficient similarity measure, [MV] *measure_option*
angle of text, [G-4] *anglestyle*
anglestyle, [G-4] *anglestyle*
angular similarity measure, [MV] *measure_option*
ANOVA, [R] **anova**, [R] **contrast**, [R] **loneway**, [R] **oneway**
    Kruskal–Wallis, [R] **kwallis**
    plots, [R] **marginsplot**
    repeated measures, [R] **anova**
anova command, [R] **anova**, [R] **anova postestimation**
anova, estat subcommand, [MV] **discrim lda postestimation**
anti, estat subcommand, [MV] **factor postestimation**, [MV] **pca postestimation**
anti-image
    correlation matrix, [MV] **factor postestimation**, [MV] **pca postestimation**, [MV] **Glossary**
    covariance matrix, [MV] **factor postestimation**, [MV] **pca postestimation**, [MV] **Glossary**
any( ) function, [M-5] **all( )**
anycount( ), egen function, [D] **egen**
anymatch( ), egen function, [D] **egen**
anyof( ) function, [M-5] **all( )**
anyvalue( ), egen function, [D] **egen**
A-PARCH, [TS] **arch**
append command, [D] **append**, [U] **22 Combining datasets**
append, mi subcommand, [MI] **mi append**
_append variable, [D] **append**
appending data, [D] **append**, [MI] **mi append**, [U] **22 Combining datasets**
appending rows and columns to matrix, [P] **matrix define**
apply recording, [G-2] **graph play**
approximating Euclidean distances, [MV] **mds postestimation**
AR, [TS] **arch**, [TS] **arfima**, [TS] **arima**, [TS] **dfactor**, [TS] **sspace**, [TS] **ucm**
arbitrary pattern of missing values, [MI] **mi impute chained**, [MI] **mi impute mvn**, [MI] **Glossary**, *also see* pattern of missingness
arccosine, arcsine, and arctangent functions, [D] **functions**

ARCH, *also see* multivariate GARCH
  effects, [TS] **arch**
    testing for, [R] **regress postestimation time series**
  model, [TS] **arch**, [TS] **Glossary**
  postestimation, [TS] **arch postestimation**
arch command, [TS] **arch**, [TS] **arch postestimation**
archlm, estat subcommand, [R] **regress postestimation time series**
area, graph twoway subcommand, [G-2] **graph twoway area**
areas, [G-4] *colorstyle*, *also see* fill, areas, dimming and brightening, *also see* fill, color, setting
areastyle, [G-4] *areastyle*
areg command, [R] **areg**, [R] **areg postestimation**
Arellano–Bond, [XT] **xtdpd**, [XT] **xtdpdsys**
  estimator, [XT] **xtabond**, [XT] **Glossary**
Arellano–Bover estimator, [XT] **xtdpd**, [XT] **xtdpdsys**
ARFIMA
  model, [TS] **arfima**, [TS] **Glossary**
  postestimation, [TS] **arfima postestimation**, [TS] **psdensity**
arfima command, [TS] **arfima**, [TS] **arfima postestimation**
arg() function, [M-5] **sin( )**
args command, [P] **syntax**
args() function, [M-5] **args( )**
arguments,
  program, [M-2] **declarations**, [M-6] **Glossary**
  values returned in, [M-1] **returnedargs**
  varying number, [M-2] **optargs**, [M-5] **args( )**
ARIMA
  model, [TS] **arima**, [TS] **Glossary**
  postestimation, [TS] **arima postestimation**, [TS] **psdensity**
arima command, [TS] **arima**, [TS] **arima postestimation**
arithmetic operators, [M-2] **op_arith**, [M-2] **op_colon** [M-5] **_negate( )** [P] **matrix define**, [U] **13.2.1 Arithmetic operators**
ARMA, [TS] **arch**, [TS] **arfima**, [TS] **arima**, [TS] **sspace**, [TS] **ucm**, [TS] **Glossary**
ARMAX, [TS] **arfima**, [TS] **arima**, [TS] **dfactor**, [TS] **sspace**, [TS] **ucm**, [TS] **Glossary**
array, [M-6] **Glossary**
arrays, class, [P] **class**
.Arrdropall built in class modifier, [P] **class**
.Arrdropel built-in class modifier, [P] **class**
.arrindexof built-in class function, [P] **class**
.arrnels built-in class function, [P] **class**
arrows, [G-2] **graph twoway pcarrow**
.Arrpop built-in class modifier, [P] **class**
.Arrpush built-in class modifier, [P] **class**
as error, display directive, [P] **display**
as input, display directive, [P] **display**
as result, display directive, [P] **display**
as text, display directive, [P] **display**
as txt, display directive, [P] **display**

asarray() function, [M-5] **asarray( )**
asarray_contains() function, [M-5] **asarray( )**
asarray_contents() function, [M-5] **asarray( )**
asarray_create() function, [M-5] **asarray( )**
asarray_elements() function, [M-5] **asarray( )**
asarray_first() function, [M-5] **asarray( )**
asarray_key() function, [M-5] **asarray( )**
asarray_keys() function, [M-5] **asarray( )**
asarray_next() function, [M-5] **asarray( )**
asarray_notfound() function, [M-5] **asarray( )**
asarray_remove() function, [M-5] **asarray( )**
ascategory() option, [G-2] **graph bar**, [G-2] **graph box**, [G-2] **graph dot**
ASCII
  codes, [M-5] **ascii( )**
  text files, writing and reading, [P] **file**
ascii() function, [M-5] **ascii( )**
asclogit command, [R] **asclogit**, [R] **asclogit postestimation**
asin() function, [D] **functions**, [M-5] **sin( )**
asinh() function, [D] **functions**, [M-5] **sin( )**
_asis, display directive, [P] **display**
asis print color mapping, [G-2] **set printcolor**
asmprobit command, [R] **asmprobit**, [R] **asmprobit postestimation**
aspect ratio, [G-3] *aspect_option*
  changing, [G-2] **graph display**
  controlling, [G-2] **graph combine**
asroprobit command, [R] **asroprobit**, [R] **asroprobit postestimation**
assert command, [D] **assert**
assert() function, [M-5] **assert( )**
asserteq() function, [M-5] **assert( )**
assignment, class, [P] **class**
assignment operator, [M-2] **op_assignment**
association tests, *see* tests, association
association, measures of, [R] **tabulate twoway**
associative arrays, [M-5] **asarray( )**
asymmetry, *see* skewness
asymptotic distribution free, [SEM] **intro 3**, [SEM] **methods and formulas**, [SEM] **Glossary**
asyvars option, [G-2] **graph bar**, [G-2] **graph box**, [G-2] **graph dot**
at risk, [ST] **Glossary**
atan() function, [D] **functions**, [M-5] **sin( )**
atan2() function, [D] **functions**, [M-5] **sin( )**
atanh() function, [D] **functions**, [M-5] **sin( )**
at-risk table, [ST] **sts graph**
attributable fraction, [ST] **epitab**, [ST] **Glossary**
attributable proportion, [ST] **epitab**
AUC, [R] **logistic postestimation**, *also see* pk (pharmacokinetic data), *also see* ROC analysis
augmented
  component-plus-residual plot, [R] **regress postestimation**
  partial residual plot, [R] **regress postestimation**
  regression, *see* imputation, perfect prediction

Author Support Program, [U] **3.8.2 For authors**
auto.dta, [U] **1.2.2 Example datasets**
autocode() function, [D] **functions**,
    [U] **25.1.2 Converting continuous variables to
    categorical variables**
autocorrelation, [R] **regress postestimation time
    series**, [TS] **arch**, [TS] **arfima**, [TS] **arima**,
    [TS] **corrgram**, [TS] **dfactor**, [TS] **newey**,
    [TS] **prais**, [TS] **psdensity**, [TS] **sspace**,
    [TS] **ucm**, [TS] **var**, [TS] **varlmar**,
    [TS] **Glossary**, *also see* HAC variance estimate
    dynamic model, [XT] **xtabond**, [XT] **xtdpd**,
        [XT] **xtdpdsys**
    residual, [XT] **xtgee**, [XT] **xtgls**, [XT] **xtpcse**,
        [XT] **xtregar**
    test, [XT] **xtabond**, [XT] **xtabond postestimation**,
        [XT] **xtdpd postestimation**, [XT] **xtdpdsys**,
        [XT] **xtdpdsys postestimation**
automatic print color mapping, [G-2] **set printcolor**
Automation, [P] **automation**
autoregressive
    conditional heteroskedasticity, *see* ARCH
        testing for, [R] **regress postestimation time
        series**
    fractionally integrated moving-average model, *see*
        ARFIMA
    integrated moving average, *see* ARIMA
    model, [TS] **dfactor**, [TS] **psdensity**, [TS] **sspace**,
        [TS] **ucm**
    moving average, *see* ARMA
    process, [TS] **Glossary**, [XT] **xtabond**, [XT] **xtdpd**,
        [XT] **xtdpdsys**, [XT] **Glossary**
autotabgraphs, set subcommand, [R] **set**
available area, [G-3] *region_options*
available-case analysis, [MI] **intro substantive**
average
    marginal effects, [R] **margins**, [R] **marginsplot**
    partial effects (APEs), [R] **margins**, [R] **marginsplot**
    predictions, [R] **margins**, [R] **marginsplot**
    RVI, [MI] **mi estimate**, [MI] **Glossary**
averagelinkage,
    clustermat subcommand, [MV] **cluster linkage**
    cluster subcommand, [MV] **cluster linkage**
average-linkage clustering, [MV] **cluster**,
    [MV] **clustermat**, [MV] **cluster linkage**,
    [MV] **Glossary**
averages, *see* means
avplot and avplots commands, [R] **regress
    postestimation**
[aweight=*exp*] modifier, [U] **11.1.6 weight**,
    [U] **20.22.2 Analytic weights**
axes
    multiple scales, [G-3] *axis_choice_options*
    setting offset between and plot region,
        [G-3] *region_options*
    suppressing, [G-3] *axis_scale_options*
axis
    labeling, [G-3] *axis_label_options*,
        [G-3] *axis_options*

axis, *continued*
    line, look of, [G-3] *axis_scale_options*,
        [G-3] *cat_axis_label_options*,
        [G-3] *cat_axis_line_options*
    log, [G-3] *axis_scale_options*
    overall look, [G-4] *axisstyle*
    range, [G-3] *axis_scale_options*
    reversed, [G-3] *axis_scale_options*
    scale, [G-3] *axis_options*,
        [G-3] *axis_scale_options*
    selection of, [G-3] *axis_choice_options*
    suppressing, [G-3] *axis_scale_options*
    ticking, [G-3] *axis_label_options*
    titling, [G-3] *axis_options*,
        [G-3] *axis_title_options*
        suppressing, [G-3] *axis_title_options*
axisstyle, [G-4] *axisstyle*

B

_b[], [U] **13.5 Accessing coefficients and standard
    errors**
b() function, [D] **functions**
b1title() option, [G-3] *title_options*
b2title() option, [G-3] *title_options*
background color, [G-4] **schemes intro**
    setting, [G-3] *region_options*
balanced data, [XT] **Glossary**
balanced repeated replication, [SVY] *brr_options*,
    [SVY] **svy brr**, [SVY] **variance estimation**,
    [SVY] **Glossary**
band-pass filters, [TS] **tsfilter bk**, [TS] **tsfilter cf**,
    [TS] **Glossary**
bar
    graph subcommand, [G-2] **graph bar**
    graph twoway subcommand, [G-2] **graph twoway
        bar**
bar charts, [G-2] **graph bar**
barbsize option, [G-2] **graph twoway pcarrow**
barlook options, [G-3] *barlook_options*
bars
    labeling, [G-3] *blabel_option*
    look of, [G-3] *barlook_options*
Bartlett's
    bands, [TS] **corrgram**
    periodogram test, [TS] **wntestb**
    scoring, [MV] **factor postestimation**
    test for equal variances, [R] **oneway**
base conversion, [M-5] **inbase( )**
BASE directory, [P] **sysdir**, [U] **17.5 Where does Stata
    look for ado-files?**
base, fvset subcommand, [R] **fvset**
base level, [U] **11.4.3 Factor variables**
base plottypes, [G-3] *advanced_options*
baseline, [ST] **Glossary**
    comparisons, [SEM] **estat gof**, [SEM] **example 4**
    dataset, [ST] **stbase**

baseline, *continued*
  hazard and survivor functions, [ST] **stcox**, [ST] **stcox PH-assumption tests**, [ST] **stcrreg**
  model, [SEM] **estat gof**, [SEM] **example 4**, [SEM] **methods and formulas**, [SEM] **Glossary**
baseline suboption, [G-4] *alignmentstyle*
baseopts option, *see* sem option baseopts( )
basis, orthonormal, [P] **matrix svd**
Battese–Coelli parameterization, [XT] **xtfrontier**
Baxter–King filter, [TS] **tsfilter**, [TS] **tsfilter bk**
Bayes' theorem, [MV] **Glossary**
Bayesian concepts, [MI] **intro substantive**
Bayesian information criterion, [R] **BIC note**, [R] **estat**, [R] **estimates stats**, [R] **glm**, [SEM] **estat gof**, [SEM] **example 4**, [SEM] **methods and formulas**, [TS] **varsoc**
bcal
  check command, [D] **bcal**
  describe command, [D] **bcal**
  dir command, [D] **bcal**
  load command, [D] **bcal**
bcskew0 command, [R] **lnskew0**
Bentler–Raykov squared multiple-correlation coefficient, [SEM] **estat eqgof**
Bentler–Weeks matrices, [SEM] **intro 6**, [SEM] **estat framework**, [SEM] **example 11**, [SEM] **Glossary**
Bentler's invariant pattern simplicity rotation, [MV] **rotate**, [MV] **rotatemat**, [MV] **Glossary**
Berndt–Hall–Hall–Hausman algorithm, [M-5] **moptimize( )**, [M-5] **optimize( )**, [R] **ml**
beta
  coefficients, [R] **regress**
  density,
    central, [D] **functions**
    noncentral, [D] **functions**
  distribution,
    cumulative, [D] **functions**
    cumulative noncentral, [D] **functions**
    inverse cumulative, [D] **functions**
    inverse cumulative noncentral, [D] **functions**
    inverse reverse cumulative, [D] **functions**
    reverse cumulative, [D] **functions**
  function
    complement to incomplete, [D] **functions**
    incomplete, [D] **functions**, [M-5] **normal( )**
betaden( ) function, [D] **functions**, [M-5] **normal( )**
between-cell means and variances, [XT] **xtdescribe**, [XT] **xtsum**
between estimators, [XT] **xtivreg**, [XT] **xtreg**, [XT] **Glossary**
between-imputation variability, [MI] **mi estimate**, [MI] **mi predict**
between matrix, [MV] **Glossary**
BFGS algorithm, [R] **ml**
bgodfrey, estat subcommand, [R] **regress postestimation time series**
BHHH algorithm, [R] **ml**

bias corrected and accelerated, [R] **bootstrap postestimation**, [R] **bstat**
BIC, *see* Bayesian information criterion
bin( ) option, [G-2] **graph twoway histogram**
binary
  files, writing and reading, [P] **file**
  I/O, [M-5] **bufio( )**
  operator, [M-6] **Glossary**
  outcome model, *see* outcomes, binary
  variable imputation, *see* imputation, binary
binomial
  distribution,
    confidence intervals, [R] **ci**
    cumulative, [D] **functions**
    inverse cumulative, [D] **functions**
    inverse reverse cumulative, [D] **functions**
    reverse cumulative, [D] **functions**
  family regression, [R] **binreg**
  probability mass function, [D] **functions**
  probability test, [R] **bitest**
binomial( ) function, [D] **functions**, [M-5] **normal( )**
binomialp( ) function, [D] **functions**, [M-5] **normal( )**
binomialtail( ) function, [D] **functions**, [M-5] **normal( )**
binormal( ) function, [D] **functions**, [M-5] **normal( )**
binreg command, [R] **binreg**, [R] **binreg postestimation**
bioequivalence tests, [R] **pk**, [R] **pkequiv**
biopharmaceutical data, *see* pk (pharmacokinetic data)
biplot command, [MV] **biplot**
biplots, [MV] **biplot**, [MV] **ca postestimation**, [MV] **Glossary**
biprobit command, [R] **biprobit**, [R] **biprobit postestimation**
biquartimax rotation, [MV] **rotate**, [MV] **rotatemat**, [MV] **Glossary**
biquartimin rotation, [MV] **rotate**, [MV] **rotatemat**, [MV] **Glossary**
bitest and bitesti commands, [R] **bitest**
bitmap, [G-3] *png_options*, [G-3] *tif_options*
bivariate normal function, [D] **functions**
bivariate probit regression, [R] **biprobit**, [SVY] **svy estimation**
biweight regression estimates, [R] **rreg**
biyearly( ) function, [U] **25 Working with categorical data and factor variables**
bk, tsfilter subcommand, [TS] **tsfilter bk**
blanks, removing from strings, [D] **functions**
block diagonal covariance, [MV] **mvtest covariances**
block diagonal matrix, [M-5] **blockdiag( )**
block exogeneity, [TS] **vargranger**
blockdiag( ) function, [M-5] **blockdiag( )**
blog, *see* Stata Blog
blogit command, [R] **glogit**, [R] **glogit postestimation**
Blundell–Bond estimator, [XT] **xtdpd**, [XT] **xtdpdsys**
BLUPs, [XT] **Glossary**
bofd( ) function, [D] **datetime business calendars**, [D] **functions**, [M-5] **date( )**

bold font, [G-4] *text*

Bonferroni's multiple-comparison adjustment, *see*
    multiple comparisons, Bonferroni's method

bootstrap
    sampling and estimation, [P] **postfile**, [R] **bootstrap**,
      [R] **bsample**, [R] **bstat**, [R] **qreg**, [R] **rocreg**,
      [R] **simulate**, [SVY] *bootstrap_options*,
      [SVY] **svy bootstrap**, [SVY] **variance**
      estimation, [SVY] **Glossary**
    standard errors, [R] *vce_option*, [XT] *vce_options*

bootstrap_options, [SVY] *bootstrap_options*

bootstrap prefix command, [R] **bootstrap**,
    [R] **bootstrap postestimation**

bootstrap, estat subcommand, [R] **bootstrap**
    postestimation

border around plot region, suppressing,
    [G-3] *region_options*

borders
    misplacement of, [G-3] *added_text_options*
    suppressing, [G-4] *linestyle*
    suppressing around plot region,
      [G-3] *region_options*

Boston College archive, *see* SSC archive

bottom suboption, [G-4] *alignmentstyle*

boundary kernel, [ST] **Glossary**

boundary solution, [MV] **Glossary**

Box–Cox
    power transformations, [R] **lnskew0**
    regression, [R] **boxcox**

box, graph subcommand, [G-2] **graph box**

Box M test, [MV] **mvtest covariances**

box plots, [G-2] **graph box**

boxcox command, [R] **boxcox**, [R] **boxcox**
    postestimation

Box's conservative epsilon, [R] **anova**

bprobit command, [R] **glogit**, [R] **glogit**
    postestimation

break command, [M-2] **break**, [P] **break**

Break key, [U] **9 The Break key**, [U] **16.1.4 Error**
    handling in do-files
    interception, [P] **break**, [P] **capture**

break key processing, [M-5] **setbreakintr( )**

breakkey( ) function, [M-5] **setbreakintr( )**

breakkeyreset( ) function, [M-5] **setbreakintr( )**

Breitung test, [XT] **xtunitroot**

breitung, xtunitroot subcommand, [XT] **xtunitroot**

Breusch–Godfrey test, [R] **regress postestimation time**
    series

Breusch–Pagan/Cook–Weisberg test for
    heteroskedasticity, [R] **regress postestimation**

Breusch–Pagan Lagrange multiplier test, [XT] **xtreg**
    postestimation

Breusch–Pagan test of independence, [R] **mvreg**,
    [R] **sureg**

brier command, [R] **brier**

Brier score decomposition, [R] **brier**

broad type, [M-6] **Glossary**

browse command, [D] **edit**

browse, view subcommand, [R] **view**

Broyden–Fletcher–Goldfarb–Shanno algorithm,
    [M-5] **moptimize( )**, [M-5] **optimize( )**, [R] **ml**

BRR, *see* balanced repeated replication

brr_options, [SVY] *brr_options*

bsample command, [R] **bsample**

bsqreg command, [R] **qreg**, [R] **qreg postestimation**

bstat command, [R] **bstat**

bstyle( ) option, [G-3] *barlook_options*

bufbfmtisnum( ) function, [M-5] **bufio( )**

bufbfmtlen( ) function, [M-5] **bufio( )**

bufbyteorder( ) function, [M-5] **bufio( )**

buffered I/O, [M-5] **bufio( )**

bufget( ) function, [M-5] **bufio( )**

bufio( ) function, [M-5] **bufio( )**

bufmissingvalue( ) function, [M-5] **bufio( )**

bufput( ) function, [M-5] **bufio( )**

build, ssd subcommand, [SEM] **ssd**

building a graph, [G-1] **graph intro**

built-in, class, [P] **class**

built-in variables, [U] **11.3 Naming conventions**,
    [U] **13.4 System variables (_variables)**

bullet symbol, [G-4] *text*

burn-between period, [MI] **mi impute**, [MI] **mi impute**
    chained, [MI] **mi impute mvn**, [MI] **Glossary**

burn-in period, [MI] **mi impute**, [MI] **mi impute**
    chained, [MI] **mi impute mvn**, [MI] **Glossary**

business calendars, [D] **bcal**, [D] **datetime**
    business calendars, [D] **datetime business**
    calendars creation, [M-5] **date( )**, [TS] **intro**,
    [U] **24.7 Business dates and calendars**

business dates, *see* business calendars

Butterworth filter, [TS] **tsfilter**, [TS] **tsfilter bw**

bw, tsfilter subcommand, [TS] **tsfilter bw**

_by( ) function, [P] **byable**

by( ) option, [G-2] **graph bar**, [G-3] *by_option*
    use of legends with, [G-3] *by_option*,
      [G-3] *clegend_option*, [G-3] *legend_options*

by *varlist*: prefix, [D] **by**, [P] **byable**, [U] **11.5 by**
    varlist: construct, [U] **13.7 Explicit**
    subscripting, [U] **27.2 The by construct**

by-graphs, look of, [G-4] *bystyle*

by-groups, [D] **by**, [D] **statsby**, [P] **byable**, [U] **11.5 by**
    varlist: construct

byable( ), [P] **byable**

_byindex( ) function, [P] **byable**

_bylastcall( ) function, [P] **byable**

_byn1( ) function, [P] **byable**

_byn2( ) function, [P] **byable**

bysort *varlist*: prefix, [D] **by**

bystyle, [G-4] *bystyle*

byte (storage type), [D] **data types**, [I] **data types**
    [U] **12.2.2 Numeric storage types**

byteorder( ) function, [D] **functions**,
    [M-5] **byteorder( )**

C

C() function, [M-5] **C( )**
c() function, [M-5] **c( )**
c() pseudofunction, [D] **functions**
c(adopath) c-class value, [P] **creturn**, [P] **sysdir**
c(adosize) c-class value, [P] **creturn**, [P] **sysdir**
c(ALPHA) c-class value, [P] **creturn**
c(alpha) c-class value, [P] **creturn**
c(autotabgraphs) c-class value, [P] **creturn**
c(bit) c-class value, [P] **creturn**
c(born_date) c-class value, [P] **creturn**
c(byteorder) c-class value, [P] **creturn**
c(cformat) c-class value, [P] **creturn**, [R] **set cformat**
c(changed) c-class value, [P] **creturn**
c(checksum) c-class value, [D] **checksum**, [P] **creturn**
c(cmdlen) c-class value, [P] **creturn**
c(console) c-class value, [P] **creturn**
c(copycolor) c-class value, [P] **creturn**
c(current_date) c-class value, [P] **creturn**
c(current_time) c-class value, [P] **creturn**
c(dirsep) c-class value, [P] **creturn**
c(dockable) c-class value, [P] **creturn**
c(dockingguides) c-class value, [P] **creturn**
c(doublebuffer) c-class value, [P] **creturn**
c(dp) c-class value, [D] **format**, [P] **creturn**
c(emptycells) c-class value, [P] **creturn**
c(eolchar) c-class value, [P] **creturn**
c(epsdouble) c-class value, [P] **creturn**
c(epsfloat) c-class value, [P] **creturn**
c(eqlen) c-class value, [P] **creturn**
c(fastscroll) c-class value, [P] **creturn**
c(filedate) c-class value, [P] **creturn**
c(filename) c-class value, [P] **creturn**
c(flavor) c-class value, [P] **creturn**
c(graphics) c-class value, [P] **creturn**
c(httpproxy) c-class value, [P] **creturn**
c(httpproxyauth) c-class value, [P] **creturn**
c(httpproxyhost) c-class value, [P] **creturn**
c(httpproxyport) c-class value, [P] **creturn**
c(httpproxypw) c-class value, [P] **creturn**
c(httpproxyuser) c-class value, [P] **creturn**
c(include_bitmap) c-class value, [P] **creturn**
c(k) c-class value, [P] **creturn**
c(level) c-class value, [P] **creturn**
c(linegap) c-class value, [P] **creturn**
c(linesize) c-class value, [P] **creturn**
c(locksplitters) c-class value, [P] **creturn**
c(logtype) c-class value, [P] **creturn**
c(lstretch) c-class value, [P] **creturn**
c(machine_type) c-class value, [P] **creturn**
c(macrolen) c-class value, [P] **creturn**
c(matacache) c-class value, [P] **creturn**
c(matafavor) c-class value, [P] **creturn**
c(matalibs) c-class value, [P] **creturn**
c(matalnum) c-class value, [P] **creturn**
c(matamofirst) c-class value, [P] **creturn**

c(mataoptimize) c-class value, [P] **creturn**
c(matastrict) c-class value, [P] **creturn**
c(matsize) c-class value, [P] **creturn**
c(max_cmdlen) c-class value, [P] **creturn**
c(max_k_theory) c-class value, [P] **creturn**
c(max_macrolen) c-class value, [P] **creturn**
c(max_matsize) c-class value, [P] **creturn**
c(max_memory) c-class value, [D] **memory**,
   [P] **creturn**
c(max_N_theory) c-class value, [P] **creturn**
c(max_width_theory) c-class value, [P] **creturn**
c(maxbyte) c-class value, [P] **creturn**
c(maxdb) c-class value, [P] **creturn**
c(maxdouble) c-class value, [P] **creturn**
c(maxfloat) c-class value, [P] **creturn**
c(maxint) c-class value, [P] **creturn**
c(maxiter) c-class value, [P] **creturn**
c(maxlong) c-class value, [P] **creturn**
c(maxstrvarlen) c-class value, [P] **creturn**
c(maxvar) c-class value, [D] **memory**, [P] **creturn**
c(memory) c-class value, [P] **creturn**
c(min_matsize) c-class value, [P] **creturn**
c(min_memory) c-class value, [D] **memory**,
   [P] **creturn**
c(minbyte) c-class value, [P] **creturn**
c(mindouble) c-class value, [P] **creturn**
c(minfloat) c-class value, [P] **creturn**
c(minint) c-class value, [P] **creturn**
c(minlong) c-class value, [P] **creturn**
c(mode) c-class value, [P] **creturn**
c(Mons) c-class value, [P] **creturn**
c(Months) c-class value, [P] **creturn**
c(more) c-class value, [P] **creturn**, [P] **more**
c(MP) c-class value, [P] **creturn**
c(N) c-class value, [P] **creturn**
c(namelen) c-class value, [P] **creturn**
c(niceness) c-class value, [D] **memory**, [P] **creturn**
c(noisily) c-class value, [P] **creturn**
c(notifyuser) c-class value, [P] **creturn**
c(odbcmgr) c-class value, [P] **creturn**
c(os) c-class value, [P] **creturn**
c(osdtl) c-class value, [P] **creturn**
c(pagesize) c-class value, [P] **creturn**
c(pformat) c-class value, [P] **creturn**, [R] **set cformat**
c(pi) c-class value, [P] **creturn**
c(pinnable) c-class value, [P] **creturn**
c(playsnd) c-class value, [P] **creturn**
c(printcolor) c-class value, [P] **creturn**
c(processors) c-class value, [P] **creturn**
c(processors_lic) c-class value, [P] **creturn**
c(processors_mach) c-class value, [P] **creturn**
c(processors_max) c-class value, [P] **creturn**
c(pwd) c-class value, [P] **creturn**
c(rc) c-class value, [P] **capture**, [P] **creturn**
c(reventries) c-class value, [P] **creturn**
c(revkeyboard) c-class value, [P] **creturn**
c(rmsg) c-class value, [P] **creturn**, [P] **rmsg**

categorical, *continued*
> variables, [U] **25.1.2 Converting continuous
> variables to categorical variables**

cause-specific hazard, [ST] **Glossary**

cc command, [ST] **epitab**

ccc, mgarch subcommand, [TS] **mgarch ccc**

cchart command, [R] **qc**

cci command, [ST] **epitab**

CD, *see* coefficient of determination

cd command, [D] **cd**

cd, net subcommand, [R] **net**

Cdhms( ) function, [D] **datetime**, [D] **functions**,
> [M-5] **date( )**

cdir, classutil subcommand, [P] **classutil**

ceil( ) function, [D] **functions**, [M-5] **trunc( )**

ceiling function, [D] **functions**

censored, [ST] **Glossary**

censored-normal regression, *see* interval regression

censoring, *see* imputation, interval-censored data

census, [SVY] **Glossary**
> data, [SVY] **survey**, [SVY] **direct standardization**,
> [SVY] **variance estimation**

center suboption, [G-4] *justificationstyle*

centered data, [MV] **Glossary**

centile command, [R] **centile**

centiles, *see* percentiles, displaying, *see* percentiles

central tendency, measures of, *see* means, *see* medians

centroidlinkage,
> clustermat subcommand, [MV] **cluster linkage**
> cluster subcommand, [MV] **cluster linkage**

centroid-linkage clustering, [MV] **cluster**,
> [MV] **clustermat**, [MV] **cluster linkage**,
> [MV] **Glossary**

certainty
> strata, [SVY] **estat**
> units, [SVY] **variance estimation**

certifying data, [D] **assert**, [D] **checksum**,
> [D] **count**, [D] **datasignature**, [D] **inspect**,
> [P] **_datasignature**, [P] **signestimationsample**

certifying mi data are consistent, [MI] **mi update**

cf command, [D] **cf**

cf, tsfilter subcommand, [TS] **tsfilter cf**

CFA, *see* models, confirmatory factor analysis

CFI, *see* comparative fit index

cformat, set subcommand, [R] **set**, [R] **set cformat**

cgraph, irf subcommand, [TS] **irf cgraph**

chained equations, *see* imputation, chained equations

changeeol command, [D] **changeeol**

changing
> data, *see* editing data
> directories, [D] **cd**

_char(#), display directive, [P] **display**

char, [U] **12.8 Characteristics**
> define command, [P] **char**
> list command, [P] **char**
> macro extended function, [P] **macro**
> rename command, [P] **char**

char( ) function, [D] **functions**, [M-5] **ascii( )**

character
> data, *see* string variables
> variables, [D] **infile (free format)**

characteristic roots, [M-5] **eigensystem( )**

characteristics, [P] **char**, [U] **12.8 Characteristics**,
> [U] **18.3.6 Extended macro functions**,
> [U] **18.3.13 Referring to characteristics**

charset, set subcommand, [P] **smcl**

chdir command, [D] **cd**

_chdir( ) function, [M-5] **chdir( )**

chdir( ) function, [M-5] **chdir( )**

check,
> bcal subcommand, [D] **bcal**
> icd9 subcommand, [D] **icd9**
> icd9p subcommand, [D] **icd9**
> ml subcommand, [R] **ml**

checkestimationsample command,
> [P] **signestimationsample**

checking data, [D] **assert**

checkpoint, [D] **snapshot**

checksum, set subcommand, [D] **checksum**, [R] **set**

checksum command, [D] **checksum**

checksums of data, [D] **checksum**, [D] **datasignature**,
> [P] **_datasignature**, [P] **signestimationsample**

cholp command, [R] **help**

chi2( ) function, [D] **functions**, [M-5] **normal( )**

chi2tail( ) function, [D] **functions**, [M-5] **normal( )**

chi-squared
> distribution,
> > cumulative, [D] **functions**
> > cumulative noncentral, [D] **functions**
> > inverse cumulative, [D] **functions**
> > inverse cumulative noncentral, [D] **functions**
> > inverse reverse cumulative, [D] **functions**
> > reverse cumulative, [D] **functions**
> hypothesis test, [R] **hausman**, [R] **lrtest**, [R] **sdtest**,
> > [R] **tabulate twoway**, [R] **test**, [R] **testnl**
> probability plot and quantile plots, [G-2] **graph
> > other**, [R] **diagnostic plots**
> quantile plot, [R] **diagnostic plots**
> test of independence, [R] **tabulate twoway**,
> > [SEM] **methods and formulas**, [ST] **epitab**,
> > [SVY] **svy: tabulate twoway**

Chms( ) function, [D] **datetime**, [D] **functions**,
> [M-5] **date( )**

choice models, [R] **asclogit**, [R] **asmprobit**,
> [R] **asroprobit**, [R] **clogit**, [R] **cloglog**,
> [R] **exlogistic**, [R] **glm**, [R] **glogit**, [R] **heckprob**,
> [R] **hetprob**, [R] **ivprobit**, [R] **logistic**, [R] **logit**,
> [R] **mlogit**, [R] **mprobit**, [R] **nlogit**, [R] **ologit**,
> [R] **oprobit**, [R] **probit**, [R] **rologit**, [R] **scobit**,
> [R] **slogit**, [R] **suest**, [XT] **xtgee**, [XT] **xtlogit**,
> [XT] **xtmelogit**, [XT] **xtprobit**

Cholesky decomposition, [M-5] **cholesky( )**, [P] **matrix
> define**

_cholesky( ) function, [M-5] **cholesky( )**

cholesky( ) function, [D] **functions**, [M-5] **cholesky( )**,
  [P] **matrix define**
Cholesky ordering, [TS] **Glossary**
_cholinv( ) function, [M-5] **cholinv( )**
cholinv( ) function, [M-5] **cholinv( )**
_cholsolve( ) function, [M-5] **cholsolve( )**
cholsolve( ) function, [M-5] **cholsolve( )**
chop( ) function, [D] **functions**
Chow test, [R] **anova**, [R] **contrast**, [R] **lrtest**
Christiano–Fitzgerald filter, [TS] **tsfilter**, [TS] **tsfilter cf**
CI, *see* confidence intervals
ci and cii commands, [R] **ci**
CIF, [ST] **Glossary**, *also see* cumulative incidence
  function
class
  definition, [P] **class**
  instance, [P] **class**
  programming, [P] **class**
  programming utilities, [P] **classutil**
class, [M-2] **class**
class exit command, [P] **class exit**
class programming, [M-6] **Glossary**
classes, [M-2] **class**
classfunctions, estat subcommand, [MV] **discrim
  lda postestimation**
classical scaling, [MV] **Glossary**
classification, *see* cluster analysis, *see* discriminant
  analysis
  data, *see* ROC analysis
  function, [MV] **Glossary**
  interrater agreement, [R] **kappa**
  table, [MV] **Glossary**, [R] **logistic postestimation**
classification, estat subcommand, [R] **logistic
  postestimation**
.classmv built-in class function, [P] **class**
.classname built-in class function, [P] **class**
classtable, estat subcommand, [MV] **discrim
  estat**, [MV] **discrim knn postestimation**,
  [MV] **discrim lda postestimation**, [MV] **discrim
  logistic postestimation**, [MV] **discrim qda
  postestimation**
classutil
  cdir command, [P] **classutil**
  describe command, [P] **classutil**
  dir command, [P] **classutil**
  drop command, [P] **classutil**
  which command, [P] **classutil**
classwide variable, [P] **class**
clean,
  icd9 subcommand, [D] **icd9**
  icd9p subcommand, [D] **icd9**
clear
  * command, [D] **clear**
  ado command, [D] **clear**
  all command, [D] **clear**
  command, [D] **clear**, [M-3] **mata clear**
  mata command, [D] **clear**
  matrix command, [D] **clear**

clear, *continued*
  programs command, [D] **clear**
  results command, [D] **clear**
clear,
  datasignature subcommand, [D] **datasignature**
  ereturn subcommand, [P] **return**, [P] **ereturn**
  _estimates subcommand, [P] **_estimates**
  estimates subcommand, [R] **estimates store**
  fvset subcommand, [R] **fvset**
  ml subcommand, [R] **ml**
  postutil subcommand, [P] **postfile**
  return subcommand, [P] **return**
  serset subcommand, [P] **serset**
  sreturn subcommand, [P] **program**, [P] **return**
  timer subcommand, [P] **timer**
clear option, [U] **11.2 Abbreviation rules**
clearing estimation results, [P] **ereturn**, [P] **_estimates**,
  [R] **estimates store**
clearing memory, [D] **clear**
clegend( ) option, [G-3] *clegend_option*,
  [G-3] *legend_options*
clinical trial, [ST] **stpower**
clip( ) function, [D] **functions**
Clock( ) function, [D] **datetime**, [D] **datetime
  translation**, [D] **functions**, [M-5] **date( )**
clock( ) function, [D] **datetime**, [D] **datetime
  translation**, [D] **functions**, [M-5] **date( )**
clock position, [G-4] *clockposstyle*
clock time, [TS] **tsset**
clogit command, [R] **bootstrap**, [R] **clogit**, [R] **clogit
  postestimation**, [R] **exlogistic**, [R] **rologit**
cloglog command, [R] **cloglog**, [R] **cloglog
  postestimation**
cloglog( ) function, [D] **functions**, [M-5] **logit( )**
clonevar command, [D] **clonevar**
close,
  cmdlog subcommand, [R] **log**
  file subcommand, [P] **file**
  log subcommand, [R] **log**
clstyle( ) option, [G-3] *connect_options*
cluster, [SVY] **survey**, [SVY] **svy estimation**,
  [SVY] **svyset**, [SVY] **variance estimation**,
  [SVY] **Glossary**
cluster, [MV] **cluster**, *also see* sem option vce( )
  averagelinkage command, [MV] **cluster linkage**
  centroidlinkage command, [MV] **cluster linkage**
  completelinkage command, [MV] **cluster linkage**
  delete command, [MV] **cluster programming
    utilities**
  dendrogram command, [MV] **cluster dendrogram**
  dir command, [MV] **cluster utility**
  drop command, [MV] **cluster utility**
  generate command, [MV] **cluster generate**
  kmeans command, [MV] **cluster kmeans and
    kmedians**
  kmedians command, [MV] **cluster kmeans and
    kmedians**
  list command, [MV] **cluster utility**

cluster, *continued*
  measures command, [MV] **cluster programming utilities**
  medianlinkage command, [MV] **cluster linkage**
  notes command, [MV] **cluster notes**
  parsedistance command, [MV] **cluster programming utilities**
  query command, [MV] **cluster programming utilities**
  rename command, [MV] **cluster utility**
  renamevar command, [MV] **cluster utility**
  set command, [MV] **cluster programming utilities**
  singlelinkage command, [MV] **cluster linkage**
  stop command, [MV] **cluster stop**
  use command, [MV] **cluster utility**
  wardslinkage command, [MV] **cluster linkage**
  waveragelinkage command, [MV] **cluster linkage**
cluster analysis, [MV] **cluster**, [MV] **cluster dendrogram**, [MV] **cluster generate**, [MV] **cluster kmeans and kmedians**, [MV] **cluster linkage**, [MV] **cluster stop**, [MV] **cluster utility**, [MV] **Glossary**, [U] **26.23 Multivariate and cluster analysis**
  dendrograms, [MV] **cluster dendrogram**
  dropping, [MV] **cluster utility**
  hierarchical, [MV] **cluster**, [MV] **clustermat**, [MV] **cluster linkage**
  kmeans, [MV] **cluster kmeans and kmedians**
  kmedians, [MV] **cluster kmeans and kmedians**
  listing, [MV] **cluster utility**
  notes, [MV] **cluster notes**
  programming, [MV] **cluster programming subroutines**, [MV] **cluster programming utilities**
  renaming, [MV] **cluster utility**
  stopping rules, [MV] **cluster**, [MV] **cluster stop**
  tree, [MV] **cluster dendrogram**, [MV] **Glossary**
  using, [MV] **cluster utility**
cluster estimator of variance, [P] **_robust**, [R] **vce_option**, [XT] **vce_options**
  alternative-specific
    conditional logit model, [R] **asclogit**
    multinomial probit regression, [R] **asmprobit**
    rank-ordered probit regression, [R] **asroprobit**
  competing-risks regression, [ST] **stcrreg**
  complementary log-log regression, [R] **cloglog**
  Cox proportional hazards model, [ST] **stcox**
  fixed-effects models,
    linear, [XT] **xtreg**
    Poisson, [XT] **xtpoisson**
  generalized linear models, [R] **glm**
    for binomial family, [R] **binreg**
  generalized methods of moments, [R] **gmm**
  heckman selection model, [R] **heckman**
  instrumental-variables regression, [R] **ivregress**
  interval regression, [R] **intreg**
  linear dynamic panel-data models, [XT] **xtabond**, [XT] **xtdpd**, [XT] **xtdpdsys**

cluster estimator of variance, *continued*
  linear regression, [R] **regress**
    constrained, [R] **cnsreg**
    truncated, [R] **truncreg**
    with dummy-variable set, [R] **areg**
  logistic regression, [R] **logistic**, [R] **logit**, *also see* logit regression subentry
    conditional, [R] **clogit**
    multinomial, [R] **mlogit**
    ordered, [R] **ologit**
    rank-ordered, [R] **rologit**
    skewed, [R] **scobit**
    stereotype, [R] **slogit**
  logit regression, [R] **logit**, *also see* logistic regression subentry
    for grouped data, [R] **glogit**
    nested, [R] **nlogit**
  maximum likelihood estimation, [R] **ml**
  multilevel mixed-effects models, [XT] **xtmixed**
  multinomial
    logistic regression, [R] **mlogit**
    probit regression, [R] **mprobit**
  negative binomial regression
    truncated, [R] **nbreg**
    zero-inflated, [R] **zinb**
  nonlinear
    least-squares estimation, [R] **nl**
    systems of equations, [R] **nlsur**
  parametric survival models, [ST] **streg**
  Poisson regression, [R] **poisson**
    truncated, [R] **tpoisson**
    zero-inflated, [R] **zip**
  population-averaged models, [XT] **xtgee**
    cloglog, [XT] **xtcloglog**
    logit, [XT] **xtlogit**
    negative binomial, [XT] **xtnbreg**
    Poisson, [XT] **xtpoisson**
    probit, [XT] **xtprobit**
  Prais–Winsten and Cochrane–Orcutt regression, [TS] **prais**
  probit regression, [R] **probit**
    bivariate, [R] **biprobit**
    for grouped data, [R] **glogit**
    heteroskedastic, [R] **hetprob**
    multinomial, [R] **mprobit**
    ordered, [R] **oprobit**
    with endogenous regressors, [R] **ivprobit**
    with sample selection, [R] **heckprob**
  random-effects models, linear, [XT] **xtreg**
  structural equation modeling, [SEM] **intro 7**, [SEM] **sem option method( )**
  summary statistics,
    mean, [R] **mean**
    proportion, [R] **proportion**
    ratio, [R] **ratio**
    total, [R] **total**

cluster estimator of variance, *continued*
    tobit model, [R] **tobit**
        with endogenous regressors, [R] **ivtobit**
    treatment-effects model, [R] **treatreg**
    truncated
        negative binomial regression, [R] **tnbreg**
        Poisson regression, [R] **tpoisson**
        regression, [R] **truncreg**
    with endogenous regressors,
        instrumental-variables regression, [R] **ivregress**
        probit model, [R] **ivprobit**
        tobit model, [R] **ivtobit**
    zero-inflated
        negative binomial regression, [R] **zinb**
        Poisson regression, [R] **zip**
cluster sampling, [P] **_robust**, [R] **bootstrap**,
    [R] **bsample**, [R] **jackknife**
clustered, [SEM] **Glossary**
clustering, *see* cluster analysis
clustermat, [MV] **clustermat**
    averagelinkage command, [MV] **cluster linkage**
    centroidlinkage command, [MV] **cluster linkage**
    completelinkage command, [MV] **cluster linkage**
    medianlinkage command, [MV] **cluster linkage**
    singlelinkage command, [MV] **cluster linkage**
    stop command, [MV] **cluster stop**
    wardslinkage command, [MV] **cluster linkage**
    waveragelinkage command, [MV] **cluster linkage**
clusters, duplicating, [D] **expandcl**
cmdlog
    close command, [R] **log**
    command, [R] **log**, [U] **15 Saving and printing
        output—log files**
    off command, [R] **log**
    on command, [R] **log**
    using command, [R] **log**
Cmdyhms() function, [D] **datetime**, [D] **functions**,
    [M-5] **date( )**
cmissing() option, [G-3] *cline_options*,
    [G-3] *connect_options*
cnsreg command, [R] **cnsreg**, [R] **cnsreg
    postestimation**
Cochrane–Orcutt regression, [TS] **prais**, [TS] **Glossary**
code, timing, [P] **timer**
codebook command, [D] **codebook**
_coef[], [U] **13.5 Accessing coefficients and
    standard errors**
coefficient
    alpha, [R] **alpha**
    of determination, [SEM] **estat eqgof**, [SEM] **estat
        ggof**, [SEM] **estat gof**, [SEM] **example 4**,
        [SEM] **example 21**, [SEM] **methods and
        formulas**, [SEM] **Glossary**
    of variation, [R] **tabstat**, [SVY] **estat**
coefficients (from estimation),
    accessing, [P] **ereturn**, [P] **matrix get**,
        [U] **13.5 Accessing coefficients and standard
        errors**

coefficients (from estimation), *continued*
    cataloging, [R] **estimates**
    estimated linear combinations, *see* linear
        combinations of estimators
    linear combinations of, *see* linear combinations of
        estimators
    nonlinear combinations of, *see* nonlinear
        combinations of estimators
    testing equality of, [R] **test**, [R] **testnl**
coeflegend option, *see* sem option coeflegend
Cofc() function, [D] **datetime**, [D] **functions**,
    [M-5] **date( )**
cofC() function, [D] **datetime**, [D] **functions**,
    [M-5] **date( )**
Cofd() function, [D] **datetime**, [D] **functions**,
    [M-5] **date( )**
cofd() function, [D] **datetime**, [D] **functions**,
    [M-5] **date( )**
cohort studies, [ST] **Glossary**
cointegration, [TS] **fcast compute**, [TS] **fcast graph**,
    [TS] **vec intro**, [TS] **vec**, [TS] **veclmar**,
    [TS] **vecnorm**, [TS] **vecrank**, [TS] **vecstable**,
    [TS] **Glossary**
coleq macro extended function, [P] **macro**
coleq, matrix subcommand, [P] **matrix rownames**
colfullnames macro extended function, [P] **macro**
collapse command, [D] **collapse**
_collate() function, [M-5] **sort( )**
collect statistics, [D] **statsby**
collinear variables, removing, [P] **_rmcoll**
collinearity,
    display of omitted variables, [R] **set showbaselevels**
    handling by regress, [R] **regress**
    retaining collinear variables, [R] **estimation options**,
        [R] **orthog**
    variance inflation factors, [R] **regress postestimation**
colmax() function, [M-5] **minmax( )**
colmaxabs() function, [M-5] **minmax( )**
colmin() function, [M-5] **minmax( )**
colminmax() function, [M-5] **minmax( )**
colmissing() function, [M-5] **missing( )**
colnames macro extended function, [P] **macro**
colnames, matrix subcommand, [P] **matrix
    rownames**
colnonmissing() function, [M-5] **missing( )**
colnumb() function, [D] **functions**, [P] **matrix define**
colon operators, [M-2] **op_colon**, [M-6] **Glossary**
color, [G-2] **palette**, [G-4] *colorstyle*
    background, [G-4] **schemes intro**
    dimming and brightening, [G-2] **graph twoway
        histogram**, [G-2] **graph twoway kdensity**,
        [G-4] *colorstyle*
    foreground, [G-4] **schemes intro**
    intensity adjustment, [G-2] **graph twoway
        histogram**, [G-2] **graph twoway kdensity**,
        [G-4] *colorstyle*
    of bars, [G-3] *barlook_options*
    of connecting lines, [G-3] *connect_options*

color, *continued*
　of markers, [G-3] *marker_options*
　of pie slices, [G-2] **graph pie**
　of text, [G-3] *textbox_options*
　setting background and fill, [G-3] *region_options*
　specifying in programs, [P] **display**
color() option, [G-2] **graph twoway histogram**,
　　[G-2] **graph twoway kdensity**
color, palette subcommand, [G-2] **palette**
colorstyle, [G-4] *colorstyle*
cols() function, [M-5] **rows( )**
colscalefactors() function, [M-5] **_equilrc( )**
colshape() function, [M-5] **rowshape( )**
colsof() function, [D] **functions**, [P] **matrix define**
colsum() function, [M-5] **sum( )**
_column(#), display directive, [P] **display**
column-join operator, [M-2] **op_join**
column-major order, [M-6] **Glossary**
columns of matrix,
　appending to, [P] **matrix define**
　names of, [P] **ereturn**, [P] **matrix define**, [P] **matrix
　　rownames**, [U] **14.2 Row and column names**
　operators on, [P] **matrix define**
　selecting, [M-5] **select( )**
colvector, [M-2] **declarations**, [M-6] **Glossary**
comb() function, [D] **functions**, [M-5] **comb( )**
combination step, [MI] **intro substantive**, [MI] **mi
　　estimate**, [MI] **mi estimate using**, [MI] **mi
　　predict**
combinatorial function, [M-5] **comb( )**
combinatorials, calculating, [D] **functions**
combine, graph subcommand, [G-2] **graph combine**
combining datasets, [D] **append**, [D] **cross**, [D] **joinby**,
　　[D] **merge**, [MI] **mi add**, [MI] **mi append**,
　　[MI] **mi merge**, [U] **22 Combining datasets**
combining graphs, [G-2] **graph combine**
command
　arguments, [P] **gettoken**, [P] **syntax**, [P] **tokenize**,
　　[U] **18.4 Program arguments**
　language, [SEM] **Glossary**
　line, launching dialog box from, [R] **db**
　parsing, [P] **gettoken**, [P] **syntax**, [P] **tokenize**,
　　[U] **18.4 Program arguments**
　timings, [U] **8 Error messages and return codes**
commands,
　abbreviating, [U] **11.2 Abbreviation rules**
　aborting, [P] **continue**, [U] **9 The Break key**,
　　[U] **10 Keyboard use**
　editing and repeating, [U] **10 Keyboard use**
　immediate, [U] **19 Immediate commands**
　repeating automatically, [D] **by**, [P] **byable**,
　　[P] **continue**, [P] **foreach**, [P] **forvalues**,
　　[P] **while**
　reviewing, [R] **#review**
　unabbreviating names of, [P] **unabcmd**
commas, reading data separated by, [D] **infile (fixed
　　format)**, [D] **infile (free format)**, [D] **insheet**

comments, [M-2] **comments**
　adding to programs, [P] **comments**
　in programs, do-files, etc., [U] **16.1.2 Comments
　　and blank lines in
　　do-files**, [U] **18.11.2 Comments and long lines
　　in ado-files**
　with data, [D] **notes**
common, estat subcommand, [MV] **factor
　　postestimation**
common factors, [MV] **Glossary**
communalities, [MV] **factor**, [MV] **factor
　　postestimation**
communality, [MV] **Glossary**
commutation matrix, [M-5] **Kmatrix( )**
comparative fit index, [SEM] **estat gof**, [SEM] **methods
　　and formulas**
comparative scatterplot, [R] **dotplot**
compare command, [D] **compare**
compare, estat subcommand, [MV] **procrustes
　　postestimation**
comparing two
　files, [D] **cf**, [D] **checksum**
　variables, [D] **compare**
compassdirstyle, [G-4] *compassdirstyle*
compatibility of Stata programs across releases,
　　[P] **version**
competing risks, [ST] **stcrreg**, [ST] **Glossary**
complementary log-log regression, [R] **cloglog**,
　　[R] **glm**, [SVY] **svy estimation**, [XT] **xtcloglog**,
　　[XT] **xtgee**
complete-cases analysis, [MI] **Glossary**
complete data, [MI] **Glossary**
complete-data analysis, [MI] **Glossary**
complete degrees of freedom for coefficients, [MI] **mi
　　estimate**, [MI] **Glossary**
complete-linkage clustering, [MV] **cluster**,
　　[MV] **clustermat**, [MV] **cluster linkage**,
　　[MV] **Glossary**
complete observations, [MI] **Glossary**
completed data, [MI] **Glossary**
completed-data analysis, [MI] **intro substantive**,
　　[MI] **mi estimate**, [MI] **Glossary**
completelinkage,
　clustermat subcommand, [MV] **cluster linkage**
　cluster subcommand, [MV] **cluster linkage**
completely determined outcomes, [R] **logit**
complex, [M-2] **declarations**, [M-6] **Glossary**
component
　analysis, [MV] **factor**, [MV] **pca**, [MV] **rotate**,
　　[MV] **rotatemat**
　loading plot, [MV] **scoreplot**
　plot, [MV] **scoreplot**
　scores, [MV] **Glossary**
component-plus-residual plots, [G-2] **graph other**,
　　[R] **regress postestimation**
compound double quotes, [P] **macro**

compound symmetric
  correlation matrix, [MV] **mvtest correlations**
  covariance matrix, [MV] **mvtest covariances**
compress command, [D] **compress**
compress files, [D] **zipfile**
compute, fcast subcommand, [TS] **fcast compute**
Comrey's tandem 1 and 2 rotations, [MV] **rotate**,
  [MV] **rotatemat**, [MV] **Glossary**
concat(), egen function, [D] **egen**
concatenating strings, [U] **13.2.2 String operators**
concordance, estat subcommand, [ST] **stcox
  postestimation**
concordance measures, [ST] **stcox postestimation**
cond() function, [D] **functions**, [M-5] **cond( )**
condition number, [M-5] **cond( )**, [M-6] **Glossary**
condition statement, [P] **if**
conditional
  (fixed-effects) logistic regression, [SVY] **svy
    estimation**
  fixed-effects model, [XT] **Glossary**
  imputation, [MI] **mi impute**, [MI] **Glossary**, *see*
    imputation, conditional
  logistic regression, [R] **asclogit**, [R] **clogit**,
    [R] **rologit**, [R] **slogit**, [XT] **xtlogit**
  marginal effects, [R] **margins**, [R] **marginsplot**
  margins, [R] **margins**, [R] **marginsplot**
  normality, *see* normality, conditional
  operator, [M-2] **op_conditional**
  variance, [TS] **arch**, [TS] **Glossary**
confidence intervals, [SEM] **Glossary**, [SVY] **variance
  estimation**, [U] **20.7 Specifying the width of
  confidence intervals**
  for bootstrap statistics, [R] **bootstrap
    postestimation**, [R] **rocreg**, [R] **rocreg
    postestimation**
  for combinations of coefficients,
    linear, [R] **lincom**
    nonlinear, [R] **nlcom**
  for counts, [R] **ci**
  for incidence-rate ratios, [R] **glm**, [R] **nbreg**,
    [R] **poisson**, [R] **tnbreg**, [R] **tpoisson**, [R] **zinb**,
    [R] **zip**, [R] **nlcom**
  for intragroup correlations, [R] **loneway**
  for means, [R] **ci**, [R] **ameans**, [R] **mean**, [R] **ttest**
  for medians and percentiles, [R] **centile**
  for odds ratios, [R] **glm**, [R] **logistic**, [R] **logit**,
    [R] **ologit**, [R] **nlcom**, [ST] **epitab**, [ST] **stci**
  for proportions, [R] **ci**, [R] **proportion**
  for ratios, [R] **ratio**
  for relative-risk ratios, [R] **mlogit**, [R] **nlcom**
  for standardized mortality ratios, [R] **dstdize**
  for tabulated proportions, [SVY] **svy: tabulate
    twoway**
  for totals, [R] **total**
  linear combinations, [SVY] **svy postestimation**
  set default, [R] **level**
config, estat subcommand, [MV] **mds
  postestimation**

configuration, [MV] **Glossary**
configuration plot, [MV] **mds postestimation**,
  [MV] **Glossary**
confirm
  existence command, [P] **confirm**
  file command, [P] **confirm**
  format command, [P] **confirm**
  matrix command, [P] **confirm**
  names command, [P] **confirm**
  number command, [P] **confirm**
  scalar command, [P] **confirm**
  variable command, [P] **confirm**
confirm, datasignature subcommand,
  [D] **datasignature**
confirmatory factor analysis, [MV] **intro**, *see* models,
  confirmatory factor analysis
conformability, [M-2] **void**, [M-6] **Glossary**, *also see*
  c-conformability, *also see* p-conformability, *also
  see* r-conformability
confounding, [ST] **Glossary**
confusion matrix, [MV] **Glossary**
_conj() function, [M-5] **conj( )**
conj() function, [M-5] **conj( )**
conjoint analysis, [R] **rologit**
conjugate, [M-5] **conj( )**, [M-6] **Glossary**
conjugate transpose, [M-2] **op_transpose**, [M-5] **conj( )**,
  [M-6] **Glossary**
connect() option, [G-3] *cline_options*,
  [G-3] *connect_options*, [G-4] *connectstyle*
connected, graph twoway subcommand, [G-2] **graph
  twoway connected**
connectstyle, [G-4] *connectstyle*
conren, set subcommand, [R] **set**
console,
  controlling scrolling of output, [P] **more**, [R] **more**
  obtaining input from, [P] **display**
constant conditional-correlation model, [TS] **mgarch**,
  [TS] **mgarch ccc**
constrained estimation, [R] **constraint**, [R] **estimation
  options**
  alternative-specific
    conditional logistic model, [R] **asclogit**
    multinomial probit regression, [R] **asmprobit**
    rank-ordered probit regression, [R] **asroprobit**
  ARCH, [TS] **arch**
  ARFIMA, [TS] **arfima**
  ARIMA and ARMAX, [TS] **arima**
  competing risks, [ST] **stcrreg**
  complementary log-log regression, [R] **cloglog**
  dynamic factor model, [TS] **dfactor**
  fixed-effects models
    logit, [XT] **xtlogit**
    negative binomial, [XT] **xtnbreg**
    Poisson, [XT] **xtpoisson**
  GARCH model, [TS] **mgarch ccc**, [TS] **mgarch
    dcc**, [TS] **mgarch dvech**, [TS] **mgarch vcc**
  generalized linear models, [R] **glm**
    for binomial family, [R] **binreg**

constrained estimation, *continued*
    generalized negative binomial regression, [R] **nbreg**
    heckman selection model, [R] **heckman**
    interval regression, [R] **intreg**
    linear regression, [R] **cnsreg**
        seemingly unrelated, [R] **sureg**
        stochastic frontier, [R] **frontier**
        three-stage least squares, [R] **reg3**
        truncated, [R] **truncreg**
    logistic regression, [R] **logistic**, [R] **logit**, *also see*
        logit regression subentry
        conditional, [R] **clogit**
        multinomial, [R] **mlogit**
        ordered, [R] **ologit**
        skewed, [R] **scobit**
        stereotype, [R] **slogit**
    logit regression, [R] **logit**, *also see* logistic regression
        subentry
        for grouped data, [R] **glogit**
        nested, [R] **nlogit**
    maximum likelihood estimation, [R] **ml**
    multinomial
        logistic regression, [R] **mlogit**
        probit regression, [R] **mprobit**
    negative binomial regression, [R] **nbreg**
        truncated, [R] **tnbreg**
        zero-inflated, [R] **zinb**
    parametric survival models, [ST] **streg**
    Poisson regression, [R] **poisson**
        truncated, [R] **tpoisson**
        zero-inflated, [R] **zip**
    probit regression, [R] **probit**
        bivariate, [R] **biprobit**
        for grouped data, [R] **glogit**
        heteroskedastic, [R] **hetprob**
        multinomial, [R] **mprobit**
        ordered, [R] **oprobit**
        with endogenous regressors, [R] **ivprobit**
        with sample selection, [R] **heckprob**
    programming, [P] **makecns**
    random-effects models
        cloglog, [XT] **xtcloglog**
        interval data regression, [XT] **xtintreg**
        logit, [XT] **xtlogit**
        negative binomial, [XT] **xtnbreg**
        Poisson, [XT] **xtpoisson**
        probit, [XT] **xtprobit**
        tobit, [XT] **xttobit**
    state-space model, [TS] **sspace**
    stochastic frontier models for panel data,
        [XT] **xtfrontier**
    structural vector autoregressive models, [TS] **var**
        svar
    tobit model with endogenous regressors, [R] **ivtobit**
    treatment-effects model, [R] **treatreg**
    truncated
        negative binomial regression, [R] **tnbreg**

constrained estimation, *continued*
    Poisson regression, [R] **tpoisson**
    regression, [R] **truncreg**
    unobserved-components model, [TS] **ucm**
    vector autoregressive models, [TS] **var**
    vector error-correction models, [TS] **vec**
    with endogenous regressors
        probit regression, [R] **ivprobit**
        tobit model, [R] **ivtobit**
    zero-inflated
        negative binomial regression, [R] **zinb**
        Poisson regression, [R] **zip**
constrained linear regression, [SVY] **svy estimation**
constraint macro extended function, [P] **macro**
constraint command, [R] **constraint**
constraint matrix, creating and displaying, [P] **makecns**
constraints, [SEM] **sem option constraints( )**,
    [SEM] **Glossary**
    across groups, [SEM] **intro 5**
    normalization, [SEM] **intro 3**, [SEM] **sem**
    relaxing, [SEM] **intro 5**, [SEM] **sem path notation**
    specifying, [SEM] **intro 3**, [SEM] **intro 5**,
        [SEM] **sem option constraints( )**, [SEM] **sem**
        option covstructure( ), [SEM] **sem path**
        notation
constraints( ) option, *see* **sem option**
    constraints( )
constructor, [M-2] **class**
containers, [M-5] **asarray( )**
contents of data, [D] **codebook**, [D] **labelbook**,
    [D] **describe**, [D] **ds**
context, class, [P] **class**
contingency tables, [MV] **ca**, [R] **roctab**,
    [R] **symmetry**, [R] **table**, [R] **tabulate twoway**,
    [ST] **epitab**, [SVY] **svy: tabulate twoway**
_continue, display directive, [P] **display**
continue command, [P] **continue**
continuous variable imputation, *see* imputation,
    continuous
contour, graph twoway subcommand, [G-2] **graph**
    twoway contour
contour plot, [G-2] **graph twoway contour**,
    [G-3] *clegend_option*
contour-line plot, [G-2] **graph twoway contourline**
contourline, graph twoway subcommand,
    [G-2] **graph twoway contourline**
contract command, [D] **contract**
contrast command, [R] **anova postestimation**,
    [R] **contrast**, [R] **contrast postestimation**,
    [R] **margins, contrast**, [SVY] **svy**
    postestimation, [U] **20.18 Obtaining contrasts,**
    tests of interactions, and main effects
contrasts, [MV] **intro**, [MV] **manova postestimation**,
    [MV] **Glossary**, [R] **contrast**, [R] **margins,**
    contrast, [R] **marginsplot**, [U] **20.18 Obtaining**
    contrasts, tests of interactions, and main effects
    graphing, [U] **20.19 Graphing margins, marginal**
    effects, and contrasts

control charts, [R] **qc**

convergence, [SEM] **intro 3**, [SEM] **sem**, [SEM] **sem option from( )**

convergence criteria, [R] **maximize**

convergence of MCMC, *see* MCMC, convergence

conversion, file, [D] **changeeol**, [D] **filefilter**

convert, mi subcommand, [MI] **mi convert**

converting between styles, [MI] **mi convert**

convolve( ) function, [M-5] **fft( )**

Cook–Weisberg test for heteroskedasticity, [R] **regress postestimation**

Cook's D, [R] **glm postestimation**, [R] **regress postestimation**

coordinates, estat subcommand, [MV] **ca postestimation**, [MV] **mca postestimation**

copy and paste, [D] **edit**

.copy built-in class function, [P] **class**

copy command, [D] **copy**

copy graph, [G-2] **graph copy**

copy macro extended function, [P] **macro**

copy,
> graph subcommand, [G-2] **graph copy**
> label subcommand, [D] **label**
> mi subcommand, [MI] **mi copy**, [MI] **styles**
> ssc subcommand, [R] **ssc**

copycolor, set subcommand, [R] **set**

copying variables, [D] **clonevar**, [D] **edit**

copyright
> boost, [R] **copyright boost**
> freetype, [R] **copyright freetype**
> icu, [R] **copyright icu**
> JagPDF, [R] **copyright jagpdf**
> lapack, [R] **copyright lapack**
> libpng, [R] **copyright libpng**
> scintilla, [R] **copyright scintilla**
> symbol, [G-4] *text*
> ttf2pt1, [R] **copyright ttf2pt1**
> zlib, [R] **copyright zlib**

copyright command, [R] **copyright**

copysource, [M-1] **source**

Cornfield confidence intervals, [ST] **epitab**

Corr( ) function, [M-5] **fft( )**

_corr( ) function, [M-5] **corr( )**

corr( ) function, [D] **functions**, [M-5] **corr( )**, [P] **matrix define**

corr2data command, [D] **corr2data**

correcting data, *see* editing data

correlate command, [R] **correlate**

correlated errors, *see* robust, Huber/White/sandwich estimator of variance, *also see* autocorrelation

correlated uniqueness models, *see* models, correlated uniqueness

correlation, [M-5] **corr( )**, [M-5] **mean( )**, [M-5] **fft( )**, [R] **correlate**
> binary variables, [R] **tetrachoric**
> canonical, [MV] **canon**
> compound symmetric, [MV] **mvtest correlations**

correlation, *continued*
> continuous variables, [R] **correlate**
> data generation, [D] **corr2data**, [D] **drawnorm**
> factoring of, [MV] **factor**
> interitem, [R] **alpha**
> intracluster, [R] **loneway**
> Kendall's rank, [R] **spearman**
> matrices, [MV] **mvtest correlations**, [P] **matrix define**, [R] **correlate**, [R] **estat**
> > anti-image, [MV] **factor postestimation**, [MV] **pca postestimation**
> models, *see* models, correlation
> pairwise, [R] **correlate**
> partial and semipartial, [R] **pcorr**
> principal components of, [MV] **pca**
> serial, [R] **runtest**
> similarity measure, [MV] *measure_option*
> Spearman's rank, [R] **spearman**
> structure, [XT] **Glossary**
> testing equality, [MV] **mvtest correlations**
> tetrachoric, [R] **tetrachoric**

correlation, estat subcommand, [R] **asmprobit postestimation**, [R] **asroprobit postestimation**

correlation( ) function, [M-5] **mean( )**

correlations,
> estat subcommand, [MV] **canon postestimation**, [MV] **discrim lda postestimation**, [MV] **discrim qda postestimation**, [MV] **mds postestimation**
> mvtest subcommand, [MV] **mvtest correlations**

correlations,
> obtaining, [SEM] **example 16**
> tests of, [SEM] **estat stdize**

correlogram, [G-2] **graph other**, [TS] **corrgram**, [TS] **Glossary**

correspondence analysis, [MV] **ca**, [MV] **mca**, [MV] **Glossary**

correspondence analysis projection, [MV] **Glossary**

corrgram command, [TS] **corrgram**

cos( ) function, [D] **functions**, [M-5] **sin( )**

cosh( ) function, [D] **functions**, [M-5] **sin( )**

cosine function, [D] **functions**

cost frontier model, [R] **frontier**, [XT] **xtfrontier**

costs, [MV] **Glossary**

count command, [D] **count**

count data, *also see* imputation, count data
> confidence intervals for counts, [R] **ci**
> estimation, [R] **expoisson**, [R] **glm**, [R] **gmm**, [R] **nbreg**, [R] **poisson**, [R] **tnbreg**, [R] **tpoisson**, [R] **zinb**, [R] **zip** [U] **26.10 Count dependent-variable models**
> graphs, [R] **histogram**, [R] **kdensity**, [R] **spikeplot**
> interrater agreement, [R] **kappa**
> summary statistics of, [R] **table**, [R] **tabstat**, [R] **tabulate oneway**, [R] **tabulate twoway**, [R] **tabulate, summarize()**
> symmetry and marginal homogeneity tests, [R] **symmetry**

count( ), egen function, [D] **egen**

count, ml subcommand, [R] ml
count-time data, [ST] ct, [ST] ctset, [ST] cttost,
    [ST] sttoct, [ST] Glossary, [SVY] svy
    estimation
counts, making dataset of, [D] collapse
courses about Stata, [U] 3.7.2 NetCourses
covariance, [SEM] intro 3, [SEM] Glossary, also see
    constraints, specifying
    analysis of, [R] anova
    assumptions, [SEM] sem
    matrix, anti-image, [MV] factor postestimation,
      [MV] pca postestimation
    matrix block diagonal, [MV] mvtest covariances
    matrix of estimators, [P] ereturn, [P] matrix get,
      [R] estat, [R] estimates store
    matrix spherical, [MV] mvtest covariances
    matrix testing equality of, [MV] mvtest covariances
    of variables or coefficients, [R] correlate
    principal components of, [MV] pca
    stationarity, [TS] Glossary
covariance, estat subcommand, [MV] discrim
    lda postestimation, [MV] discrim qda
    postestimation, [R] asmprobit postestimation,
    [R] asroprobit postestimation
covariance() option, see sem option covariance()
covariances, mvtest subcommand, [MV] mvtest
    covariances
covariances, creating dataset from, see summary
    statistics data
covariate class, [D] duplicates
covariate patterns, [R] logistic postestimation, [R] logit
    postestimation, [R] probit postestimation
covariates, [ST] Glossary
covarimin rotation, [MV] rotate, [MV] rotatemat,
    [MV] Glossary
COVRATIO, [R] regress postestimation
covstructure() option, see sem option
    covstructure()
Cox proportional hazards model, [ST] stcox, [SVY] svy
    estimation
    power, [ST] stpower cox
    sample size, [ST] stpower cox
    test of assumption, [ST] stcox, [ST] stcox PH-
      assumption tests, [ST] stcox postestimation,
      [ST] stsplit
    Wald test, power, [ST] stpower cox
Cox–Snell residual, [ST] stcox postestimation,
    [ST] streg postestimation
cox, stpower subcommand, [ST] stpower cox
cprplot command, [R] regress postestimation
Cramér's V, [R] tabulate twoway
Crawford–Ferguson rotation, [MV] rotate,
    [MV] rotatemat, [MV] Glossary
create, irf subcommand, [TS] irf create
create, serset subcommand, [P] serset
create_cspline, serset subcommand, [P] serset
create_xmedians, serset subcommand, [P] serset
creturn list command, [P] creturn

crexternal() function, [M-5] findexternal()
Cronbach's alpha, [R] alpha
cross command, [D] cross
cross-correlation function, [TS] xcorr, [TS] Glossary
cross-correlogram, [G-2] graph other, [TS] xcorr
cross() function, [M-5] cross()
cross product, [M-5] cross(), [M 5] crossdev(),
    [M-5] quadcross()
cross-product matrices, [P] matrix accum
cross-sectional studies, [ST] Glossary
crossdev() function, [M-5] crossdev()
crossed-effects model, [XT] Glossary
crossed variables, [MV] Glossary
crossing variables, [MV] Glossary
crossover designs, [R] pk, [R] pkcross, [R] pkshape
cross-sectional
    data, [XT] Glossary
    time-series data, [XT] Glossary
cross-tabulations, see tables
crude estimate, [ST] epitab, [ST] Glossary
cs command, [ST] epitab
csi command, [ST] epitab
ct command, [ST] ctset
ct data, [ST] Glossary, also see count-time data
ctable, irf subcommand, [TS] irf ctable
ctset command, [ST] ctset
cttost command, [ST] cttost
cubic natural splines, [M-5] spline3()
cumsp command, [TS] cumsp, [TS] psdensity
cumul command, [R] cumul
cumulative distribution functions, [D] functions
cumulative distribution, empirical, [R] cumul
cumulative hazard function, [ST] stcurve, [ST] sts,
    [ST] sts generate, [ST] sts graph, [ST] sts list,
    [ST] Glossary
cumulative hazard ratio, see hazard ratio
cumulative incidence
    data, [R] poisson, [ST] epitab
    estimator, [ST] stcrreg, [ST] Glossary
    function, [ST] stcrreg, [ST] stcurve, [ST] Glossary
cumulative spectral distribution, empirical, [TS] cumsp,
    [TS] psdensity
cumulative subhazard function, [ST] stcrreg,
    [ST] stcurve, [ST] Glossary
curse of dimensionality, [MV] Glossary
curved path, [SEM] Glossary
custom prediction equations, [MI] mi impute chained,
    [MI] mi impute monotone
cusum command, [R] cusum
cusum plots, [G-2] graph other
cut(), egen function, [D] egen
cutil, see classutil
cv, estat subcommand, [SVY] estat
cvpermute() function, [M-5] cvpermute()
cvpermutesetup() function, [M-5] cvpermute()
cyclical component, [TS] Glossary

D

DA, *see* data augmentation

dashed lines, [G-4] *linepatternstyle*

data, [D] **data types**, [U] **12 Data**

  appending, *see* appending data

  autocorrelated, *see* autocorrelation

  case–cohort, *see* case–cohort data

  case–control, *see* case–control data

  categorical, *see* categorical data

  certifying, *see* certifying data

  characteristics of, *see* characteristics

  checksums of, *see* checksums of data

  combining, *see* combining datasets

  contents of, *see* contents of data

  count-time, *see* count-time data

  current, [P] **creturn**

  displaying, *see* displaying data

  discrete survival, *see* discrete survival data

  documenting, *see* documenting data

  editing, *see* editing data

  entering, *see* importing data, *see* inputting data
interactively

  exporting, *see* exporting data

  extended missing values, *see* missing values

  flong, *see* flong

  flongsep, *see* flongsep

  generating, *see* generating data

  importing, *see* importing data

  inputting, *see* importing data, *see* inputting data
interactively, *see* reading data from disk

  labeling, *see* labeling data

  large, dealing with, *see* memory

  listing, *see* listing data

  loading, *see* importing data, *see* inputting data
interactively, *see* loading saved data

  matched case–control, *see* matched case–control data

  missing values, *see* missing values

  mlong, *see* mlong

  multiple-record st, *see* multiple-record st data

  nested case–control, *see* nested case–control data

  preserving, [P] **preserve**

  range of, *see* range of data

  ranking, *see* ranking data

  reading, *see* importing data, *see* loading saved data,
see reading data from disk

  recoding, *see* recoding data

  rectangularizing, *see* rectangularize dataset

  reordering, *see* reordering data

  reorganizing, *see* reorganizing data

  restoring, *see* restoring data

  sampling, *see* sampling

  saving, *see* exporting data, *see* saving data

  single-record st, *see* st data

  stacking, *see* stacking data

  strings, *see* string variables

  summarizing, *see* summarizing data

  survey, *see* survey data

data, *continued*

  survival-time, *see* survival analysis

  time-series, *see* time-series analysis

  time-span, *see* time-span data

  transposing, *see* transposing data

  verifying, *see* certifying data

  wide, *see* wide

data augmentation, [MI] **mi impute**, [MI] **mi impute
mvn**, [MI] **Glossary**

Data Browser, *see* Data Editor

Data Editor, [D] **edit**

  copy and paste, [D] **edit**

data entry, *see* importing data, *see* inputting data
interactively, *see* reading data from disk

data-have-changed flag, [M-5] **st_updata( )**

data label macro extended function, [P] **macro**

data, label subcommand, [D] **label**

data management, [MI] **mi add**, [MI] **mi append**,
[MI] **mi expand**, [MI] **mi extract**, [MI] **mi
merge**, [MI] **mi rename**, [MI] **mi replace0**,
[MI] **mi reset**, [MI] **mi reshape**

data manipulation, [R] **fvrevar**, [R] **fvset**,
[TS] **tsappend**, [TS] **tsfill**, [TS] **tsreport**,
[TS] **tsrevar**, [TS] **tsset**, [XT] **xtset**

data matrix, [M-5] **st_data( )**, [M-5] **st_view( )**,
[M-6] **Glossary**

data reduction, [MV] **ca**, [MV] **canon**, [MV] **factor**,
[MV] **mds**, [MV] **pca**

data signature, [D] **datasignature**, [P] **_datasignature**,
[P] **signestimationsample**

data transfer, *see* exporting data, *see* importing data

data types, [I] **data types**

database, reading data from, [D] **odbc**,
[U] **21.4 Transfer programs**

dataset,

  adding notes to, [D] **notes**

  comparing, [D] **cf**, [D] **checksum**

  creating, [D] **corr2data**, [D] **drawnorm**

  example, [U] **1.2.2 Example datasets**

  loading, *see* importing data, *see* inputting data
interactively, *see* loading saved data

  rectangularize, [D] **fillin**

  saving, *see* exporting data, *see* saving data

dataset labels, [D] **label**, [D] **label language**, [D] **notes**

  determining, [D] **codebook**, [D] **describe**

  managing, [D] **varmanage**

datasignature

  clear command, [D] **datasignature**

  command, [D] **datasignature**, [SEM] **example 25**,
[SEM] **ssd**

  confirm command, [D] **datasignature**

  report command, [D] **datasignature**

  set command, [D] **datasignature**

_datasignature command, [P] **_datasignature**

date,

  displaying, [U] **12.5.3 Date and time formats**,
[U] **24.3 Displaying dates and times**

date, *continued*
    formats, [U] **12.5.3 Date and time formats**,
        [U] **24.3 Displaying dates and times**
    functions, [D] **datetime**, [D] **datetime translation**,
        [D] **functions**, [M-5] **date( )**, [U] **24.5 Extracting**
        components of dates and times
    inputting, [U] **24.2 Inputting dates and times**
    variables, [U] **24 Working with dates and times**
date and time stamp, [D] **describe**
date( ) function, [D] **datetime**, [D] **datetime**
    translation, [D] **functions**, [M-5] **date( )**
datelist, [U] **11.1.9 datelist**
dates,
    business, *see* business calendars
    Excel, [D] **datetime**
    OpenOffice, [D] **datetime**
    R, [D] **datetime**
    SAS, [D] **datetime**
    SPSS, [D] **datetime**
dates and times, [D] **datetime**, [D] **datetime**
    business calendars, [D] **datetime business**
    calendars creation, [D] **datetime display**
    formats, [D] **datetime translation** [M-5] **c( )**,
    [M-5] **date( )**, [P] **creturn**
datetime, [D] **datetime**, [D] **datetime business**
    calendars, [D] **datetime business calendars**
    creation, [D] **datetime display formats**,
    [D] **datetime translation**
Davidon–Fletcher–Powell algorithm,
    [M-5] **moptimize( )**, [M-5] **optimize( )**, [R] **ml**
day( ) function, [D] **datetime**, [D] **functions**,
    [M-5] **date( )**, [U] **24.5 Extracting components**
    of dates and times
db command, [R] **db**
dBASE, reading data from, [U] **21.4 Transfer**
    programs
dcc, mgarch subcommand, [TS] **mgarch dcc**
.dct filename suffix, [D] **import**, [D] **infile (fixed**
    format), [D] **infix (fixed format)**, [D] **outfile**,
    [U] **11.6 Filenaming conventions**
debugging, [P] **discard**, [P] **pause**, [P] **trace**
decimal symbol, setting, [D] **format**
declarations, [M-2] **declarations**, [M-6] **Glossary**
.Declare built-in class modifier, [P] **class**
declare, class, [P] **class**
decode command, [D] **encode**
decomposition, [M-5] **cholesky( )**, [M-5] **fullsvd( )**,
    [M-5] **ghessenbergd( )**, [M-5] **gschurd( )**,
    [M-5] **hessenbergd( )**, [M-5] **lud( )**, [M-5] **qrd( )**,
    [M-5] **schurd( )**, [M-5] **svd( )**
deconvolve( ) function, [M-5] **fft( )**
decrement operator, [M-2] **op_increment**
default settings of system parameters, [R] **query**,
    [R] **set_defaults**
defective matrix, [M-6] **Glossary**
DEFF, *see* design effects
define,
    char subcommand, [P] **char**
    label subcommand, [D] **label**

define, *continued*
    matrix subcommand, [P] **matrix define**
    program subcommand, [P] **program**, [P] **program**
    properties
    scalar subcommand, [P] **scalar**
    transmap subcommand, [R] **translate**
DEFT, *see* design effects
degrees of freedom, [MI] **mi estimate**, [MI] **mi predict**
    adjustment, [SEM] **Glossary**
    for coefficients, complete, *see* complete degrees of
        freedom for coefficients, *also see* estimation,
        degrees of freedom for coefficients
degree-to-radian conversion, [D] **functions**
delete, [M-5] **unlink( )**
delete, cluster subcommand, [MV] **cluster**
    programming utilities
deleting
    casewise, [D] **egen**
    files, [D] **erase**
    variables or observations, [D] **drop**
#delimit command, [M-2] **semicolons**, [P] **#delimit**
delimiter
    for comments, [P] **comments**
    for lines, [P] **#delimit**
delta beta influence statistic, [R] **clogit postestimation**,
    [R] **logistic postestimation**, [R] **logit**
    postestimation
delta chi-squared influence statistic, [R] **clogit**
    postestimation, [R] **logistic postestimation**,
    [R] **logit postestimation**
delta deviance influence statistic, [R] **clogit**
    postestimation, [R] **logistic postestimation**,
    [R] **logit postestimation**
delta method, [R] **margins**, [R] **nlcom**, [R] **predictnl**,
    [R] **testnl**, [SEM] **estat residuals**, [SEM] **estat**
    teffects, [SVY] **variance estimation**,
    [SVY] **Glossary**
dendrogram, cluster subcommand, [MV] **cluster**
    dendrogram
dendrograms, [G-2] **graph other**, [MV] **cluster**,
    [MV] **cluster dendrogram**, [MV] **Glossary**
density-distribution sunflower plot, [R] **sunflower**
density estimation, kernel, [R] **kdensity**
density functions, [M-5] **normal( )**
density option, [G-2] **graph twoway histogram**
density smoothing, [G-2] **graph other**
dereference, [M-6] **Glossary**
dereferencing, [M-2] **ftof**, [M-2] **pointers**
_deriv( ) function, [M-5] **deriv( )**
deriv( ) function, [M-5] **deriv( )**
derivative of incomplete gamma function, [D] **functions**
derivatives, [M-5] **deriv( )**
derivatives, numeric, [R] **dydx**, [R] **testnl**
derived plottypes, [G-3] *advanced_options*
deriv_init( ) functions, [M-5] **deriv( )**
deriv_init_*( ) functions, [M-5] **deriv( )**
deriv_query( ) function, [M-5] **deriv( )**
deriv_result_*( ) functions, [M-5] **deriv( )**

describe command, [D] **describe**, [M-3] **mata describe**, [U] **12.6 Dataset, variable, and value labels**

describe,
  ado subcommand, [R] **net**
  bcal subcommand, [D] **bcal**
  classutil subcommand, [P] **classutil**
  estimates subcommand, [R] **estimates describe**
  graph subcommand, [G-2] **graph describe**
  irf subcommand, [TS] **irf describe**
  mi subcommand, [MI] **mi describe**
  net subcommand, [R] **net**
  ssc subcommand, [R] **ssc**
  ssd subcommand, [SEM] **ssd**
describing graph, [G-2] **graph describe**
describing mi data, [MI] **mi describe**
descriptive statistics, [R] **summarize**, [SEM] **estat summarize**
  CIs for means, proportions, and counts, [R] **ci**
  correlations, [R] **correlate**, [R] **pcorr**, [R] **tetrachoric**
  creating dataset containing, [D] **collapse**
  creating variables containing, [D] **egen**
  displays, [D] **codebook**, [D] **pctile**, [R] **grmeanby**, [R] **lv**, [XT] **xtsum**, [XT] **xttab**
  estimation, [R] **mean**, [R] **proportion**, [R] **ratio**, [R] **total**
  means, [R] **ameans**, [R] **summarize**
  percentiles, [R] **centile**
  pharmacokinetic data,
    make dataset of, [R] **pkcollapse**
    summarize, [R] **pksumm**
  tables, [R] **table**, [R] **tabstat**, [R] **tabulate oneway**, [R] **tabulate twoway**, [R] **tabulate, summarize()**
design, fvset subcommand, [R] **fvset**
design effects, [R] **loneway**, [SVY] **estat**, [SVY] **svy: tabulate oneway**, [SVY] **svy: tabulate twoway**, [SVY] **Glossary**
design matrix, [M-5] **designmatrix()**, [M-5] **I()**
designmatrix() function, [M-5] **designmatrix()**
destring command, [D] **destring**
destroy() class function, [M-2] **class**
destructor, [M-2] **class**
destructors, class, [P] **class**
det() function, [D] **functions**, [M-5] **det()**, [P] **matrix define**
determinant of matrix, [M-5] **det()**, [P] **matrix define**
deterministic trend, [TS] **Glossary**
dettriangular() function, [M-5] **det()**
deviance residual, [R] **binreg postestimation**, [R] **fracpoly postestimation**, [R] **glm postestimation**, [R] **logistic postestimation**, [R] **logit postestimation**, [R] **probit postestimation**, [ST] **stcox postestimation**, [ST] **streg postestimation**
deviation cross product, [M-5] **crossdev()**, [M-5] **quadcross()**

dexponential, tssmooth subcommand, [TS] **tssmooth dexponential**
dfactor command, [TS] **dfactor**
DFBETA, [R] **regress postestimation**, [ST] **stcox postestimation**, [ST] **stcrreg postestimation**, [ST] **Glossary**
dfbeta command, [R] **regress postestimation**
dfgls command, [TS] **dfgls**
DFITS, [R] **regress postestimation**
DFP algorithm, [R] **ml**
dfuller command, [TS] **dfuller**
dgammapda() function, [D] **functions**, [M-5] **normal()**
dgammapdada() function, [D] **functions**, [M-5] **normal()**
dgammapdadx() function, [D] **functions**, [M-5] **normal()**
dgammapdx() function, [D] **functions**, [M-5] **normal()**
dgammapdxdx() function, [D] **functions**, [M-5] **normal()**
dhms() function, [D] **datetime**, [D] **functions**, [M-5] **date()**
_diag() function, [M-5] **_diag()**
diag() function, [D] **functions**, [M-5] **diag()**, [P] **matrix define**
diag0cnt() function, [D] **functions**, [M-5] **diag0cnt()**, [P] **matrix define**
diagnostic codes, [D] **icd9**
diagnostic plots, [G-2] **graph other**, [R] **diagnostic plots**, [R] **logistic postestimation**, [R] **regress postestimation**
diagnostics, regression, [R] **regress postestimation**
diagonal, [M-5] **diagonal()**, [M-6] **Glossary**
diagonal() function, [M-5] **diagonal()**
diagonal matrix, [M-5] **_diag()**, [M-5] **diag()**, [M-5] **diagonal()**, [M-5] **isdiagonal()**, [M-6] **Glossary**
diagonal vech model, [TS] **mgarch**, [TS] **mgarch dvech**
diagonals of matrices, [P] **matrix define**
dialog
  box, [P] **dialog programming**, [P] **window programming**, [R] **db**
  programming, [P] **dialog programming**, [P] **window programming**
Dice coefficient similarity measure, [MV] *measure_option*
dichotomous outcome model, [XT] **xtcloglog**, [XT] **xtgee**, [XT] **xtlogit**, [XT] **xtmelogit**, [XT] **xtprobit**, see outcomes, binary
Dickey–Fuller test, [TS] **dfgls**, [TS] **dfuller**
dictionaries, [D] **export**, [D] **import**, [D] **infile (fixed format)**, [D] **infix (fixed format)**, [D] **outfile**, [M-5] **asarray()**
diff(), egen function, [D] **egen**
difference of estimated coefficients, see linear combinations of estimators
difference operator, [TS] **Glossary**, [U] **11.4.4 Time-series varlists**
differentiation, [M-5] **deriv()**

difficult option, [R] **maximize**
digamma( ) function, [D] **functions**, [M-5] **factorial( )**
digitally signing data, *see* datasignature command
digits, controlling the number displayed, [D] **format**,
　　　[U] **12.5 Formats: Controlling how data are
　　　displayed**
dilation, [MV] **Glossary**
dimension, [MV] **Glossary**
dir,
　　classutil subcommand, [P] **classutil**
　　ado subcommand, [R] **net**
　　bcal subcommand, [D] **bcal**
　　cluster subcommand, [MV] **cluster utility**
　　_estimates subcommand, [P] **_estimates**
　　estimates subcommand, [R] **estimates store**
　　graph subcommand, [G-2] **graph dir**
　　label subcommand, [D] **label**
　　macro subcommand, [P] **macro**
　　matrix subcommand, [P] **matrix utility**
　　postutil subcommand, [P] **postfile**
　　program subcommand, [P] **program**
　　_return subcommand, [P] **_return**
　　scalar subcommand, [P] **scalar**
　　serset subcommand, [P] **serset**
　　sysuse subcommand, [D] **sysuse**
dir command, [D] **dir**
dir( ) function, [M-5] **dir( )**
dir macro extended function, [P] **macro**
direct standardization, [R] **dstdize**, [R] **mean**,
　　　[R] **proportion**, [R] **ratio**, [SVY] **direct
　　　standardization**, [SVY] **Glossary**
directories, [M-5] **chdir( )**, [M-5] **dir( )**,
　　　[M-5] **direxists( )**, [P] **creturn**,
　　　[U] **11.6 Filenaming conventions**,
　　　[U] **18.3.11 Constructing Windows filenames
　　　by using macros**
　　location of ado-files, [U] **17.5 Where does Stata
　　　look for ado-files?**
　　changing, [D] **cd**
　　creating, [D] **mkdir**
　　listing, [D] **dir**
　　removing, [D] **rmdir**
directory, class, [P] **classutil**
direxists( ) function, [M-5] **direxists( )**
direxternal( ) function, [M-5] **direxternal( )**
discard command, [P] **discard**,
　　　[U] **18.11.3 Debugging ado-files**
discard, relationship to graph drop, [G-2] **graph
　　drop**
discrete option, [G-2] **graph twoway histogram**
discrete survival data, [ST] **discrete**
discrim
　　knn command, [MV] **discrim**, [MV] **discrim knn**
　　lda command, [MV] **discrim**, [MV] **discrim lda**
　　logistic command, [MV] **discrim**, [MV] **discrim
　　　logistic**
　　qda command, [MV] **discrim**, [MV] **discrim qda**

discriminant analysis, [MV] **candisc**, [MV] **discrim**,
　　　[MV] **discrim knn**, [MV] **discrim lda**,
　　　[MV] **discrim logistic**, [MV] **discrim qda**,
　　　[MV] **Glossary**
　　loading plot, [MV] **scoreplot**
　　score plot, [MV] **scoreplot**
discriminant function, [MV] **Glossary**
discriminating variables, [MV] **Glossary**
disparity, [MV] **Glossary**
dispersion, measures of, [D] **pctile**, [XT] **xtsum**, *see*
　　　standard deviations, displaying, *see* variance,
　　　displaying, *see* percentiles, displaying, *see* range
　　　of data
display
　　as error, [M-5] **displayas( )**, [M-5] **errprintf( )**
　　as text, as result, etc., [M-5] **displayas( )**
　　formats, [D] **describe**, [D] **format**, [P] **macro**,
　　　[U] **12.5 Formats: Controlling how data are
　　　displayed**, [U] **24.3 Displaying dates and times**
　　graph, [G-2] **graph display**
　　settings, [R] **set showbaselevels**
　　width and length, [R] **log**
display,
　　ereturn subcommand, [P] **ereturn**
　　graph subcommand, [G-2] **graph display**
　　ml subcommand, [R] **ml**
display command, [P] **macro**, [P] **display**,
　　　[U] **19.1.2 A list of the immediate commands**
　　as a calculator, [R] **display**
display( ) function, [M-5] **display( )**
display macro extended function, [P] **display**
displayas( ) function, [M-5] **displayas( )**
displayflush( ) function, [M-5] **displayflush( )**
displaying, *also see* printing, logs (output)
　　contents, [D] **describe**
　　data, [D] **edit**, [D] **list**
　　files, [D] **type**
　　macros, [P] **macro**
　　matrix, [P] **matrix utility**
　　output, [P] **display**, [P] **quietly**, [P] **smcl**,
　　　[P] **tabdisp**
　　named graphs, [G-2] **graph display**, [G-2] **graph
　　　use**
　　previously typed lines, [R] **#review**
　　saved results, [R] **saved results**
　　scalar expressions, [P] **display**, [P] **scalar**
dissimilarity, [MV] **Glossary**
　　matrices, [MV] **matrix dissimilarity**,
　　　[MV] **Glossary**, [P] **matrix dissimilarity**
　　measures,
　　　[MV] **cluster**, [MV] **cluster programming
　　　utilities**, [MV] **matrix dissimilarity**, [MV] **mds**,
　　　[MV] *measure_option*, [P] **matrix dissimilarity**
　　　absolute value, [MV] *measure_option*
　　　Bray and Curtis, [MV] **clustermat**
　　　Canberra, [MV] *measure_option*
　　　Euclidean, [MV] *measure_option*
　　　Gower, [MV] *measure_option*

dissimilarity, measures, *continued*
    maximum value, [MV] *measure_option*
    Minkowski, [MV] *measure_option*
dissimilarity, matrix subcommand, [MV] **matrix dissimilarity**, [P] **matrix dissimilarity**
distance matrices, [MV] **matrix dissimilarity**, [P] **matrix dissimilarity**
distances, *see* dissimilarity measures
distances, estat subcommand, [MV] **ca postestimation**
distribution functions, [M-5] **normal( )**
distributional diagnostic plots, [G-2] **graph other**
distributions,
    examining, [D] **pctile**, [R] **ameans**, [R] **centile**, [R] **kdensity**, [R] **mean**, [R] **pksumm**, [R] **summarize**, [R] **total**
    income, [R] **inequality**
    plots, [R] **cumul**, [R] **cusum**, [R] **diagnostic plots**, [R] **dotplot**, [R] **histogram**, [R] **kdensity**, [R] **ladder**, [R] **lv**, [R] **spikeplot**, [R] **stem**
    standard population, [R] **dstdize**
    testing equality of, [R] **ksmirnov**, [R] **kwallis**, [R] **ranksum**, [R] **signrank**
    testing for normality, [MV] **mvtest normality**, [R] **sktest**, [R] **swilk**
    transformations
        to achieve normality, [R] **boxcox**, [R] **ladder**
        to achieve zero skewness, [R] **lnskew0**
disturbance term, [XT] **Glossary**
division, [M-2] **op_arith**, [M-2] **op_colon**
division operator, *see* arithmetic operators
divisive hierarchical clustering methods, [MV] **Glossary**
DLL, [P] **plugin**
Dmatrix( ) function, [M-5] **Dmatrix( )**
do command, [R] **do**, [U] **16 Do-files**
.do filename suffix, [U] **11.6 Filenaming conventions**
do ... while, [M-2] **do**, [M-2] **continue**, [M-2] **break**
dockable, set subcommand, [R] **set**
dockingguides, set subcommand, [R] **set**
documentation, [U] **1 Read this—it will help**
    keyword search on, [R] **search**, [U] **4 Stata's help and search facilities**
documenting data, [D] **codebook**, [D] **labelbook**, [D] **notes**
doedit command, [R] **doedit**
dofb( ) function, [D] **datetime business calendars**, [D] **functions**, [M-5] **date( )**
dofC( ) function, [D] **datetime**, [D] **functions**, [M-5] **date( )**
dofc( ) function, [D] **datetime**, [D] **functions**, [M-5] **date( )**
dofh( ) function, [D] **datetime**, [D] **functions**, [M-5] **date( )**
do-files, [P] **break**, [P] **include**, [P] **version**, [R] **do**, [U] **16 Do-files**, [U] **18.2 Relationship between a program and a do-file**
    adding comments to, [P] **comments**
    editing, [R] **doedit**

do-files, *continued*
    long lines, [P] **#delimit**, [U] **18.11.2 Comments and long lines in ado-files**
dofm( ) function, [D] **datetime**, [D] **functions**, [M-5] **date( )**
dofq( ) function, [D] **datetime**, [D] **functions**, [M-5] **date( )**
dofw( ) function, [D] **datetime**, [D] **functions**, [M-5] **date( )**
dofy( ) function, [D] **datetime**, [D] **functions**, [M-5] **date( )**
domain sampling, [R] **alpha**
Doornik–Hansen normality test, [MV] **mvtest normality**
dose–response models, [R] **binreg**, [R] **glm**, [R] **logistic**
dot,
    graph subcommand, [G-2] **graph dot**
    graph twoway subcommand, [G-2] **graph twoway dot**
dot plots, [G-2] **graph dot**, [G-2] **graph twoway dot**, [G-3] *area_options*, [G-3] *line_options*
dotplot command, [R] **dotplot**
dotted lines, [G-4] *linepatternstyle*
double (storage type), [D] **data types**, [I] **data types**, [U] **12.2.2 Numeric storage types**
double-exponential smoothing, [TS] **tssmooth dexponential**
double-precision floating point number, [U] **12.2.2 Numeric storage types**
double quotes, [P] **macro**
doublebuffer, set subcommand, [R] **set**
dow( ) function, [D] **datetime**, [D] **functions**, [M-5] **date( )** [U] **24.5 Extracting components of dates and times**
doy( ) function, [D] **datetime**, [D] **functions**, [M-5] **date( )**
dp, set subcommand, [D] **format**, [R] **set**
drawnorm command, [D] **drawnorm**
drift, [TS] **Glossary**
drop,
    classutil subcommand, [P] **classutil**
    cluster subcommand, [MV] **cluster utility**
    duplicates subcommand, [D] **duplicates**
    _estimates subcommand, [P] **_estimates**
    estimates subcommand, [R] **estimates store**
    graph subcommand, [G-2] **graph drop**
    irf subcommand, [TS] **irf drop**
    label subcommand, [D] **label**
    notes subcommand, [D] **notes**
    macro subcommand, [P] **macro**
    matrix subcommand, [P] **matrix utility**
    program subcommand, [P] **program**
    _return subcommand, [P] **_return**
    scalar subcommand, [P] **scalar**
    serset subcommand, [P] **serset**
drop command, [D] **drop**, [M-3] **mata drop**
dropline, graph twoway subcommand, [G-2] **graph twoway dropline**

dropping graphs, [G-2] **graph drop**

dropping programs, [P] **discard**

dropping variables and observations, [D] **drop**

ds command, [D] **ds**

dsign( ) function, [M-5] **dsign( )**, [M-5] **sign( )**

dstdize command, [R] **dstdize**

.dta file extension, [U] **11.6 Filenaming conventions**
technical description, [P] **file formats .dta**

.dtasig filename suffix, [U] **11.6 Filenaming conventions**

dual scaling, [MV] **ca**

Duda and Hart index stopping rules, [MV] **cluster stop**

dummy variables, *see* indicator variables

Duncan's multiple-comparison adjustment, *see* multiple comparisons, Duncan's method

Dunnett's multiple-comparison adjustment, *see* multiple comparisons, Dunnett's method

dunnettprob( ) function, [D] **functions**, [M-5] **normal( )**

Dunnett's multiple range, [D] **functions**

_dup(#), display directive, [P] **display**

duplicate observations,
    dropping, [D] **duplicates**
    identifying, [D] **duplicates**

duplicates
    drop command, [D] **duplicates**
    examples command, [D] **duplicates**
    list command, [D] **duplicates**
    report command, [D] **duplicates**
    tag command, [D] **duplicates**

duplicating
    clustered observations, [D] **expandcl**
    observations, [D] **expand**

duplication matrix, [M-5] **Dmatrix( )**

Durbin–Watson statistic, [R] **regress postestimation time series**, [TS] **prais**

durbinalt, estat subcommand, [R] **regress postestimation time series**

Durbin's alternative test, [R] **regress postestimation time series**

dvech, mgarch subcommand, [TS] **mgarch dvech**

dwatson, estat subcommand, [R] **regress postestimation time series**

dyadic operator, [M-2] **syntax**, [M-6] **Glossary**

dydx command, [R] **dydx**

dynamic
    conditional-correlation model, [TS] **mgarch**, [TS] **mgarch dcc**
    factor model, [TS] **dfactor**, [TS] **dfactor postestimation**, *also see* state-space model
    forecast, [TS] **arch**, [TS] **arfima**, [TS] **fcast compute**, [TS] **fcast graph**, [TS] **mgarch**, [TS] **Glossary**
    model, [XT] **Glossary**
    multiplier function, [TS] **irf**, [TS] **irf cgraph**, [TS] **irf create**, [TS] **irf ctable**, [TS] **irf ograph**, [TS] **irf table**, [TS] **var intro**, [TS] **Glossary**

dynamic, *continued*
    panel-data regression, [XT] **xtabond**, [XT] **xtdpd**, [XT] **xtdpdsys**
    regression model, [TS] **arfima**, [TS] **arima**, [TS] **var**
    structural simultaneous equations, [TS] **var svar**

.dynamicmv built-in class function, [P] **class**

E

e( ) scalars, macros, matrices, functions, [D] **functions**, [M-5] **e( )**, [P] **ereturn**, [P] **_estimates**, [P] **return**, [R] **saved results**, [U] **18.8 Accessing results calculated by other programs**, [U] **18.9 Accessing results calculated by estimation commands**, [U] **18.10.2 Saving results in e( )**

e(functions) macro extended function, [P] **macro**

e(macros) macro extended function, [P] **macro**

e(matrices) macro extended function, [P] **macro**

e(sample) function, [P] **ereturn**, [P] **return**

e(scalars) macro extended function, [P] **macro**

e(sample) function, [D] **functions**
    resetting, [R] **estimates save**

e-class command, [R] **saved results**, [P] **program**, [P] **return**, [U] **18.8 Accessing results calculated by other programs**

EBCDIC files, [D] **filefilter**, [D] **infile (fixed format)**, [U] **21.2.9 If you have EBCDIC data**

economist scheme, [G-3] *axis_options*, [G-4] **scheme economist**

edit command, [D] **edit**

edit, graphs, [G-1] **graph editor**

editing
    ado-files and do-files, [R] **doedit**
    commands, [U] **10 Keyboard use**
    data, [D] **edit**, [D] **generate**, [D] **merge**, [D] **recode**
    files while in Stata, [R] **doedit**
    output, [U] **15 Saving and printing output—log files**

_editmissing( ) function, [M-5] **editmissing( )**

editmissing( ) function, [M-5] **editmissing( )**

_edittoint( ) function, [M-5] **edittoint( )**

edittoint( ) function, [M-5] **edittoint( )**

_edittointtol( ) function, [M-5] **edittoint( )**

edittointtol( ) function, [M-5] **edittoint( )**

_edittozero( ) function, [M-5] **edittozero( )**

edittozero( ) function, [M-5] **edittozero( )**

_edittozerotol( ) function, [M-5] **edittozero( )**

edittozerotol( ) function, [M-5] **edittozero( )**

_editvalue( ) function, [M-5] **editvalue( )**

editvalue( ) function, [M-5] **editvalue( )**

effect size, [ST] **stpower**, [ST] **stpower cox**, [ST] **stpower exponential**, [ST] **stpower logrank**, [ST] **Glossary**

effects,
    direct, [SEM] **estat teffects**, [SEM] **example 7**, [SEM] **methods and formulas**, [SEM] **Glossary**

effects, *continued*
  indirect, [SEM] **estat teffects**, [SEM] **example 7**,
      [SEM] **methods and formulas**, [SEM] **Glossary**
  total, [SEM] **estat teffects**, [SEM] **example 7**,
      [SEM] **methods and formulas**, [SEM] **Glossary**
effects, estat subcommand, [SVY] **estat**
eform_option, [R] ***eform_option***
EGARCH, [TS] **arch**
egen command, [D] **egen**, [MI] **mi passive**, [MI] **mi
      xeq**
_eigen_la( ) function, [M-5] **eigensystem( )**
_eigensystem( ) function, [M-5] **eigensystem( )**
eigensystem( ) function, [M-5] **eigensystem( )**
_eigensystem_select( ) functions,
      [M-5] **eigensystemselect( )**
_eigensystemselect*( ) functions,
      [M-5] **eigensystemselect( )**
eigensystemselect*( ) functions,
      [M-5] **eigensystemselect( )**
eigenvalue stability condition, [TS] **varstable**,
      [TS] **vecstable**
eigenvalue stability index, [SEM] **estat stable**
eigenvalues, [M-5] **eigensystem( )**, [M-6] **Glossary**,
      [MV] **factor**, [MV] **factor postestimation**,
      [MV] **pca**, [MV] **rotate**, [MV] **rotatemat**,
      [MV] **screeplot**, [MV] **Glossary**, [P] **matrix
      eigenvalues**, [P] **matrix svd**, [P] **matrix
      symeigen**
_eigenvalues( ) function, [M-5] **eigensystem( )**
eigenvalues( ) function, [M-5] **eigensystem( )**
eigenvalues, matrix subcommand, [P] **matrix
      eigenvalues**
eigenvectors, [M-5] **eigensystem( )**, [M-6] **Glossary**,
      [MV] **factor**, [MV] **factor postestimation**,
      [MV] **pca**, [MV] **rotate**, [MV] **rotatemat**,
      [MV] **scoreplot**, [MV] **Glossary**, [P] **matrix svd**,
      [P] **matrix symeigen**
EIM, *see* expected information matrix
eim, *see* sem option vce( )
eivreg command, [R] **eivreg**, [R] **eivreg
      postestimation**
el( ) function, [D] **functions**, [P] **matrix define**
elimination matrix, [M-5] **Lmatrix( )**
ellipsis, [G-4] *text*
else command, [P] **if**
eltype, [M-2] **declarations**, [M-6] **Glossary**
eltype( ) function, [M-5] **eltype( )**
EM, [MI] **mi impute mvn**, [MI] **Glossary**
  parameter trace files, [MI] **mi ptrace**
empirical cumulative distribution function, [R] **cumul**
emptycells, set subcommand, [R] **set**, [R] **set
      emptycells**
Encapsulated PostScript, [G-3] *eps_options*,
      [G-2] **graph export**, [G-4] *text*
encode command, [D] **encode**, [U] **23.2 Categorical
      string variables**
end, [M-3] **end**
end-of-line characters, [D] **changeeol**
ending a Stata session, [P] **exit**, [R] **exit**

endless loop, *see* loop, endless
endogeneity test, [R] **ivregress postestimation**
endogenous
  covariates, [R] **gmm**, [R] **ivprobit**, [R] **ivregress**,
      [R] **ivtobit**, [R] **reg3**, [XT] **xtdpd**,
      [XT] **xtdpdsys**, [XT] **xthtaylor**, [XT] **xtivreg**
  treatment, [R] **treatreg**
  variable, [SEM] **intro 3**, [SEM] **Glossary**,
      [TS] **Glossary**, [SVY] **svy estimation**
      [XT] **Glossary**
endogenous, estat subcommand, [R] **ivregress
      postestimation**
ends( ), egen function, [D] **egen**
Engle's LM test, [R] **regress postestimation time series**
Enhanced Metafile, [G-2] **graph export**
ensuring mi data are consistent, [MI] **mi update**
entering data, *see* importing data, *see* inputting data
      interactively, *see* reading data from disk
environment macro extended function, [P] **macro**
environment variables (Unix), [P] **macro**
eolchar, set subcommand, [R] **set**
Epanechnikov kernel density function, [R] **kdensity**
epidemiological tables, [ST] **epitab**
epidemiology and related,
  Brier score decomposition, [R] **brier**
  interrater agreement, [R] **kappa**
  pharmacokinetic data, *see* pk (pharmacokinetic data)
  ROC analysis, *see* ROC analysis
  standardization, [R] **dstdize**
  symmetry and marginal homogeneity tests,
      [R] **symmetry**
  tables, [R] **tabulate twoway**
epsdouble( ) function, [D] **functions**
epsfloat( ) function, [D] **functions**
epsilon( ) function, [M-5] **epsilon( )**, [M-6] **Glossary**
eqgof, estat subcommand, [SEM] **estat eqgof**
eqtest, estat subcommand, [SEM] **estat eqtest**
equal FMI test, [MI] **mi estimate**, [MI] **mi test**,
      [MI] **Glossary**
equality of means tests, [MV] **hotelling**, [MV] **manova**
equality operator, [U] **13.2.3 Relational operators**
equality test, survivor functions, [ST] **sts test**
equality tests of
  binomial proportions, [R] **bitest**
  coefficients, [R] **pwcompare**, [R] **test**, [R] **testnl**
  distributions, [R] **ksmirnov**, [R] **kwallis**,
      [R] **ranksum**, [R] **signrank**
  margins, [R] **margins**, [R] **pwcompare**
  means, [MV] **hotelling**, [MV] **manova**,
      [MV] **mvtest means**, [R] **contrast**, [R] **pwmean**,
      [R] **sampsi**, [R] **ttest**
  medians, [R] **ranksum**
  proportions, [R] **bitest**, [R] **prtest**, [R] **sampsi**
  ROC areas, [R] **roccomp**, [R] **rocreg**
  variances, [R] **sdtest**
equamax rotation, [MV] **rotate**, [MV] **rotatemat**,
      [MV] **Glossary**

equation names of matrix, [P] **ereturn**, [P] **matrix define**, [P] **matrix rownames**, [U] **14.2 Row and column names**
_equilc( ) function, [M-5] _equilrc( )
equilibration, [M-5] _equilrc( )
_equilr( ) function, [M-5] _equilrc( )
_equilrc( ) function, [M-5] _equilrc( )
equivalence tests, [R] **pk**, [R] **pkequiv**
erase, [M-5] **unlink( )**
erase, mi subcommand, [MI] **mi erase**, [MI] **styles**
erase, snapshot subcommand, [D] **snapshot**
erase command, [D] **erase**
erasing files, [D] **erase**
erasing graph files, [G-2] **graph drop**
ereturn
    clear command, [P] **ereturn**, [P] **return**
    display command, [P] **ereturn**
    list command, [P] **ereturn**, [P] **return**, [R] **saved results**
    local command, [P] **ereturn**, [P] **return**
    matrix command, [P] **ereturn**, [P] **return**
    post command, [P] **ereturn**, [P] **makecns**, [P] **return**
    repost command, [P] **ereturn**, [P] **return**
    scalar command, [P] **ereturn**, [P] **return**
error, [SEM] **Glossary**
    checking, [D] **assert**
    handling, [P] **capture**, [P] **confirm**, [P] **error**, [U] **16.1.4 Error handling in do-files**
    messages and return codes, [M-2] **errors**, [P] **error**, [P] **rmsg**, [R] **error messages**, [U] **4.9.5 Return codes**, [U] **8 Error messages and return codes**, *also see* error handling
    searching, [R] **search**
    variable, [SEM] **intro 3**, [SEM] **Glossary**
error-bar charts, [R] **serrbar**
error command, [P] **error**
_error( ) function, [M-5] **error( )**
error( ) function, [M-5] **error( )**
error, reshape subcommand, [D] **reshape**
error-components model, [XT] **xthtaylor**, [XT] **Glossary**
errorrate, estat subcommand, [MV] **discrim estat**, [MV] **discrim knn postestimation**, [MV] **discrim lda postestimation**, [MV] **discrim logistic postestimation**, [MV] **discrim qda postestimation**
errors-in-variables regression, [R] **eivreg**
errprintf( ) function, [M-5] **errprintf( )**
esample, estimates subcommand, [R] **estimates save**
estat, [P] **estat programming**
    abond command, [XT] **xtabond postestimation**, [XT] **xtdpd postestimation**, [XT] **xtdpdsys postestimation**
    alternatives command, [R] **asclogit postestimation**, [R] **asmprobit postestimation**, [R] **asroprobit postestimation**
    anova command, [MV] **discrim lda postestimation**

estat, *continued*
    anti command, [MV] **factor postestimation**, [MV] **pca postestimation**
    archlm command, [R] **regress postestimation time series**
    bgodfrey command, [R] **regress postestimation time series**
    bootstrap command, [R] **bootstrap postestimation**
    canontest command, [MV] **discrim lda postestimation**
    classfunctions command, [MV] **discrim lda postestimation**
    classification command, [R] **logistic postestimation**
    classtable command, [MV] **discrim estat**, [MV] **discrim knn postestimation**, [MV] **discrim lda postestimation**, [MV] **discrim logistic postestimation**, [MV] **discrim qda postestimation**
    common command, [MV] **factor postestimation**
    compare command, [MV] **procrustes postestimation**
    concordance command, [ST] **stcox postestimation**
    config command, [MV] **mds postestimation**
    coordinates command, [MV] **ca postestimation**, [MV] **mca postestimation**
    correlation command, [R] **asmprobit postestimation**, [R] **asroprobit postestimation**
    correlations command, [MV] **canon postestimation**, [MV] **discrim lda postestimation**, [MV] **discrim qda postestimation**, [MV] **mds postestimation**
    covariance command, [MV] **discrim lda postestimation**, [MV] **discrim qda postestimation**, [R] **asmprobit postestimation**, [R] **asroprobit postestimation**
    cv command, [SVY] **estat**
    distances command, [MV] **ca postestimation**
    durbinalt command, [R] **regress postestimation time series**
    dwatson command, [R] **regress postestimation time series**
    effects command, [SVY] **estat**
    endogenous command, [R] **ivregress postestimation**
    eqgof command, [SEM] **intro 6**, [SEM] **estat eqgof**, [SEM] **example 3**
    eqtest command, [SEM] **intro 6**, [SEM] **estat eqtest**, [SEM] **example 13**
    errorrate command, [MV] **discrim estat**, [MV] **discrim knn postestimation**, [MV] **discrim lda postestimation**, [MV] **discrim logistic postestimation**, [MV] **discrim qda postestimation**
    factors command, [MV] **factor postestimation**
    facweights command, [R] **asmprobit postestimation**, [R] **asroprobit postestimation**
    firststage command, [R] **ivregress postestimation**

estat, *continued*

framework command, [SEM] **intro 6**, [SEM] **estat framework**, [SEM] **example 11**

ggof command, [SEM] **intro 6**, [SEM] **estat ggof**, [SEM] **example 21**

ginvariant command, [SEM] **intro 6**, [SEM] **estat ginvariant**, [SEM] **example 22**

gof command, [R] **logistic postestimation**, [R] **poisson postestimation**, [SEM] **estat gof**, [SEM] **example 4**, [SVY] **estat**

grdistances command, [MV] **discrim lda postestimation**, [MV] **discrim qda postestimation**

grmeans command, [MV] **discrim lda postestimation**

group command, [XT] **xtmelogit**, [XT] **xtmelogit postestimation**, [XT] **xtmepoisson**, [XT] **xtmepoisson postestimation**, [XT] **xtmixed**, [XT] **xtmixed postestimation**

grsummarize command, [MV] **discrim estat**, [MV] **discrim knn postestimation**, [MV] **discrim lda postestimation**, [MV] **discrim logistic postestimation**, [MV] **discrim qda postestimation**

hettest command, [R] **regress postestimation**

ic command, [R] **estat**

imtest command, [R] **regress postestimation**

inertia command, [MV] **ca postestimation**

kmo command, [MV] **factor postestimation**, [MV] **pca postestimation**

lceffects command, [SVY] **estat**

list command, [MV] **discrim estat**, [MV] **discrim knn postestimation**, [MV] **discrim lda postestimation**, [MV] **discrim logistic postestimation**, [MV] **discrim qda postestimation**

loadings command, [MV] **ca postestimation**, [MV] **canon postestimation**, [MV] **discrim lda**, [MV] **discrim lda postestimation**, [MV] **pca postestimation**

manova command, [MV] **discrim lda postestimation**

mfx command, [R] **asclogit postestimation**, [R] **asmprobit postestimation**, [R] **asroprobit postestimation**

mindices command, [SEM] **intro 6**, [SEM] **estat mindices**, [SEM] **example 5**, [SEM] **example 9**

mvreg command, [MV] **procrustes postestimation**

nproc command, [R] **rocreg postestimation**

overid command, [R] **gmm postestimation**, [R] **ivregress postestimation**

ovtest command, [R] **regress postestimation**

pairwise command, [MV] **mds postestimation**

phtest command, [ST] **stcox PH-assumption tests**

predict command, [R] **exlogistic postestimation**

profiles command, [MV] **ca postestimation**

quantiles command, [MV] **mds postestimation**

estat, *continued*

recovariance command, [XT] **xtmelogit**, [XT] **xtmelogit postestimation**, [XT] **xtmepoisson**, [XT] **xtmepoisson postestimation**, [XT] **xtmixed**, [XT] **xtmixed postestimation**

residuals command, [MV] **factor postestimation**, [MV] **pca postestimation**, [SEM] **intro 6**, [SEM] **estat residuals**, [SEM] **example 10**

rotate command, [MV] **canon postestimation**

rotatecompare command, [MV] **canon postestimation**, [MV] **factor postestimation**, [MV] **pca postestimation**

sargan command, [XT] **xtabond postestimation**, [XT] **xtdpd postestimation**, [XT] **xtdpdsys postestimation**

scoretests command, [SEM] **intro 6**, [SEM] **estat scoretests**, [SEM] **example 8**

sd command, [SVY] **estat**

se command, [R] **exlogistic postestimation**, [R] **expoisson postestimation**

size command, [SVY] **estat**

smc command, [MV] **factor postestimation**, [MV] **pca postestimation**

stable command, [SEM] **intro 6**, [SEM] **estat stable**, [SEM] **example 7**

stdize: prefix command, [SEM] **estat stdize**, [SEM] **example 16**

strata command, [SVY] **estat**

stress command, [MV] **mds postestimation**

structure command, [MV] **discrim lda postestimation**, [MV] **factor postestimation**

subinertia command, [MV] **mca postestimation**

summarize command, [MV] **ca postestimation**, [MV] **discrim estat**, [MV] **discrim knn postestimation**, [MV] **discrim lda postestimation**, [MV] **discrim logistic postestimation**, [MV] **discrim qda postestimation**, [MV] **factor postestimation**, [MV] **mca postestimation**, [MV] **mds postestimation**, [MV] **pca postestimation**, [MV] **procrustes postestimation**, [R] **estat**, [SEM] **estat summarize**

svyset command, [SVY] **estat**

szroeter command, [R] **regress postestimation**

table command, [MV] **ca postestimation**

teffects command, [SEM] **estat teffects**, [SEM] **example 7**

vce command, [R] **estat**, [SVY] **estat**

vif command, [R] **regress postestimation**

wcorrelation command, [XT] **xtgee postestimation**

estimate, mi subcommand, [MI] **mi estimate**, [MI] **mi estimate using**

estimated generalized least squares (EGLS), [XT] **xtgls**, [XT] **xtivreg**, [XT] **xtreg**

_estimates
    clear command, [P] _estimates
    dir command, [P] _estimates
    drop command, [P] _estimates
    hold command, [P] _estimates
    unhold command, [P] _estimates
estimates
    clear command, [R] estimates store
    command, [R] suest, [SVY] svy postestimation
        introduction, [R] estimates
    describe command, [R] estimates describe
    dir command, [R] estimates store
    drop command, [R] estimates store
    esample command, [R] estimates save
    for command, [R] estimates for
    notes command, [R] estimates notes
    query command, [R] estimates store
    replay command, [R] estimates replay
    restore command, [R] estimates store
    save command, [R] estimates save
    stats command, [R] estimates stats
    store command, [R] estimates store
    table command, [R] estimates table
    title command, [R] estimates title
    use command, [R] estimates save
estimation
    allowed estimation commands, [MI] estimation
    commands, [I] estimation commands, [P] ereturn,
        [P] _estimates, [U] 18.9 Accessing results
        calculated by estimation commands,
        [U] 26 Overview of Stata estimation commands
    accessing stored information from, [P] matrix get
    allowing constraints in, [P] makecns
    eliminating stored information from, [P] discard
    obtaining predictions after, [P] _predict
    obtaining robust estimates, [P] _robust
    saving results from, [P] _estimates
    degrees of freedom for coefficients, [MI] mi
        estimate
    method, [SEM] Glossary
    options, [R] estimation options, [SEM] sem
        estimation options
    posting VCE, [MI] mi estimate
    predictions after, [MI] mi predict
    results,
        clearing, [P] ereturn, [P] _estimates,
            [R] estimates store
        listing, [P] ereturn, [P] _estimates
        saving, [P] ereturn, [P] _estimates
        storing and restoring, [R] estimates store
        tables of, [R] estimates table
    sample, summarizing, [R] estat
    tests after, [MI] mi estimate, [MI] mi test
estimators,
    covariance matrix of, [P] ereturn, [P] matrix get,
        [R] correlate, [R] estat, [U] 20.9 Obtaining the
        variance–covariance matrix

estimators, continued
    linear combinations of, [R] lincom,
        [U] 20.13 Obtaining linear combinations of
        coefficients
    nonlinear combinations of, [R] nlcom
etiologic fraction, [ST] epitab
Euclidean dissimilarity measure,
        [MV] measure_option
Euclidean distance, [MV] Glossary
event, [ST] stpower logrank, [ST] Glossary
event of interest, [ST] Glossary
_Ex, [SEM] sem option covstructure( )
exact statistics, [U] 26.11 Exact estimators
    binary confidence intervals, [R] ci, [R] exlogistic,
        [R] roctab
    centiles, [R] centile
    indirect standardization, [R] dstdize
    one-way anova, [R] loneway
    regression, [R] exlogistic, [R] expoisson
    test,
        binomial probability, [R] bitest
        equality of distributions, [R] ksmirnov
        equality of medians, [R] ranksum
        Fisher's, [R] tabulate twoway
        symmetry and marginal homogeneity,
            [R] symmetry
    tetrachoric correlations, [R] tetrachoric
example datasets, [U] 1.2.2 Example datasets
examples, duplicates subcommand, [D] duplicates
Excel, [U] 21 Inputting and importing data
excel,
    export subcommand, [D] import excel
    import subcommand, [D] import excel
Excel dates, [D] datetime
Excel, Microsoft, reading data from, [D] import excel,
        [D] odbc, [D] xmlsave, also see spreadsheets,
        transferring
excess fraction, [ST] epitab
exec( ), odbc subcommand, [D] odbc
existence, confirm subcommand, [P] confirm
exit class program, [P] class exit
exit, class subcommand, [P] class exit
exit command, [P] capture, [P] exit, [R] exit,
        [U] 16.1.4 Error handling in do-files
exit( ) function, [M-5] exit( )
exit Mata, [M-3] end
exiting Stata, see exit command
exlogistic command, [R] exlogistic, [R] exlogistic
        postestimation
exogeneity test, see endogeneity test
exogenous variable, [SEM] intro 3, [SEM] Glossary,
        [TS] Glossary, [XT] Glossary
exp, [M-2] exp, [M-6] Glossary
=exp, [U] 11 Language syntax
exp( ) function, [D] functions, [M-5] exp( )
exp_list, [SVY] svy bootstrap, [SVY] svy brr,
        [SVY] svy jackknife, [SVY] svy sdr,
        [TS] rolling

expand command, [D] **expand**

expand factor varlists, [P] **fvexpand**

expand for mi data, [MI] **mi expand**

expand, mi subcommand, [MI] **mi expand**

expandcl command, [D] **expandcl**

expectation-maximization algorithm, *see* EM

expected information matrix, [SEM] **Glossary**

exploded logit model, [R] **rologit**

expoisson command, [R] **expoisson**, [R] **expoisson postestimation**

exponential

distribution, [ST] **streg**

function, [D] **functions**

notation, [U] **12.2 Numbers**

smoothing, [TS] **tssmooth**, [TS] **tssmooth exponential**, [TS] **Glossary**

survival

power, [ST] **stpower exponential**

regression, [ST] **streg**

sample size, [ST] **stpower exponential**

test, [ST] **Glossary**

power, [ST] **stpower exponential**

sample size, [ST] **stpower exponential**

exponential,

stpower subcommand, [ST] **stpower exponential**

tssmooth subcommand, [TS] **tssmooth exponential**

exponentiated coefficients, [R] *eform_option*

exponentiation, [M-5] **exp( )**, [M-5] **matexpsym( )**

export

excel command, [D] **import excel**

sasxport command, [D] **import sasxport**

export,

graph subcommand, [G-2] **graph export**

mi subcommand, [MI] **mi export**, [MI] **mi export ice**, [MI] **mi export nhanes1**

exporting data, [D] **export**, [D] **import excel**, [D] **import sasxport**, [D] **odbc**, [D] **outfile**, [D] **outsheet**, [D] **xmlsave**, [I] **import and export data**, [MI] **mi export**, [MI] **mi export ice**, [MI] **mi export nhanes1**, [U] **21.4 Transfer programs**

exporting graphs, [G-2] **graph export**, [G-2] **graph set**, [G-3] *eps_options*, [G-3] *png_options*, [G-3] *ps_options*, [G-3] *tif_options*, [G-4] *text*

exposure variable, [ST] **Glossary**

expressions, [M-2] **exp**, [P] **matrix define**, [U] **13 Functions and expressions**

extended macro functions, [P] **char**, [P] **display**, [P] **macro**, [P] **macro lists**, [P] **serset**

external, [M-2] **declarations**

externals, [M-2] **declarations**, [M-5] **findexternal( )**, [M-5] **valofexternal( )**, [M-5] **direxternal( )**, [M-6] **Glossary**

extract diagonal, [M-5] **diagonal( )**, [M-5] **diag( )**

extract, mi subcommand, [MI] **mi extract**, [MI] **mi replace0**

extracting $m=\#$ data from mi data, [MI] **mi extract**, [MI] **mi select**

extracting original data from mi data, [MI] **mi extract**

extrapolation, [D] **ipolate**

F

F( ) function, [D] **functions**, [M-5] **normal( )**

F density,

central, [D] **functions**

noncentral, [D] **functions**

F distribution,

cumulative, [D] **functions**

inverse cumulative, [D] **functions**

inverse reverse cumulative, [D] **functions**

inverse reverse cumulative noncentral, [D] **functions**

reverse cumulative, [D] **functions**

reverse cumulative noncentral, [D] **functions**

Facebook, *see* Stata on Facebook

factor, [MV] **Glossary**

analysis, [MV] **canon**, [MV] **factor**, [MV] **factor postestimation**, [MV] **Glossary**, [R] **alpha** *also see* models, confirmatory factor analysis

loading plot, [MV] **scoreplot**, [MV] **Glossary**

loadings, [MV] **Glossary**

[TS] **dfactor**

parsimony rotation, [MV] **rotate**, [MV] **rotatemat**, [MV] **Glossary**

score plot, [MV] **scoreplot**

scores, [MV] **factor postestimation**, [MV] **Glossary**, [SEM] **intro 6**, [SEM] **example 14**, [SEM] **methods and formulas**, [SEM] **predict**

variables, [P] **fvexpand**, [P] **matrix rownames**, [P] **_rmcoll**, [P] **syntax**, [P] **unab**, [R] **fvrevar**, [R] **fvset**, [U] **11.4.3 Factor variables**, [U] **13.8 Indicator values for levels of factor variables**, [U] **14.2.2 Two-part names**, [U] **20.11 Accessing estimated coefficients**, [U] **25 Working with categorical data and factor variables**

factor command, [MV] **factor**

factor-variable settings, [R] **fvset**

factorial, [U] **11.4.3 Factor variables**

design, [MV] **manova**, [R] **anova**

function, [D] **functions**

factorial( ) function, [M-5] **factorial( )**

factormat command, [MV] **factor**

factors, estat subcommand, [MV] **factor postestimation**

facweights, estat subcommand, [R] **asmprobit postestimation**, [R] **asroprobit postestimation**

failure event, [ST] **Glossary**

failure tables, [ST] **ltable**

failure-time models, *see* survival analysis

false-negative result, *see* type II error

false-positive result, *see* type I error

FAQs, [U] **3.2.1 The Stata website (www.stata.com)**

search, [R] **search**, [U] **4.9.4 FAQ searches**

fastscroll, set subcommand, [R] **set**

favorspeed( ) function, [M-5] **favorspeed( )**

fbufget() function, [M-5] **bufio( )**
fbufput() function, [M-5] **bufio( )**
fcast compute command, [TS] **fcast compute**
fcast graph command, [TS] **fcast graph**
_fclose() function, [M-5] **fopen( )**
fclose() function, [M-5] **fopen( )**
FCS, *see* fully conditional specification
Fden() function, [D] **functions**, [M-5] **normal( )**
feasible generalized least squares (FGLS), [R] **reg3**,
    [TS] **dfgls**, [TS] **prais**, [TS] **var**, [XT] **xtgls**,
    [XT] **xtivreg**, [XT] **xtreg**
feedback loops, [SEM] **estat stable**, [SEM] **estat
    teffects**
fences, [R] **lv**
ferrortext() function, [M-5] **ferrortext( )**
FEVD, [G-2] **graph other**, [TS] **irf**, [TS] **irf create**,
    [TS] **irf ograph**, [TS] **irf table**, [TS] **var intro**,
    [TS] **varbasic**, [TS] **vec intro**, [TS] **Glossary**
_fft() function, [M-5] **fft( )**
fft() function, [M-5] **fft( )**
_fget() function, [M-5] **fopen( )**
fget() function, [M-5] **fopen( )**
_fgetmatrix() function, [M-5] **fopen( )**
fgetmatrix() function, [M-5] **fopen( )**
_fgetnl() function, [M-5] **fopen( )**
fgetnl() function, [M-5] **fopen( )**
FGLS, *see* feasible generalized least squares (FGLS)
fictional data, [SEM] **Glossary**
file
    conversion, [D] **changeeol**, [D] **filefilter**
    format, Stata, [P] **file formats .dta**
    extensions, [I] **file extensions**
    modification, [D] **changeeol**, [D] **filefilter**
    processing, [M-4] **io**, [M-5] **bufio( )**, [M-5] **cat( )**,
        [M-5] **ferrortext( )**, [M-5] **fileexists( )**,
        [M-5] **findfile( )**, [M-5] **fopen( )**, [M-5] **unlink( )**
    translation, [D] **changeeol**, [D] **filefilter**
file
    close command, [P] **file**
    open command, [P] **file**
    query command, [P] **file**
    read command, [P] **file**
    seek command, [P] **file**
    sersetread command, [P] **serset**
    sersetwrite command, [P] **serset**
    set command, [P] **file**
    write command, [P] **file**
file, confirm subcommand, [P] **confirm**
file, find in path, [P] **findfile**
fileexists() function, [M-5] **fileexists( )**
filefilter command, [D] **filefilter**
filename manipulation, [M-5] **adosubdir( )**,
    [M-5] **pathjoin( )**
filenames, displaying, [D] **dir**
files,
    checksum of, [D] **checksum**
    comparison, [D] **cf**

files, *continued*
    compressing, [D] **zipfile**
    copying and appending, [D] **copy**
    display contents of, [D] **type**
    downloading, [D] **checksum**, [D] **copy**,
        [R] **adoupdate**, [R] **net**, [R] **sj**, [R] **ssc**,
        [R] **update**, [U] **28 Using the Internet to keep
        up to date**
    erasing, [D] **erase**
    exporting, *see* exporting data
    extensions, [U] **11.6 Filenaming conventions**
    importing, *see* importing data
    loading, [D] **use**
    names, [U] **11.6 Filenaming conventions**,
        [U] **18.3.11 Constructing Windows filenames
        by using macros**
    opening, [P] **window programming**
    reading ASCII text or binary, [P] **file**
    saving, [D] **save**, [P] **window programming**
    temporary, [P] **macro**, [P] **preserve**, [P] **scalar**
    uncompressing, [D] **zipfile**
    writing ASCII text or binary, [P] **file**
fill
    areas, dimming and brightening, [G-2] **graph
        twoway histogram**, [G-2] **graph twoway
        kdensity**, [G-4] *colorstyle*
    color, setting, [G-3] *region_options*
fill(), egen function, [D] **egen**
fillin command, [D] **fillin**
filling in values, [ST] **stfill**
_fillmissing() function, [M-5] **_fillmissing( )**
filters, [TS] **tsfilter**, *also see* smoothers
    Baxter–King, [TS] **tsfilter bk**
    Butterworth, [TS] **tsfilter bw**
    Christiano–Fitzgerald, [TS] **tsfilter cf**
    Hodrick–Prescott, [TS] **tsfilter hp**
final, [M-2] **class**
findexternal() function, [M-5] **findexternal( )**
findfile command, [P] **findfile**
findfile() function, [M-5] **findfile( )**
finding file in path, [P] **findfile**
finding variables, [D] **lookfor**
findit command, [R] **search**
finite population correction, [SVY] **survey**, [SVY] **svy
    estimation**, [SVY] **svyset**, [SVY] **variance
    estimation**, [SVY] **Glossary**
first-order latent variables, [SEM] **Glossary**
first-differenced estimator, [XT] **xtabond**, [XT] **xtdpd**,
    [XT] **xtdpdsys**, [XT] **xtivreg**
firststage, estat subcommand, [R] **ivregress
    postestimation**
Fisher-type tests, [XT] **xtunitroot**
fisher, xtunitroot subcommand, [XT] **xtunitroot**
Fisher's exact test, [R] **tabulate twoway**, [ST] **epitab**
fits, adding, [G-2] **graph twoway fpfit**, [G-2] **graph
    twoway fpfitci**, [G-2] **graph twoway lfit**,
    [G-2] **graph twoway lfitci**, [G-2] **graph twoway
    qfit**, [G-2] **graph twoway qfitci**

fixed-effects model, [R] **anova**, [R] **areg**, [R] **asclogit**,
  [R] **clogit**, [XT] **xtabond**, [XT] **xtdpd**,
  [XT] **xtdpdsys**, [XT] **xtivreg**, [XT] **xtlogit**,
  [XT] **xtnbreg**, [XT] **xtpoisson**, [XT] **xtreg**,
  [XT] **xtregar**, [XT] **Glossary**

F-keys, [U] **10 Keyboard use**

flexible functional form, [R] **boxcox**, [R] **fracpoly**,
  [R] **mfp**

flist command, [D] **list**

float() function, [D] **functions**, [U] **13.11 Precision
  and problems therein**

float (storage type), [D] **data types**, [I] **data types**,
  [M-5] **floatround( )**, [U] **12.2.2 Numeric storage
  types**, [U] **13.11 Precision and problems therein**

floatresults, set subcommand, [R] **set**

floatround() function, [M-5] **floatround( )**

floatwindows, set subcommand, [R] **set**

flong
  data style, [MI] **styles**, [MI] **Glossary**
  technical description, [MI] **technical**

flongsep
  data style, [MI] **mi xeq**, [MI] **styles**, [MI] **Glossary**
  estimating memory requirements, [MI] **mi convert**
  style, [MI] **mi copy**, [MI] **mi erase**
  technical description, [MI] **technical**

floor() function, [D] **functions**, [M-5] **trunc( )**

_flopin function, [M-5] **lapack( )**

_flopout function, [M-5] **lapack( )**

FMI, *see* fraction missing information

%fmts, [D] **format**, [U] **12.5 Formats: Controlling
  how data are displayed**

fmtwidth() function, [D] **functions**, [M-5] **fmtwidth( )**

folders, *see* directories

follow-up period, [ST] **Glossary**

follow-up studies, *see* incidence studies

fonts, in graphs, [G-4] *text*

footnote, ml subcommand, [R] **ml**

_fopen() function, [M-5] **fopen( )**

fopen() function, [M-5] **fopen( )**

fopen, window subcommand, [P] **window
  programming**

for, [M-2] **for**, [M-2] **continue**, [M-2] **break**,
  [M-2] **semicolons**

for, estimates subcommand, [R] **estimates for**

forcecorrelations option, *see* sem option
  forcecorrelations

forcenoanchor option, *see* sem option
  forcenoanchor

foreach command, [P] **foreach**

forecast,
  ARCH model, [TS] **arch postestimation**
  ARFIMA model, [TS] **arfima postestimation**
  ARIMA model, [TS] **arima postestimation**
  dynamic-factor model, [TS] **dfactor postestimation**
  MGARCH model, *see* multivariate GARCH
    postestimation
  standard error of, [R] **regress postestimation**
  state-space model, [TS] **sspace postestimation**

forecast, *continued*
  structural vector autoregressive model, [TS] **var svar
    postestimation**
  unobserved-components model, [TS] **ucm
    postestimation**
  vector autoregressive model, [TS] **var
    postestimation**
  vector error-correction model, [TS] **vec
    postestimation**

forecast-error variance decomposition, *see* FEVD

forecasting, [TS] **arch**, [TS] **arfima**, [TS] **arima**,
  [TS] **fcast compute**, [TS] **fcast graph**,
  [TS] **irf create**, [TS] **mgarch**, [TS] **tsappend**,
  [TS] **tssmooth**, [TS] **tssmooth dexponential**,
  [TS] **tssmooth exponential**, [TS] **tssmooth
  hwinters**, [TS] **tssmooth ma**, [TS] **tssmooth
  shwinters**, [TS] **ucm**, [TS] **var intro**, [TS] **var**,
  [TS] **vec intro**, [TS] **vec**

forecasts, [G-2] **graph other**

foreground color, [G-4] **schemes intro**

format command, [D] **format**

format, confirm subcommand, [P] **confirm**

format macro extended function, [P] **macro**

format settings, [R] **set cformat**

format width, [M-5] **fmtwidth( )**

formats, [U] **24.3 Displaying dates and times**,
  [D] **datetime**, [D] **describe**, [D] **format**,
  [D] **varmanage**, [U] **12.5 Formats: Controlling
  how data are displayed**, [U] **20.8 Formatting
  the coefficient table**

formatted data, reading, *see* importing data

formatting contents of macros, [P] **macro**

formatting statistical output, [D] **format**

FORTRAN, [M-2] **goto**, [M-5] **dsign( )**

forvalues command, [P] **forvalues**

forward operator, [TS] **Glossary**

fourfold tables, [ST] **epitab**

Fourier transform, [M-5] **fft( )**

FoxPro, reading data from, [U] **21.4 Transfer programs**

FPC, *see* finite population correction

fpfit, graph twoway subcommand, [G-2] **graph
  twoway fpfit**

fpfitci, graph twoway subcommand, [G-2] **graph
  twoway fpfitci**

_fput() function, [M-5] **fopen( )**

fput() function, [M-5] **fopen( )**

_fputmatrix() function, [M-5] **fopen( )**

fputmatrix() function, [M-5] **fopen( )**

fracgen command, [R] **fracpoly**

fracplot command, [R] **fracpoly postestimation**

fracpoly prefix command, [R] **fracpoly**, [R] **fracpoly
  postestimation**

fracpred command, [R] **fracpoly postestimation**

fraction defective, [R] **qc**

fraction missing information, [MI] **mi estimate**, [MI] **mi
  predict**, [MI] **mi test**, [MI] **Glossary**

fraction option, [G-2] **graph twoway histogram**

fractional polynomial regression, [R] **fracpoly**
multivariable, [R] **mfp**
fractionally integrated autoregressive moving average
model, [TS] **psdensity**
frailty, [ST] **Glossary**
frailty model, [ST] **stcox**, [ST] **stcurve**, [ST] **streg**
framework, estat subcommand, [SEM] **estat**
framework
_fread() function, [M-5] **fopen( )**
fread() function, [M-5] **fopen( )**
frequencies,
creating dataset of, [D] **collapse**, [D] **contract**
graphical representation, [R] **histogram**,
[R] **kdensity**
table of, [R] **table**, [R] **tabstat**, [R] **tabulate**
oneway, [R] **tabulate twoway**, [R] **tabulate,**
summarize(), [SVY] **svy: tabulate oneway**,
[SVY] **svy: tabulate twoway**
frequency-domain analysis, [TS] **cumsp**, [TS] **pergram**,
[TS] **psdensity**, [TS] **Glossary**
frequency option, [G-2] **graph twoway histogram**
frequency table, *see* tables frequency
frequency weights, [U] **11.1.6 weight,**
[U] **20.22.1 Frequency weights**
[frequency=*exp*] modifier, [U] **11.1.6 weight,**
[U] **20.22.1 Frequency weights**
frequentist concepts, [MI] **Intro substantive**
freturncode() function, [M-5] **ferrortext( )**
from,
net subcommand, [R] **net**
update subcommand, [R] **update**
from() option, [R] **maximize**, *see* sem option from()
frombase() function, [M-5] **inbase( )**
frontier command, [R] **frontier**, [R] **frontier**
postestimation
frontier models, *see* stochastic frontier models
fsave, window subcommand, [P] **window**
programming
_fseek() function, [M-5] **fopen( )**
fseek() function, [M-5] **fopen( )**
fstatus() function, [M-5] **fopen( )**
Ftail() function, [D] **functions**, [M-5] **normal( )**
_ftell() function, [M-5] **fopen( )**
ftell() function, [M-5] **fopen( )**
ftfreqs() function, [M-5] **fft( )**
ftpad() function, [M-5] **fft( )**
ftperiodogram() function, [M-5] **fft( )**
ftretime() function, [M-5] **fft( )**
_ftruncate() function, [M-5] **fopen( )**
ftruncate() function, [M-5] **fopen( )**
ftunwrap() function, [M-5] **fft( )**
ftwrap() function, [M-5] **fft( )**
full factorial, [U] **11.4.3 Factor variables**
fullsdiag() function, [M-5] **fullsvd( )**
_fullsvd() function, [M-5] **fullsvd( )**
fullsvd() function, [M-5] **fullsvd( )**
fully conditional specification, [MI] **mi impute**, [MI] **mi**
impute chained, [MI] **Glossary**

function, graph twoway subcommand, [G-2] **graph**
twoway function
functions, [D] **functions**, [M-2] **declarations,**
[M-4] **intro**, [M-5] **intro**, [M-6] **Glossary,**
[U] **13.3 Functions**
aggregate, [D] **egen**
arguments, [M-1] **returnedargs**, *also see* arguments
cluster generate, adding, [MV] **cluster**
programming subroutines
combinations of estimators, [R] **lincom**, [R] **nlcom**
combinatorial, [D] **functions**
cumulative distribution, [R] **cumul**
creating dataset of, [D] **collapse**, [D] **obs**
date, [U] **24.5 Extracting components of dates and**
times
date and time, [D] **functions**
derivatives and integrals of, [R] **dydx**
estimable, [R] **margins**
evaluator program, [R] **gmm**, [R] **nl**, [R] **nlsur**
extended macro, [P] **char**, [P] **display**, [P] **macro,**
[P] **macro lists**, [P] **serset**
fractional polynomial, [R] **fracpoly**, [R] **mfp**
graphing, [D] **range**, [G-2] **graph twoway function**
index, [R] **logistic postestimation**, [R] **logit**
postestimation, [R] **probit postestimation**
kernel, [R] **kdensity**, [R] **lpoly**
link, [R] **glm**
mathematical, [D] **functions**
matrix, [D] **functions**, [P] **matrix define,**
[U] **14.8 Matrix functions**
maximizing likelihood, [R] **maximize**, [R] **ml**
naming convention, [M-1] **naming**
obtaining help for, [R] **help**
orthogonalization, [R] **orthog**
passing to functions, [M-2] **ftof**
piecewise cubic and piecewise linear, [R] **mkspline**
prediction, [R] **predict**, [R] **predictnl**
production and cost, [R] **frontier**
programming, [D] **functions**
random number, [D] **generate**
statistical, [D] **functions**
string, [D] **functions**
time-series, [D] **functions**
underscore, [M-6] **Glossary**
variance, [R] **glm**
future history, [ST] **stset**, [ST] **Glossary**
fvexpand command, [P] **fvexpand**
fvrevar command, [R] **fvrevar**
fvset
base command, [R] **fvset**
clear command, [R] **fvset**
design command, [R] **fvset**
report command, [R] **fvset**
fvset command for mi data, [MI] **mi XXXset**
fvset, mi subcommand, [MI] **mi XXXset**
fvunab command, [P] **unab**
[fweight=*exp*] modifier, [U] **11.1.6 weight,**
[U] **20.22.1 Frequency weights**

_fwrite() function, [M-5] **fopen( )**
fwrite() function, [M-5] **fopen( )**
fxsize() option, [G-2] **graph combine**
fysize() option, [G-2] **graph combine**

G

g2 inverse of matrix, [P] **matrix define**, [P] **matrix svd**
gain, [TS] **Glossary**
gamma
    density function, [D] **functions**
        incomplete, [D] **functions**
    distribution
        cumulative, [D] **functions**
        inverse cumulative, [D] **functions**
        inverse reverse cumulative, [D] **functions**
        reverse cumulative, [D] **functions**
gamma() function, [M-5] **factorial( )**
gammaden() function, [D] **functions**, [M-5] **normal( )**
gammap() function, [D] **functions**, [M-5] **normal( )**
gammaptail() function, [D] **functions**,
    [M-5] **normal( )**
gap() option, [G-2] **graph twoway histogram**
gaps, [ST] **Glossary**
GARCH, [TS] **arch**, [TS] **Glossary**, *also see*
    multivariate GARCH
Gauss–Hermite quadrature, [XT] **quadchk**
Gauss, reading data from, [U] **21.4 Transfer programs**
GEE, *see* generalized estimating equations
_geigen_la() function, [M-5] **geigensystem( )**
_geigenselect*_la() functions,
    [M-5] **geigensystem( )**
geigensystem() function, [M-5] **geigensystem( )**
_geigensystem_la() function, [M-5] **geigensystem( )**
geigensystemselect*() functions,
    [M-5] **geigensystem( )**
generalized
    autoregressive conditional heteroskedasticity, *see*
        ARCH model, *see* GARCH
    eigensystem, [M-5] **geigensystem( )**
    eigenvalues, [M-6] **Glossary**
    estimating equations (GEE), [XT] **xtgee**,
        [XT] **Glossary**
    gamma survival regression, [ST] **streg**
    Hessenberg decomposition, [M-5] **ghessenbergd( )**
    inverse, [M-5] **invsym( )**, [M-5] **pinv( )**,
        [M-5] **qrinv( )**
    inverse of matrix, [P] **matrix define**, [P] **matrix svd**
    least squares, [SEM] **intro 3**
        estimated, *see* estimated generalized least squares
            estimator, [TS] **Glossary**
        feasible, *see* feasible generalized least squares
    linear latent and mixed models, *see* GLLAMM
    linear models, *see* GLM

generalized, *continued*
    method of moments (GMM), [P] **matrix accum**,
        [SEM] **Glossary**, [U] **26.19 Generalized**
        method of moments (GMM), [XT] **xtabond**,
        [XT] **xtdpd**, [XT] **xtdpdsys**, *also see* gmm
        command
    negative binomial regression, [R] **nbreg**, [SVY] **svy**
        estimation
    Schur decomposition, [M-5] **gschurd( )**
generate,
    cluster subcommand, [MV] **cluster generate**
    icd9 subcommand, [D] **icd9**
    icd9p subcommand, [D] **icd9**
    sts subcommand, [ST] **sts generate**
generate command, [D] **generate**, [MI] **mi passive**,
    [MI] **mi xeq**
generate functions, adding, [MV] **cluster**
    programming subroutines
generating data, [D] **egen**, [D] **generate**
generating variables, [ST] **stgen**, [ST] **sts generate**
get() function, [D] **functions**, [P] **matrix define**,
    [P] **matrix get**
get, net subcommand, [R] **net**
getmata command, [D] **putmata**
getting started, [U] **1 Read this—it will help**
Getting Started with Stata manuals, [U] **1.1 Getting**
    Started with Stata
    keyword search of, [U] **4 Stata's help and search**
        facilities
gettoken command, [P] **gettoken**
Geweke–Hajivassiliou–Keane multivariate normal
    simulator, [M-5] **ghk( )**, [M-5] **ghkfast( )**
ggof, estat subcommand, [SEM] **estat ggof**
ghalton() function, [M-5] **halton( )**
_ghessenbergd() function, [M-5] **ghessenbergd( )**
ghessenbergd() function, [M-5] **ghessenbergd( )**
_ghessenbergd_la() function, [M-5] **ghessenbergd( )**
ghk() function, [M-5] **ghk( )**
ghkfast() function, [M-5] **ghkfast( )**
ghkfast_i() function, [M-5] **ghkfast( )**
ghkfast_init() function, [M-5] **ghkfast( )**
ghkfast_init_*() function, [M-5] **ghkfast( )**
ghkfast_query_*() function, [M-5] **ghkfast( )**
ghk_init() function, [M-5] **ghk( )**
ghk_init_*() function, [M-5] **ghk( )**
ghk_query_npts() function, [M-5] **ghk( )**
ginvariant, estat subcommand, [SEM] **estat**
    ginvariant
ginvariant() option, *see* sem option ginvariant()
GJR, [TS] **arch**
gladder command, [R] **ladder**
GLLAMM, [R] **gllamm**
gllamm command, [R] **gllamm**
GLM, [R] **binreg**, [R] **glm**, [SVY] **svy estimation**,
    [U] **26.5 Generalized linear models**,
    [U] **26.17.3 Generalized linear models with**
    panel data, [XT] **xtgee**, [XT] **Glossary**

glm command, [R] **glm**, [R] **glm postestimation**
Global, class prefix operator, [P] **class**
global command, [P] **macro**, [U] **18.3.2 Global macros**, [U] **18.3.10 Advanced global macro manipulation**
global variable, [M-2] **declarations**, [M-5] **direxternal( )**, [M-5] **findexternal( )**, [M-5] **valofexternal( )**, [M-6] **Glossary**
glogit command, [R] **glogit**, [R] **glogit postestimation**
glsaccum, matrix subcommand, [P] **matrix accum**
GMM, *see* generalized method of moments
gmm command, [R] **gmm**, [R] **gmm postestimation**
gnbreg command, [R] **nbreg**, [R] **nbreg postestimation**
gof, estat subcommand, [R] **logistic postestimation**, [R] **poisson postestimation**, [SEM] **estat gof**, [SVY] **estat**
Gompertz survival regression, [ST] **streg**
Gönen and Heller's K, [ST] **stcox postestimation**
Goodman and Kruskal's gamma, [R] **tabulate twoway**
goodness-of-fit statistics and tests, [R] **brier**, [R] **diagnostic plots**, [R] **ksmirnov**, [R] **logistic postestimation**, [R] **poisson postestimation**, [R] **regress postestimation**, [SEM] **intro 6**, [SEM] **estat eqgof**, [SEM] **estat ggof**, [SEM] **estat gof**, [SEM] **example 3**, [SEM] **example 4**, [SEM] **Glossary**, [SVY] **estat** *also see* deviance residual, *also see* normal distribution and normality, test for
goto, [M-2] **goto**
Gower coefficient similarity measure, [MV] *measure_option*
gph files, [G-2] **graph manipulation**, [G-4] **concept: gph files**, [U] **11.6 Filenaming conventions** describing contents, [G-2] **graph describe**
gprobit command, [R] **glogit**, [R] **glogit postestimation**
gradient option, [R] **maximize**
grammar, [M-2] **syntax**
Granger causality, [TS] **vargranger**, [TS] **Glossary**
graph
    bar command, [G-2] **graph bar**
    box command, [G-2] **graph box**
    combine command, [G-2] **graph combine**
    copy command, [G-2] **graph copy**
    describe, [G-2] **graph describe**
    dir command, [G-2] **graph dir**
    display command, [G-2] **graph display**
    dot command, [G-3] *area_options*, [G-3] *line_options*, [G-2] **graph dot**
    drop command, [G-2] **graph drop**
    export command, [G-2] **graph export**
    hbar command, [G-2] **graph bar**
    hbox command, [G-2] **graph box**
    matrix command, [G-2] **graph matrix**
    pie command, [G-2] **graph pie**
    play command, [G-2] **graph play**

graph, *continued*
    print command, [G-2] **graph set**, [G-2] **graph print**, [G-3] *pr_options*
    query command, [G-2] **graph query**
    rename command, [G-2] **graph rename**
    save command, [G-2] **graph save**
    set command, [G-2] **graph set**
    twoway area command, [G-2] **graph twoway area**
    twoway bar command, [G-2] **graph twoway bar**
    twoway command, [G-2] **graph twoway**
    twoway connected command, [G-2] **graph twoway connected**
    twoway contour command, [G-2] **graph twoway contour**
    twoway contourline command, [G-2] **graph twoway contourline**
    twoway dot command, [G-2] **graph twoway dot**
    twoway dropline command, [G-2] **graph twoway dropline**
    twoway fpfit command, [G-2] **graph twoway fpfit**
    twoway fpfitci command, [G-2] **graph twoway fpfitci**
    twoway function command, [G-2] **graph twoway function**
    twoway histogram command, [G-2] **graph twoway histogram**
    twoway kdensity command, [G-2] **graph twoway kdensity**
    twoway lfit command, [G-2] **graph twoway lfit**
    twoway lfitci command, [G-2] **graph twoway lfitci**
    twoway line command, [G-2] **graph twoway line**
    twoway lowess command, [G-2] **graph twoway lowess**
    twoway lpoly command, [G-2] **graph twoway lpoly**
    twoway lpolyci command, [G-2] **graph twoway lpolyci**
    twoway mband command, [G-2] **graph twoway mband**
    twoway mspline command, [G-2] **graph twoway mspline**
    twoway pcarrow command, [G-2] **graph twoway pcarrow**
    twoway pcarrowi command, [G-2] **graph twoway pcarrowi**
    twoway pcbarrow command, [G-2] **graph twoway pcarrow**
    twoway pccapsym command, [G-2] **graph twoway pccapsym**
    twoway pci command, [G-2] **graph twoway pci**
    twoway pcscatter command, [G-2] **graph twoway pcscatter**
    twoway pcspike command, [G-2] **graph twoway pcspike**
    twoway qfit command, [G-2] **graph twoway qfit**
    twoway qfitci command, [G-2] **graph twoway qfitci**

graph, *continued*
    twoway rarea command, [G-2] **graph twoway rarea**
    twoway rbar command, [G-2] **graph twoway rbar**
    twoway rcap command, [G-2] **graph twoway rcap**
    twoway rcapsym command, [G-2] **graph twoway rcapsym**
    twoway rconnected command, [G-2] **graph twoway rconnected**
    twoway rline command, [G-2] **graph twoway rline**
    twoway rscatter command, [G-2] **graph twoway rscatter**
    twoway rspike command, [G-2] **graph twoway rspike**
    twoway scatter command, [G-2] **graph twoway scatter**
    twoway scatteri command, [G-2] **graph twoway scatteri**
    twoway spike command, [G-2] **graph twoway spike**
    twoway tsline command, [G-2] **graph twoway tsline**
    twoway tsrline command, [G-2] **graph twoway tsline**
    use command, [G-2] **graph use**
graph,
    fcast subcommand, [TS] **fcast graph**
    irf subcommand, [TS] **irf graph**
    ml subcommand, [R] **ml**
    sts subcommand, [ST] **sts graph**
graph command, [G-2] **graph**
Graph Editor, [G-1] **graph editor**
graph region, [G-3] *region_options*
graph text, [G-4] *text*
graphical user interface, [P] **dialog programming**, *see* GUI
graphics, set subcommand, [G-2] **set graphics**, [R] **set**
graphregion() option, [G-3] *region_options*
graphs,
    added-variable plot, [R] **regress postestimation**
    adjusted Kaplan–Meier survivor curves, [ST] **sts**
    adjusted partial residual plot, [R] **regress postestimation**
    augmented component-plus-residual plot, [R] **regress postestimation**
    augmented partial residual plot, [R] **regress postestimation**
    autocorrelations, [TS] **corrgram**
    baseline hazard and survivor, [ST] **stcox**, [ST] **sts**
    binary variable cumulative sum, [R] **cusum**
    biplot, [MV] **biplot**, [MV] **ca postestimation**
    CA dimension projection, [MV] **ca postestimation**
    cluster tree, *see* graphs, dendrogram
    component-plus-residual, [R] **regress postestimation**
    contrasts, *see* subentry margins
    correlogram, [TS] **corrgram**

graphs, *continued*
    cross-correlogram, [TS] **xcorr**
    cross-sectional time-series data, [XT] **xtdata**, [XT] **xtline**
    cumulative distribution, [R] **cumul**
    cumulative hazard function, [ST] **stcurve**, [ST] **sts graph**
    cumulative spectral density, [TS] **cumsp**
    dendrogram, [MV] **clustermat**, [MV] **cluster dendrogram**, [MV] **cluster generate**, [MV] **cluster linkage**, [MV] **cluster stop**
    density, [R] **kdensity**
    density-distribution sunflower, [R] **sunflower**
    derivatives, [R] **dydx**, [R] **testnl**
    describing contents, [G-2] **graph describe**
    diagnostic, [R] **diagnostic plots**
    dotplot, [R] **dotplot**
    eigenvalue
        after discrim lda, [MV] **discrim lda postestimation**, [MV] **screeplot**
        after factor, [MV] **factor postestimation**, [MV] **screeplot**
        after manova, [MV] **screeplot**
        after mca, [MV] **screeplot**
        after mds, [MV] **screeplot**
        after pca, [MV] **pca postestimation**, [MV] **screeplot**
    error-bar charts, [R] **serrbar**
    forecasts, [TS] **fcast graph**
    fractional polynomial, [R] **fracpoly postestimation**
    functions, [D] **obs**, [D] **range**
    hazard function, [ST] **ltable**, [ST] **stcurve**, [ST] **sts graph**
    histograms, [R] **histogram**, [R] **kdensity**
    impulse–response functions, [TS] **irf**, [TS] **irf cgraph**, [TS] **irf graph**, [TS] **irf ograph**
    integrals, [R] **dydx**
    interaction plots, [R] **marginsplot**
    Kaplan–Meier survivor curves, [ST] **stcox PH-assumption tests**, [ST] **sts**
    ladder-of-power histograms, [R] **ladder**
    letter-value display, [R] **lv**
    leverage-versus-(squared)-residual, [R] **regress postestimation**
    loading
        after candisc, [MV] **candisc**, [MV] **scoreplot**
        after discrim lda, [MV] **discrim lda**, [MV] **discrim lda postestimation**, [MV] **scoreplot**
        after factor, [MV] **factor postestimation**, [MV] **scoreplot**
        after pca, [MV] **pca postestimation**, [MV] **scoreplot**
    logistic diagnostic, [R] **logistic postestimation**
    log-log curve, [ST] **stcox PH-assumption tests**
    lowess smoothing, [R] **lowess**
    margins, [U] **20.19 Graphing margins, marginal effects, and contrasts**

graphs, *continued*
 margins plots, [R] **marginsplot**
 MDS configuration, [MV] **mds postestimation**
 means and medians, [R] **grmeanby**
 normal probability, [R] **diagnostic plots**
 overall look of, [G-4] **schemes intro**
 parameterized curves, [D] **range**
 partial correlogram, [TS] **corrgram**
 partial-regression leverage, [R] **regress postestimation**
 partial residual, [R] **regress postestimation**
 periodogram, [TS] **pergram**
 procrustes overlay, [MV] **procrustes postestimation**
 profile plots, [R] **marginsplot**
 quality control, [R] **qc**
 quantile, [R] **diagnostic plots**
 quantile–normal, [R] **diagnostic plots**
 quantile–quantile, [R] **diagnostic plots**
 regression diagnostic, [R] **regress postestimation**
 residual versus fitted, [R] **regress postestimation**
 residual versus predictor, [R] **regress postestimation**
 ROC curve, [R] **logistic postestimation**,
  [R] **roccomp**, [R] **rocfit postestimation**,
  [R] **rocregplot**, [R] **roctab**
 rootograms, [R] **spikeplot**
 saving, [G-3] *saving_option*
 score
  after candisc, [MV] **candisc**, [MV] **scoreplot**
  after discrim lda, [MV] **discrim lda**,
   [MV] **discrim lda postestimation**,
   [MV] **scoreplot**
  after factor, [MV] **factor postestimation**,
   [MV] **scoreplot**
  after pca, [MV] **scoreplot**
 scree
  after canon, [MV] **screeplot**
  after ca, [MV] **screeplot**
  after discrim lda, [MV] **discrim lda**
   postestimation, [MV] **screeplot**
  after factor, [MV] **factor postestimation**,
   [MV] **screeplot**
  after manova, [MV] **screeplot**
  after mca, [MV] **screeplot**
  after mds, [MV] **screeplot**
  after pca, [MV] **pca postestimation**,
   [MV] **screeplot**
 Shepard diagram, [MV] **mds postestimation**
 smoothing, [R] **kdensity**, [R] **lowess**, [R] **lpoly**
 spike plot, [R] **spikeplot**
 stem-and-leaf, [R] **stem**
 sunflower, [R] **sunflower**
 suppressing, [G-3] *nodraw_option*
 survivor function, [ST] **stcurve**, [ST] **sts graph**
 symmetry, [R] **diagnostic plots**
 time-versus-concentration curve, [R] **pk**,
  [R] **pkexamine**
 white-noise test, [TS] **wntestb**

grdistances, estat subcommand, [MV] **discrim
  lda postestimation**, [MV] **discrim qda
  postestimation**
greater than (or equal) operator, [U] **13.2.3 Relational
  operators**
.grec files, [U] **11.6 Filenaming conventions**
Greek letters, [G-4] *text*
Greenhouse–Geisser epsilon, [R] **anova**
Greenwood confidence intervals, [ST] **sts**
grid
 definition, [G-4] *gridstyle*
 lines, [G-3] *axis_label_options*
  without ticks, [G-4] *tickstyle*
gridstyle, [G-4] *gridstyle*
grmeanby command, [R] **grmeanby**
grmeans, estat subcommand, [MV] **discrim lda
  postestimation**
group-data regression, [R] **glogit**, [R] **intreg**
group(), egen function, [D] **egen**
group, estat subcommand, [XT] **xtmelogit
  postestimation**, [XT] **xtmepoisson
  postestimation**, [XT] **xtmixed postestimation**
group invariance test, [SEM] **methods and formulas**
grouping variables, generating, [MV] **cluster generate**
groups, graphs by, [G-3] *by_option*
groups() option, *see* sem option groups()
groupvar, [U] **11.4 varlists**
grsummarize, estat subcommand, [MV] **discrim
  estat**, [MV] **discrim knn postestimation**,
  [MV] **discrim lda postestimation**, [MV] **discrim
  logistic postestimation**, [MV] **discrim qda
  postestimation**
gs1 print color mapping, [G-2] **set printcolor**
gs2 print color mapping, [G-2] **set printcolor**
gs3 print color mapping, [G-2] **set printcolor**
_gschurd() function, [M-5] **gschurd( )**
gschurd() function, [M-5] **gschurd( )**
_gschurdgroupby() function, [M-5] **gschurd( )**
gschurdgroupby() function, [M-5] **gschurd( )**
_gschurdgroupby_la() function, [M-5] **gschurd( )**
_gschurd_la() function, [M-5] **gschurd( )**
gsort command, [D] **gsort**
GUI, [SEM] **GUI**, [SEM] **Glossary**
 examples of, [U] **2 A brief description of Stata**
 programming, [P] **dialog programming**

H

HAC variance estimate, [R] **binreg**, [R] **glm**, [R] **gmm**,
  [R] **ivregress**, [R] **nl**
hadamard() function, [D] **functions**, [P] **matrix define**
Hadamard matrix, [SVY] **svy brr**, [SVY] **Glossary**
Hadri Lagrange multiplier stationarity test,
  [XT] **xtunitroot**
hadri, xtunitroot subcommand, [XT] **xtunitroot**
half option, [G-2] **graph matrix**
halfyear() function, [D] **datetime**, [D] **functions**,
  [M-5] **date( )**

halfyearly() function, [D] **datetime**, [D] **datetime translation**, [D] **functions**, [M-5] **date( )**

_halton() function, [M-5] **halton( )**

halton() function, [M-5] **halton( )**

Halton set, [M-5] **halton( )**

Hamann coefficient similarity measure, [MV] *measure_option*

Hammersley set, [M-5] **halton( )**

Hansen's J statistic, [R] **gmm**, [R] **gmm postestimation**, [R] **ivregress**

hard missing value, [MI] **mi impute**, [MI] **Glossary**

harmonic mean, [R] **ameans**

Harrell's C, [ST] **stcox postestimation**

Harris–Tzavalis test, [XT] **xtunitroot**

has_eprop() function, [D] **functions**

hash functions, [M-5] **hash1( )**, [M-6] **Glossary**

hash tables, [M-5] **asarray( )**, [M-6] **Glossary**

hash1() function, [M-5] **hash1( )**

hashing, [M-6] **Glossary**

hasmissing() function, [M-5] **missing( )**

hat matrix, *see* projection matrix, diagonal elements of

hausman command, [R] **hausman**

Hausman specification test, [R] **hausman**, [XT] **xtreg postestimation**

Hausman–Taylor estimator, [XT] **xthtaylor**

Haver Analytics, [TS] **haver**

haver command, [TS] **haver**

hazard contributions, [ST] **Glossary**

hazard function, [G-2] **graph other**, [ST] **sts**, [ST] **sts generate**, [ST] **sts list**, [ST] **Glossary**
  graph of, [ST] **ltable**, [ST] **stcurve**, [ST] **sts graph**

hazard ratio, [ST] **Glossary**
  minimal detectable difference, [ST] **stpower**
  minimal effect size, [ST] **stpower**

hazard tables, [ST] **ltable**

hbar, graph subcommand, [G-2] **graph bar**

hbox, graph subcommand, [G-2] **graph box**

headlabel option, [G-2] **graph twoway pccapsym**, [G-2] **graph twoway pcscatter**

health ratios, [R] **binreg**

heart attack data, [MI] **intro substantive**, [MI] **mi estimate**, [MI] **mi impute chained**, [MI] **mi impute intreg**, [MI] **mi impute logit**, [MI] **mi impute mlogit**, [MI] **mi impute monotone**, [MI] **mi impute mvn**, [MI] **mi impute nbreg**, [MI] **mi impute ologit**, [MI] **mi impute pmm**, [MI] **mi impute poisson**, [MI] **mi impute regress**, [MI] **mi impute truncreg**, [MI] **mi predict**

heckman command, [R] **heckman**, [R] **heckman postestimation**

Heckman selection model, [R] **heckman**, [R] **heckprob**, [SVY] **svy estimation**

heckprob command, [R] **heckprob**, [R] **heckprob postestimation**

height() textbox option, [G-3] *added_text_options*

Helmert contrasts, [R] **contrast**

help command, [M-1] **help**, [M-3] **mata help**, [R] **help**, [U] **4 Stata's help and search facilities**, [U] **7 –more– conditions**
  writing your own, [U] **18.11.6 Writing online help**

help file search, [R] **hsearch**

help—I don't know what to do, [U] **3 Resources for learning and using Stata**

help system, searching, [R] **hsearch**

help, view subcommand, [R] **view**

help_d, view subcommand, [R] **view**

Henze–Zirkler normality test, [MV] **mvtest normality**

Hermitian
  adjoin, [M-2] **op_transpose**, [M-5] **conj( )**
  matrices, [M-5] **issymmetric( )**, [M-5] **makesymmetric( )**, [M-6] **Glossary**
  transpose, [M-2] **op_transpose**, [M-5] **conj( )**

Hessenberg
  decomposition, [M-5] **hessenbergd( )**, [M-6] **Glossary**
  form, [M-6] **Glossary**

_hessenbergd() function, [M-5] **hessenbergd( )**

hessenbergd() function, [M-5] **hessenbergd( )**

_hessenbergd_la() function, [M-5] **hessenbergd( )**

hessian option, [R] **maximize**

heterogeneity test, [ST] **epitab**

heteroskedastic errors, *see* linear regression with heteroskedastic errors

heteroskedastic probit regression, [R] **hetprob**, [SVY] **svy estimation**

heteroskedasticity, [XT] **xtgls**, *also see* HAC variance estimate
  ARCH model, *see* ARCH model
  conditional, [R] **regress postestimation time series**
  GARCH model, *see* GARCH
  Newey–West estimator, *see* Newey–West regression
  robust variances, *see* robust, Huber/White/sandwich estimator of variance
  test for, [R] **hetprob**, [R] **regress postestimation**, [R] **regress postestimation time series**

hetprob command, [R] **hetprob**, [R] **hetprob postestimation**

hettest, estat subcommand, [R] **regress postestimation**

hexadecimal report, [D] **hexdump**

hexdump command, [D] **hexdump**

Heywood case, [MV] **Glossary**

Heywood solution, [MV] **Glossary**

hh() function, [D] **datetime**, [D] **functions**, [M-5] **date( )**

hhC() function, [D] **datetime**, [D] **functions**, [M-5] **date( )**

hidden saved results, [M-5] **st_global( )**, [M-5] **st_matrix( )**, [M-5] **st_numscalar( )**

hierarchical
  cluster analysis, [MV] **cluster**, [MV] **clustermat**, [MV] **cluster linkage**
  clustering, [MV] **Glossary**

hierarchical, *continued*
  model, [XT] **xtmelogit**, [XT] **xtmepoisson**,
    [XT] **xtmixed**, [XT] **Glossary** *also see* models,
    confirmatory factor analysis
  regression, [R] **nestreg**, [R] **stepwise**
  samples, [R] **anova**, [R] **gllamm**, [R] **loneway**,
    [R] **areg**
high-low charts, [G-2] **graph twoway rbar**,
    [G-2] **graph twoway rcap**, [G-2] **graph twoway
    rspike**
high-pass filter, [TS] **tsfilter bw**, [TS] **tsfilter hp**,
    [TS] **Glossary**
higher-order models, *see* models, confirmatory factor
    analysis
Hilbert() function, [M-5] **Hilbert( )**
Hildreth–Lu regression, [TS] **prais**
HILO, [M-5] **byteorder( )**
histogram command, [R] **histogram**
histogram, graph twoway subcommand, [G-2] **graph
    twoway histogram**
histograms, [G-2] **graph twoway histogram**,
    [R] **histogram**
  dotplots, [R] **dotplot**
  kernel density estimator, [R] **kdensity**
  ladder-of-powers, [R] **ladder**
  of categorical variables, [R] **histogram**
  rootograms, [R] **spikeplot**
  stem-and-leaf, [R] **stem**
historical saved results, [M-5] **st_global( )**,
    [M-5] **st_matrix( )**, [M-5] **st_numscalar( )**
histories, [G-2] **graph bar**, [G-2] **graph box**,
    [G-2] **graph matrix**, [G-2] **graph
    pic**, [G-2] **graph twoway histogram**,
    [G-3] *by_option*
hms() function, [D] **datetime**, [D] **functions**,
    [M-5] **date( )**
Hodrick–Prescott filter, [TS] **tsfilter**, [TS] **tsfilter hp**
hofd() function, [D] **datetime**, [D] **functions**,
    [M-5] **date( )**
hold,
  _estimates subcommand, [P] **_estimates**
  _return subcommand, [P] **_return**
Holm's multiple-comparison adjustment, *see* multiple
    comparisons, Holm's method
Holt–Winters smoothing, [TS] **tssmooth**, [TS] **tssmooth
    dexponential**, [TS] **tssmooth exponential**,
    [TS] **tssmooth hwinters**, [TS] **tssmooth
    shwinters**, [TS] **Glossary**
home resale-price data, [MI] **mi estimate**, [MI] **mi
    estimate using**, [MI] **mi predict**, [MI] **mi test**
homogeneity of variances, [R] **oneway**, [R] **sdtest**
homogeneity test, [ST] **epitab**
homoskedasticity tests, [R] **regress postestimation**
Horst normalization, *see* Kaiser normalization
Hosmer and Lemeshow
  delta chi-squared influence statistic, *see* delta chi-
    squared influence statistic

Hosmer and Lemeshow, *continued*
  delta deviance influence statistic, *see* delta deviance
    influence statistic
  goodness-of-fit test, [R] **logistic postestimation**,
    [SVY] **estat**
hot, ssc subcommand, [R] **ssc**
hotelling command, [MV] **hotelling**
Hotelling's
  generalized T-squared statistic, [MV] **manova**
  T-squared, [MV] **hotelling**, [MV] **mvtest means**,
    [MV] **Glossary**
hours() function, [D] **datetime**, [D] **functions**,
    [M-5] **date( )**
hp, tsfilter subcommand, [TS] **tsfilter hp**
_hqrd() function, [M-5] **qrd( )**
hqrd() function, [M-5] **qrd( )**
hqrdmultq() function, [M-5] **qrd( )**
hqrdmultq1t() function, [M-5] **qrd( )**
_hqrdp() function, [M-5] **qrd( )**
hqrdp() function, [M-5] **qrd( )**
_hqrdp_la() function, [M-5] **qrd( )**
hqrdq() function, [M-5] **qrd( )**
hqrdq1() function, [M-5] **qrd( )**
hqrdr() function, [M-5] **qrd( )**
hqrdr1() function, [M-5] **qrd( )**
HRF, *see* human readable form
hsearch command, [R] **hsearch**, [U] **4 Stata's help
    and search facilities**
ht, xtunitroot subcommand, [XT] **xtunitroot**
http://www.stata.com, [U] **3.2.1 The Stata website
    (www.stata.com)**
httpproxy, set subcommand, [R] **netio**, [R] **set**
httpproxyauth, set subcommand, [R] **netio**, [R] **set**
httpproxyhost, set subcommand, [R] **netio**, [R] **set**
httpproxyport, set subcommand, [R] **netio**, [R] **set**
httpproxypw, set subcommand, [R] **netio**, [R] **set**
httpproxyuser, set subcommand, [R] **netio**, [R] **set**
Huber weighting, [R] **rreg**
Huber/White/sandwich estimator of variance, *see* robust,
    Huber/White/sandwich estimator of variance
human readable form, [D] **datetime**, [D] **datetime
    display formats**, [D] **datetime translation**
Huynh–Feldt epsilon, [R] **anova**
hwinters, tssmooth subcommand, [TS] **tssmooth
    hwinters**
hyperbolic functions, [D] **functions**, [M-5] **sin( )**
hypergeometric() function, [D] **functions**,
    [M-5] **normal( )**
hypergeometric,
  cumulative distribution, [D] **functions**
  probability mass function, [D] **functions**
hypergeometricp() function, [D] **functions**,
    [M-5] **normal( )**
hypertext help, [R] **help**, [U] **4 Stata's help and search
    facilities**, [U] **18.11.6 Writing online help**
hypothesis tests, [SEM] **test**, [SEM] **testnl**, *see* tests

I

I ( ) function, [D] **functions**, [M-5] **I ( )**, [P] **matrix define**
ibeta() function, [D] **functions**, [M-5] **normal( )**
ibetatail() function, [D] **functions**, [M-5] **normal( )**
ic, estat subcommand, [R] **estat**
icd9
    check command, [D] **icd9**
    clean command, [D] **icd9**
    generate command, [D] **icd9**
    lookup command, [D] **icd9**
    query command, [D] **icd9**
    search command, [D] **icd9**
icd9p
    check command, [D] **icd9**
    clean command, [D] **icd9**
    generate command, [D] **icd9**
    lookup command, [D] **icd9**
    query command, [D] **icd9**
    search command, [D] **icd9**
ICE, *see* imputation, chained equations
ice command, [MI] **mi export ice**, [MI] **mi import ice**
ID variable, [ST] **Glossary**
identification, [SEM] **Glossary**, *see* model identification
identifier, class, [P] **class**
identifier, unique, [D] **isid**
identity matrix, [M-5] **I ( )**, [P] **matrix define**
idiosyncratic error term, [XT] **Glossary**
if *exp*, [M-2] **if**, [P] **syntax**, [U] **11 Language syntax**
if programming command, [P] **if**
ignorable missing-data mechanism, [MI] **intro substantive**, [MI] **Glossary**
IIA,
    assumption, [R] **clogit**, [R] **nlogit**
    relaxing assumption, [R] **asclogit**, [R] **asmprobit**, [R] **asroprobit**
    test for, [R] **hausman**, [R] **suest**
Im( ) function, [M-5] **Re( )**
Im–Pesaran–Shin test, [XT] **xtunitroot**
imaginary part, [M-5] **Re( )**
immediate commands, [U] **18.4.5 Parsing immediate commands**, [I] **immediate commands**, [I] **prefix commands**, [P] **display**, [R] **bitest**, [R] **ci**, [R] **prtest**, [R] **sampsi**, [R] **sdtest**, [R] **symmetry**, [R] **tabulate twoway**, [R] **ttest**, [U] **19 Immediate commands**
implied context, class, [P] **class**
import excel command, [D] **import excel**
import, mi subcommand, [MI] **mi import**, [MI] **mi import flong**, [MI] **mi import flongsep**, [MI] **mi import ice**, [MI] **mi import nhanes1**, [MI] **mi import wide**
import sasxport command, [D] **import sasxport**
importance weights, [U] **11.1.6 weight**, [U] **20.22.4 Importance weights**

importing data, [D] **import**, [D] **import excel**, [D] **import sasxport**, [D] **infile (fixed format)**, [D] **infile (free format)**, [D] **infix (fixed format)**, [D] **insheet**, [D] **odbc**, [D] **xmlsave**, [I] **import and export data**, [MI] **mi import**, [MI] **mi import flong**, [MI] **mi import flongsep**, [MI] **mi import ice**, [MI] **mi import nhanes1**, [MI] **mi import wide**, [U] **21 Inputting and importing data**, [U] **21.4 Transfer programs**, *also see* combining datasets, *also see* inputting data interactively
impulse–response functions, *see* IRF
imputation diagnostics, *see* imputation, diagnostics
imputation method, [MI] **mi impute**
    iterative, [MI] **mi impute**, [MI] **mi impute chained**, [MI] **mi impute mvn**
    monotone, [MI] **mi impute monotone**
    multivariate, [MI] **mi impute chained**, [MI] **mi impute monotone**, [MI] **mi impute mvn**
    proper, [MI] **intro substantive**
    univariate, [MI] **mi impute intreg**, [MI] **mi impute logit**, [MI] **mi impute mlogit**, [MI] **mi impute nbreg**, [MI] **mi impute ologit**, [MI] **mi impute pmm**, [MI] **mi impute poisson**, [MI] **mi impute regress**, [MI] **mi impute truncreg**
imputation using chained equations, *see* imputation, chained equations
imputation,
    binary, [MI] **mi impute logit**
    by groups, [MI] **mi impute**
    categorical, [MI] **mi impute mlogit**, [MI] **mi impute ologit**
    chained equations, [MI] **mi impute**, [MI] **mi impute chained**, [MI] **mi impute intreg**, [MI] **mi impute logit**, [MI] **mi impute mlogit**, [MI] **mi impute nbreg**, [MI] **mi impute ologit**, [MI] **mi impute pmm**, [MI] **mi impute poisson**, [MI] **mi impute regress**, [MI] **mi impute truncreg**
    conditional, [MI] **mi impute**, [MI] **mi impute chained**, [MI] **mi impute intreg**, [MI] **mi impute logit**, [MI] **mi impute mlogit**, [MI] **mi impute monotone**, [MI] **mi impute nbreg**, [MI] **mi impute ologit**, [MI] **mi impute pmm**, [MI] **mi impute poisson**, [MI] **mi impute regress**, [MI] **mi impute truncreg**
    continuous, [MI] **mi impute pmm**, [MI] **mi impute regress**
        with a limited range, [MI] **mi impute intreg**, [MI] **mi impute truncreg**
    count data, [MI] **mi impute nbreg**, [MI] **mi impute poisson**
    diagnostics, [MI] **mi impute**
    interval-censored data, [MI] **mi impute intreg**
    interval regression, [MI] **mi impute intreg**
    linear regression, [MI] **mi impute regress**
    logistic regression, [MI] **mi impute logit**
    modeling, [MI] **mi impute**
    monotone, [MI] **mi impute**, [MI] **mi impute chained**, [MI] **mi impute monotone**

imputation, *continued*
   multinomial logistic regression, [MI] **mi impute mlogit**
   multiple, [MI] **intro substantive**
   multivariate, [MI] **mi impute chained**, [MI] **mi impute monotone**, [MI] **mi impute mvn**
   monotone, [MI] **mi impute**, [MI] **mi impute intreg**, [MI] **mi impute logit**, [MI] **mi impute mlogit**, [MI] **mi impute nbreg**, [MI] **mi impute ologit**, [MI] **mi impute pmm**, [MI] **mi impute poisson**, [MI] **mi impute regress**, [MI] **mi impute truncreg**
   normal, [MI] **mi impute**, [MI] **mi impute mvn**
   negative binomial regression, [MI] **mi impute nbreg**
   on subsamples, [MI] **mi impute**
   ordered logistic regression, [MI] **mi impute ologit**
   overdispersed count data, [MI] **mi impute nbreg**
   passive, [MI] **mi impute**, [MI] **mi impute chained**
   passive variables, [MI] **mi impute regress**
   perfect prediction, [MI] **mi impute**
   Poisson regression, [MI] **mi impute poisson**
   predictive mean matching, [MI] **mi impute**, [MI] **mi impute pmm**
   regression, [MI] **mi impute**, [MI] **mi impute regress**
   semiparametric, [MI] **mi impute pmm**
   step, [MI] **intro substantive**, [MI] **mi estimate**
   transformations, [MI] **mi impute**
   truncated data, [MI] **mi impute truncreg**
   truncated regression, [MI] **mi impute truncreg**
   univariate, [MI] **mi impute intreg**, [MI] **mi impute logit**, [MI] **mi impute mlogit**, [MI] **mi impute nbreg**, [MI] **mi impute ologit**, [MI] **mi impute pmm**, [MI] **mi impute poisson**, [MI] **mi impute regress**, [MI] **mi impute truncreg**
imputations, recommended number of, [MI] **intro substantive**, [MI] **mi estimate**
impute, mi subcommand, [MI] **mi impute**, [MI] **mi impute chained**, [MI] **mi impute intreg**, [MI] **mi impute logit**, [MI] **mi impute mlogit**, [MI] **mi impute monotone**, [MI] **mi impute mvn**, [MI] **mi impute nbreg**, [MI] **mi impute ologit**, [MI] **mi impute pmm**, [MI] **mi impute poisson**, [MI] **mi impute regress**, [MI] **mi impute truncreg**
imputed data, [MI] **Glossary**
imputed variables, *see* variables, imputed
imtest, estat subcommand, [R] **regress postestimation**
in *range* modifier, [P] **syntax**, [U] **11 Language syntax**
in smcl, display directive, [P] **display**
inbase( ) function, [M-5] **inbase( )**
incidence, [ST] **Glossary**
incidence rate, [ST] **Glossary**
incidence studies, [ST] **Glossary**
incidence-rate ratio, [R] **poisson**, [R] **zip**, [ST] **epitab**, [ST] **stci**, [ST] **stir**, [ST] **stptime**, [ST] **stsum**, [XT] **xtgee**, [XT] **xtmepoisson**, [XT] **xtnbreg**, [XT] **xtpoisson**
   differences, [R] **lincom**, [R] **nlcom**

include_bitmap, set subcommand, [R] **set**
include command, [P] **include**
income distributions, [R] **inequality**
income tax rate function, [D] **egen**
incomplete
   beta function, [D] **functions**, [M-5] **normal( )**
   gamma function, [D] **functions**, [M-5] **normal( )**
incomplete observations, [MI] **Glossary**
increment operator, [M-2] **op_increment**
independence of irrelevant alternatives, *see* IIA
independence tests, *see* tests
independent and identically distributed, [TS] **Glossary**
index of probit and logit, [R] **logit postestimation**, [R] **predict**, [R] **probit postestimation**
index search, [R] **search**, [U] **4 Stata's help and search facilities**
index,
   mathematical functions, [M-4] **statistical**
   matrix functions, [M-4] **utility**
   statistical functions, [M-4] **statistical**
   stopping rules, *see* stopping rules
   utility functions, [M-4] **utility**
indexnot( ) function, [D] **functions**, [M-5] **indexnot( )**
indicator variables, [R] **tabulate oneway**, [R] **xi**, [SEM] **Glossary**, [U] **11.4.3 Factor variables**, *also see* factor variables
indirect standardization, [R] **dstdize**
ineligible missing values, [MI] **mi impute**, [MI] **Glossary**
inequality measures, [R] **inequality**
inertia, [MV] **Glossary**
inertia, estat subcommand, [MV] **ca postestimation**
infile command, [D] **infile (fixed format)**, [D] **infile (free format)**
infix command, [D] **infix (fixed format)**
influence statistics, *see* delta beta influence statistic, *see* delta chi-squared influence statistic, *see* delta deviance influence statistic, *see* DFBETA
%infmt, [D] **infile (fixed format)**
information
   criteria, *see* Akaike information criterion, *see* Bayesian information criterion
   matrix, [P] **matrix get**, [R] **correlate**, [R] **maximize**
   matrix test, [R] **regress postestimation**
Informix, reading data from, [U] **21.4 Transfer programs**
inheritance, [M-2] **class**, [P] **class**
init,
   ml subcommand, [R] **ml**
   ssd subcommand, [SEM] **ssd**
initial values, [SEM] **Glossary**, *see* starting values
initialization, class, [P] **class**
inlist( ) function, [D] **functions**
inner fence, [R] **lv**
innovation accounting, [TS] **irf**
input command, [D] **input**
input, matrix subcommand, [P] **matrix define**

input, obtaining from console in programs, *see* console, obtaining input from

input/output functions, [M-4] **io**

inputting data

from a file, *see* importing data, *see* reading data from disk

interactively, [D] **edit**, [D] **input**, *also see* editing data, *also see* importing data

inrange() function, [D] **functions**

insert, odbc subcommand, [D] **odbc**

insheet command, [D] **insheet**

inspect command, [D] **inspect**

install,

net subcommand, [R] **net**

ssc subcommand, [R] **ssc**

installation

of official updates, [R] **update**, [U] **28 Using the Internet to keep up to date**

of SJ and STB, [R] **net**, [R] **sj**, [U] **3.6 Updating and adding features from the web**, [U] **17.6 How do I install an addition?**

of user-written commands (updating), [R] **adoupdate**

instance, [M-6] **Glossary**

class, [P] **class**

.instancemv built-in class function, [P] **class**

instance-specific variable, [P] **class**

instrumental-variables, [XT] **Glossary**

estimator, [XT] **Glossary**

regression, [R] **gmm**, [R] **ivprobit**, [R] **ivregress**, [R] **ivtobit**, [R] **nlsur**, [XT] **xtabond**, [XT] **xtdpd**, [XT] **xtdpdsys**, [XT] **xthtaylor**, [XT] **xtivreg**, [XT] **xtivreg postestimation**, [SVY] **svy estimation**

int (storage type), [D] **data types**, [I] **data types** [U] **12.2.2 Numeric storage types**

int() function, [D] **functions**

integ command, [R] **dydx**

integer truncation function, [D] **functions**

integers, [M-5] **trunc( )**

integrals, numeric, [R] **dydx**

integrated autoregressive moving average model, [TS] **psdensity**

integrated process, [TS] **Glossary**

intensity, color, adjustment, [G-2] **graph twoway histogram**, [G-2] **graph twoway kdensity**, [G-4] *colorstyle*

intensitystyle, [G-4] *intensitystyle*

interaction, [R] **anova**, [R] **contrast**, [R] **fvrevar**, [R] **margins**, [R] **margins, contrast**, [R] **margins, pwcompare**, [R] **marginsplot**, [R] **pwcompare**, [R] **set emptycells**, [R] **xi** [U] **11.4.3 Factor variables**, [U] **13.5.3 Factor variables and time-series operators**, [U] **20.18 Obtaining contrasts, tests of interactions, and main effects**, [U] **25.2 Estimation with factor variables**

expansion, [R] **xi**

plots, [R] **marginsplot**

intercept, [SEM] **intro 3**, [SEM] **Glossary**, *also see* constraints, specifying

internal consistency, test for, [R] **alpha**

Internet, [U] **3.2.1 The Stata website (www.stata.com)**

commands to control connections to, [R] **netio**

installation of updates from, [R] **adoupdate**, [R] **net**, [R] **sj**, [R] **update** [U] **28 Using the Internet to keep up to date**

search, [R] **net search**

interpolation, [D] **ipolate**

interquantile range, [R] **qreg**

interquartile range, [R] **lv**

generating variable containing, [D] **egen**

making dataset of, [D] **collapse**

reporting, [R] **table**, [R] **tabstat**

summarizing, [D] **pctile**

interrater agreement, [R] **kappa**

interrupting command execution, [U] **10 Keyboard use**

interval censoring, *see* imputation, interval-censored data

interval data, [XT] **Glossary**

interval regression, [R] **intreg**, [SVY] **svy estimation**

random-effects, [XT] **xtintreg**

intracluster correlation, [R] **loneway**

intreg command, [R] **intreg**, [R] **intreg postestimation**

inv() function, [D] **functions**, [P] **matrix define**

invbinomial() function, [D] **functions**, [M-5] **normal( )**

invbinomialtail() function, [D] **functions**, [M-5] **normal( )**

invchi2() function, [D] **functions**, [M-5] **normal( )**

invchi2tail() function, [D] **functions**, [M-5] **normal( )**

invcloglog() function, [D] **functions**, [M-5] **logit( )**

invdunnettprob() function, [D] **functions**, [M-5] **normal( )**

inverse

cumulative

beta distribution, [D] **functions**

binomial function, [D] **functions**

chi-squared distribution function, [D] **functions**

F distribution function, [D] **functions**

incomplete gamma function, [D] **functions**

matrix, [M-4] **solvers**, [M-5] **invsym( )**, [M-5] **cholinv( )**, [M-5] **luinv( )**, [M-5] **qrinv( )**, [M-5] **pinv( )**, [M-5] **solve_tol( )**

noncentral

beta distribution, [D] **functions**

chi-squared distribution function, [D] **functions**

F distribution, [D] **functions**

normal distribution function, [D] **functions**

reverse cumulative

beta distribution, [D] **functions**

binomial function, [D] **functions**

chi-squared distribution function, [D] **functions**

F distribution function, [D] **functions**

inverse, reverse cumulative, *continued*
  incomplete gamma function, [D] **functions**
  t distribution function, [D] **functions**
inverse of matrix, [P] **matrix define**, [P] **matrix svd**
invF() function, [D] **functions**, [M-5] **normal( )**
_invfft() function, [M-5] **fft( )**
invfft() function, [M-5] **fft( )**
invFtail() function, [D] **functions**, [M-5] **normal( )**
invgammap() function, [D] **functions**, [M-5] **normal( )**
invgammaptail() function, [D] **functions**,
  [M-5] **normal( )**
invHilbert() function, [M-5] **Hilbert( )**
invibeta() function, [D] **functions**, [M-5] **normal( )**
invibetatail() function, [D] **functions**,
  [M-5] **normal( )**
invlogit() function, [D] **functions**, [M-5] **logit( )**
invnbinomial() function, [D] **functions**,
  [M-5] **normal( )**
invnbinomialtail() function, [D] **functions**,
  [M-5] **normal( )**
invnchi2() function, [D] **functions**, [M-5] **normal( )**
invnFtail() function, [D] **functions**, [M-5] **normal( )**
invnibeta() function, [D] **functions**, [M-5] **normal( )**
invnormal() function, [D] **functions**, [M-5] **normal( )**
invorder() function, [M-5] **invorder( )**
invpoisson() function, [D] **functions**,
  [M-5] **normal( )**
invpoissontail() function, [D] **functions**,
  [M-5] **normal( )**
_invsym() function, [M-5] **invsym( )**
invsym() function, [D] **functions**, [M-5] **invsym( )**,
  [P] **matrix define**
invtokens() function, [M-5] **invtokens( )**
invttail() function, [D] **functions**, [M-5] **normal( )**
invtukeyprob() function, [D] **functions**,
  [M-5] **normal( )**
invvech() function, [M-5] **vec( )**
I/O functions, [M-4] **io**
ipolate command, [D] **ipolate**
ips, xtunitroot subcommand, [XT] **xtunitroot**
IQR, *see* interquartile range
iqr(), egen function, [D] **egen**
iqreg command, [R] **qreg**, [R] **qreg postestimation**
ir command, [ST] **epitab**
irecode() function, [D] **functions**
IRF, [G-2] **graph other**, [TS] **irf**, [TS] **irf add**, [TS] **irf**
  cgraph, [TS] **irf create**, [TS] **irf ctable**, [TS] **irf**
  describe, [TS] **irf drop**, [TS] **irf graph**, [TS] **irf**
  ograph, [TS] **irf rename**, [TS] **irf set**, [TS] **irf**
  table, [TS] **var intro**, [TS] **varbasic**, [TS] **vec**
  intro, [TS] **Glossary**
irf, [TS] **irf**
 add command, [TS] **irf add**
 cgraph command, [TS] **irf cgraph**
 create command, [TS] **irf create**
 ctable command, [TS] **irf ctable**
 describe command, [TS] **irf describe**
 drop command, [TS] **irf drop**

irf, *continued*
 graph command, [TS] **irf graph**
 ograph command, [TS] **irf ograph**
 rename command, [TS] **irf rename**
 set command, [TS] **irf set**
 table command, [TS] **irf table**
.irf files, [U] **11.6 Filenaming conventions**
iri command, [ST] **epitab**
IRLS, [R] **glm**, [R] **reg3**
IRR, *see* incidence-rate ratio
.isa built-in class function, [P] **class**
iscale() option, [G-2] **graph matrix**
iscomplex() function, [M-5] **isreal( )**
isdiagonal() function, [M-5] **isdiagonal( )**
isfleeting() function, [M-5] **isfleeting( )**
isid command, [D] **isid**
.isofclass built-in class function, [P] **class**
isolines, [G-2] **graph twoway contourline**
ispointer() function, [M-5] **isreal( )**
isreal() function, [M-5] **isreal( )**
isrealvalues() function, [M-5] **isrealvalues( )**
isstring() function, [M-5] **isreal( )**
issymmetric() function, [D] **functions**,
  [M-5] **issymmetric( )**, [P] **matrix define**
issymmetriconly() function, [M-5] **issymmetric( )**
istdize command, [R] **dstdize**
istmt, [M-1] **how**, [M-6] **Glossary**
isview() function, [M-5] **isview( )**
italics, [G-4] *text*
iterate() option, [R] **maximize**
iterated least squares, [R] **reg3**, [R] **sureg**
iterated principal-factor method, [MV] **factor**,
  [MV] **Glossary**
iterations, controlling the maximum number,
  [R] **maximize**
itrim() function, [D] **functions**
ivprobit command, [R] **ivprobit**, [R] **ivprobit**
  postestimation
ivregress command, [R] **ivregress**, [R] **ivregress**
  postestimation
ivtobit command, [R] **ivtobit**, [R] **ivtobit**
  postestimation
[iweight=*exp*] modifier, [U] **11.1.6 weight**,
  [U] **20.22.4 Importance weights**

J

J() function, [D] **functions**, [M-5] **J( )**, [M-2] **void**,
  [M-6] **Glossary**, [P] **matrix define**
Jaccard coefficient similarity measure,
  [MV] *measure_option*
jackknife
 estimation, [R] **jackknife**,
  [SVY] *jackknife_options*, [SVY] **svy jackknife**,
  [SVY] **variance estimation**, [SVY] **Glossary**
 standard errors, [R] *vce_option*, [XT] *vce_options*
jackknife_options, [SVY] *jackknife_options*

jackknife prefix command, [R] **jackknife**,
[R] **jackknife postestimation**
jackknifed residuals, [R] **predict**, [R] **regress
postestimation**
jackknifed standard error, *see* Monte Carlo error
Jarque–Bera statistic, [TS] **varnorm**, [TS] **vecnorm**
JCA, [MV] **Glossary**
Jeffreys noninformative prior, [MI] **mi impute mvn**
jitter() option, [G-2] **graph matrix**, [G-2] **graph
twoway scatter**
jitterseed() option, [G-2] **graph matrix**,
[G-2] **graph twoway scatter**
join operator, [M-2] **op_join**
joinby command, [D] **joinby**, [U] **22 Combining
datasets**
joining datasets, *see* combining datasets
joining time-span records, [ST] **stsplit**
joint correspondence analysis, [MV] **Glossary**
joint normality, *see* normality, joint
_jumble() function, [M-5] **sort( )**
jumble() function, [M-5] **sort( )**
Junior School Project data, [MI] **mi estimate**
justification of text, [G-3] *textbox_options*
justificationstyle, [G-4] *justificationstyle*

K

Kaiser–Meyer–Olkin sampling adequacy, [MV] **factor
postestimation**, [MV] **pca postestimation**,
[MV] **Glossary**
Kaiser normalization, [MV] **factor postestimation**,
[MV] **pca postestimation**, [MV] **rotate**,
[MV] **rotatemat**
Kalman
filter, [TS] **arima**, [TS] **dfactor**, [TS] **dfactor
postestimation**, [TS] **sspace**, [TS] **sspace
postestimation**, [TS] **ucm**, [TS] **ucm
postestimation**, [TS] **Glossary**
forecast, [TS] **dfactor postestimation**, [TS] **sspace
postestimation**, [TS] **ucm postestimation**
smoothing, [TS] **dfactor postestimation**,
[TS] **sspace postestimation**, [TS] **ucm
postestimation**
kap command, [R] **kappa**
Kaplan–Meier
product-limit estimate, [ST] **sts**, [ST] **sts generate**,
[ST] **sts graph**, [ST] **sts list**, [ST] **sts test**,
[ST] **Glossary**
survivor function, [ST] **ltable**, [ST] **stcox PH-
assumption tests**, [ST] **sts**
kappa command, [R] **kappa**
kapwgt command, [R] **kappa**
kdensity command, [R] **kdensity**
kdensity, graph twoway subcommand, [G-2] **graph
twoway kdensity**
keep command, [D] **drop**
keeping variables or observations, [D] **drop**
Kendall's tau, [R] **spearman**, [R] **tabulate twoway**
kernel density estimator, [R] **kdensity**

kernel-density smoothing, [G-2] **graph other**
kernel-weighted local polynomial estimator, [R] **lpoly**
keyboard
entry, [U] **10 Keyboard use**
search, [U] **4 Stata's help and search facilities**
keys, [G-3] *clegend_option*, [G-3] *legend_options*
Kish design effects, [R] **loneway**, [SVY] **estat**
Kmatrix() function, [M-5] **Kmatrix( )**
kmeans, [MV] **Glossary**
kmeans, cluster subcommand, [MV] **cluster kmeans
and kmedians**
kmeans clustering, [MV] **cluster**, [MV] **cluster kmeans
and kmedians**
kmedians, [MV] **Glossary**
kmedians, cluster subcommand, [MV] **cluster
kmeans and kmedians**
kmedians clustering, [MV] **cluster**, [MV] **cluster
kmeans and kmedians**
KMO, [MV] **Glossary**
kmo, estat subcommand, [MV] **factor postestimation**,
[MV] **pca postestimation**
KNN, [MV] **Glossary**
knn, discrim subcommand, [MV] **discrim knn**
Kolmogorov–Smirnov test, [R] **ksmirnov**
KR-20, [R] **alpha**
Kronecker direct-product operator, [D] **cross**,
[M-2] **op_kronecker**, [P] **matrix define**
Kruskal stress, [MV] **mds postestimation**,
[MV] **Glossary**
Kruskal–Wallis test, [R] **kwallis**
ksmirnov command, [R] **ksmirnov**
ktau command, [R] **spearman**
kth-nearest neighbor, [MV] **discrim knn**,
[MV] **Glossary**
Kuder–Richardson Formula 20, [R] **alpha**
Kulczynski coefficient similarity measure,
[MV] *measure_option*
kurt(), egen function, [D] **egen**
kurtosis, [MV] **mvtest normality**, [R] **lv**, [R] **pksumm**,
[R] **regress postestimation**, [R] **sktest**,
[R] **summarize**, [R] **tabstat**, [TS] **varnorm**,
[TS] **vecnorm**
kwallis command, [R] **kwallis**

L

L-R plots, [R] **regress postestimation**
L1-norm models, [R] **qreg**
l1title() option, [G-3] *title_options*
l2title() option, [G-3] *title_options*
label
copy command, [D] **label**
command, [U] **12.6 Dataset, variable, and value
labels**
data command, [D] **label**
define command, [D] **label**
dir command, [D] **label**
drop command, [D] **label**

label, *continued*
   language command, [D] **label language**
   list command, [D] **label**
   macro extended function, [P] **macro**
   save command, [D] **label**
   values command, [D] **label**
   variable command, [D] **label**
label, snapshot subcommand, [D] **snapshot**
label values, [U] **13.10 Label values**, [P] **macro**,
   [U] **12.6 Dataset, variable, and value labels**
labelbook command, [D] **labelbook**
labeling data, [D] **describe**, [D] **label**, [D] **label**
   language, [D] **edit**, [D] **notes**, [D] **varmanage**,
   [U] **12.6 Dataset, variable, and value labels**,
   [U] **12.6.4 Labels in other languages**
labels,
   creating, [D] **edit**, [D] **varmanage**
   axis, [G-3] *axis_label_options*
   editing, [D] **edit**, [D] **varmanage**
   marker, [G-3] *marker_label_options*
LAD regression, [R] **qreg**
ladder command, [R] **ladder**
ladder-of-powers plots, [G-2] **graph other**, [R] **ladder**
lag-exclusion statistics, [TS] **varwle**
lag-order selection statistics, [TS] **var**, [TS] **var intro**,
   [TS] **var svar**, [TS] **vec intro**, [TS] **varsoc**,
   [TS] **Glossary**, [U] **11.4.4 Time-series varlists**
lagged values, [U] **13.7 Explicit subscripting**,
   [U] **11.4.4 Time-series varlists**,
   [U] **13 Functions and expressions**,
   [U] **13.9.1 Generating lags, leads, and**
   differences
Lagrange-multiplier test, [R] **regress postestimation**
   time series, [SEM] **estat ginvariant**,
   [SEM] **estat mindices**, [SEM] **estat scoretests**,
   [SEM] **Glossary**, [TS] **varlmar**, [TS] **veclmar**
Lance and Williams' formula, [MV] **cluster**
language, label subcommand, [D] **label language**
language syntax, [P] **syntax**, [U] **11 Language syntax**
languages, multiple, [D] **label language**
LAPACK, [M-1] **LAPACK**, [M-5] **cholesky( )**,
   [M-5] **cholinv( )**, [M-5] **cholsolve( )**,
   [M-5] **eigensystem( )**, [M-5] **eigensystemselect( )**,
   [M-5] **fullsvd( )**, [M-5] **ghessenbergd( )**,
   [M-5] **lapack( )**, [M-5] **lud( )**, [M-5] **luinv( )**,
   [M-5] **lusolve( )**, [M-5] **qrd( )**, [M-5] **qrinv( )**,
   [M-5] **qrsolve( )**, [M-5] **svd( )**, [M-5] **svsolve( )**,
   [M-6] **Glossary**
latent( ) option, *see* sem option latent( )
latent growth models, *see* models, latent growth
latent roots, [M-5] **eigensystem( )**
latent variable, [SEM] **intro 3**, [SEM] **Glossary**
Latin-square designs, [MV] **manova**, [R] **anova**,
   [R] **pkshape**
LAV regression, [R] **qreg**
Lawley–Hotelling trace statistic, [MV] **canon**,
   [MV] **manova**, [MV] **mvtest means**,
   [MV] **Glossary**
lceffects, estat subcommand, [SVY] **estat**

lcolor( ) option, [G-3] *connect_options*,
   [G-3] *rspike_options*
LDA, [MV] **discrim lda**, [MV] **Glossary**
lda, discrim subcommand, [MV] **discrim lda**
lead
   operator, [U] **11.4.4 Time-series varlists**
   values, *see* lagged values
leap seconds, [TS] **tsset**
least absolute
   deviations, [R] **qreg**
   residuals, [R] **qreg**
   value regression, [R] **qreg**
least squared deviations, *see* linear regression
least squares, *see* linear regression
   generalized, *see* FGLS
   means, [R] **margins**, [R] **marginsplot**,
   [U] **20.15.1 Obtaining estimated marginal**
   means
left eigenvectors, [M-5] **eigensystem( )**, [M-6] **Glossary**
left suboption, [G-4] *justificationstyle*
left-censoring, [ST] **Glossary**, *see* imputation, interval-
   censored data
left-truncation, *see* imputation, truncated data, *see*
   truncation
_lefteigensystem( ) function, [M-5] **eigensystem( )**
lefteigensystem( ) function, [M-5] **eigensystem( )**
lefteigensystemselect*( ) functions,
   [M-5] **eigensystemselect( )**
leftgeigensystem( ) function, [M-5] **geigensystem( )**
leftgeigensystemselect*( ) function,
   [M-5] **geigensystem( )**
legend( ) option, [G-3] *legend_options*
legends, [G-3] *clegend_option*, [G-3] *legend_options*
   problems, [G-3] *legend_options*
   use with by( ), [G-3] *by_option*,
   [G-3] *clegend_option*, [G-3] *legend_options*
legendstyle, [G-4] *legendstyle*
length, [M-5] **abs( )**, [M-5] **rows( )**, [M-5] **strlen( )**
length macro extended function, [P] **macro**
length( ) function, [D] **functions**, [M-5] **rows( )**
length of string function, [D] **functions**
less than (or equal) operator, [U] **13.2.3 Relational**
   operators
letter values, [R] **lv**
level command and value, [P] **macro**
level( ) option, *see* sem option level( )
level, set subcommand, [R] **level**, [R] **set**
levels, [U] **11.4.3 Factor variables**
levelsof command, [P] **levelsof**
Levene's robust test statistic, [R] **sdtest**
leverage, [R] **logistic postestimation**, [R] **predict**,
   [R] **regress postestimation**
   obtaining with weighted data, [R] **predict**
leverage-versus-(squared)-residual plot, [G-2] **graph**
   other, [R] **regress postestimation**
Levin–Lin–Chu test, [XT] **xtunitroot**
_LEx, [SEM] **sem option covstructure( )**
lexis command, [ST] **stsplit**

lexis diagram, [ST] **stsplit**

lfit, graph twoway subcommand, [G-2] **graph twoway lfit**

lfitci, graph twoway subcommand, [G-2] **graph twoway lfitci**

libraries, [M-1] **how**, [M-3] **mata mlib**, [M-3] **mata which**

license, [R] **about**

life tables, [ST] **ltable**, [ST] **Glossary**

likelihood, *see* maximum likelihood estimation

likelihood displacement value, [ST] **stcox postestimation**, [ST] **Glossary**

likelihood-ratio
    chi-squared of association, [R] **tabulate twoway**
    test, [R] **lrtest** [SEM] **lrtest**, [SEM] **methods and formulas**, [U] **20.12.3 Likelihood-ratio tests**

Likert summative scales, [R] **alpha**

limited dependent variables, [R] **asclogit**,
    [R] **asmprobit**, [R] **asroprobit**, [R] **binreg**,
    [R] **biprobit**, [R] **brier**, [R] **clogit**, [R] **cloglog**,
    [R] **cusum**, [R] **exlogistic**, [R] **expoisson**,
    [R] **glm**, [R] **glogit**, [R] **heckprob**, [R] **hetprob**,
    [R] **ivprobit**, [R] **logistic**, [R] **logit**, [R] **mlogit**,
    [R] **mprobit**, [R] **nbreg**, [R] **nlogit**, [R] **ologit**,
    [R] **oprobit**, [R] **poisson**, [R] **probit**, [R] **rocfit**,
    [R] **rocreg**, [R] **rologit**, [R] **scobit**, [R] **slogit**,
    [R] **tnbreg**, [R] **tpoisson**, [R] **zinb**, [R] **zip**,
    [XT] **xtcloglog**, [XT] **xtgee**, [XT] **xtlogit**,
    [XT] **xtmelogit**, [XT] **xtmepoisson**,
    [XT] **xtnbreg**, [XT] **xtpoisson**, [XT] **xtprobit**

limits, [D] **describe**, [D] **memory**, [M-1] **limits**,
    [R] **matsize**, [U] **6 Managing memory**
    numerical and string, [P] **creturn**
    system, [P] **creturn**

lincom command, [R] **lincom**, [SEM] **intro 6**,
    [SEM] **estat stdize**, [SEM] **lincom**, [SVY] **svy postestimation**

line, definition, [G-4] *linestyle*

line, graph twoway subcommand, [G-2] **graph twoway line**

linear
    combinations, [SVY] **estat**, [SVY] **svy postestimation**
    discriminant analysis, [MV] **candisc**, [MV] **discrim lda**, [MV] **Glossary**
        forming, [P] **matrix score**
        of estimators, [R] **lincom**, [U] **20.13 Obtaining linear combinations of coefficients**
    filter, [TS] **Glossary**
    hypothesis test after estimation, [R] **contrast**,
        [R] **lrtest**, [R] **margins**, [R] **pwcompare**, [R] **test**
    interpolation and extrapolation, [D] **ipolate**
    mixed models, [XT] **xtmixed**
    prediction, *see* multiple imputation, prediction
    regression, [R] **anova**, [R] **areg**, [R] **binreg**,
        [R] **cnsreg**, [R] **eivreg**, [R] **frontier**,
        [R] **glm**, [R] **gmm**, [R] **heckman**, [R] **intreg**,
        [R] **ivregress**, [R] **ivtobit**, [R] **mvreg**, [R] **qreg**,
        [R] **reg3**, [R] **regress**, [R] **rreg**, [R] **sureg**,

linear, regression, *continued*
    [R] **tobit**, [R] **vwls**, [SEM] **Glossary**, [SVY] **svy estimation**, [TS] **newey**, [TS] **prais**, [TS] **var intro**, [TS] **var**, [TS] **var svar**, [TS] **varbasic**, [XT] **xtabond**, [XT] **xtdpd**, [XT] **xtdpdsys**, [XT] **xtfrontier**, [XT] **xtgee**, [XT] **xtgls**, [XT] **xthtaylor**, [XT] **xtintreg**, [XT] **xtivreg**, [XT] **xtmixed**, [XT] **xtpcse**, [XT] **xtrc**, [XT] **xtreg**, [XT] **xtregar**, [XT] **xttobit** *also see* generalized linear models, *also see* models, linear regression *also see* panel data, *also see* random-coefficients model,
        conditional, [U] **26.8 Conditional logistic regression**
        imputation, *see* imputation, regression
        with heteroskedastic errors, [U] **26.12 Linear regression with heteroskedastic errors**
        with simple error structures, [U] **26.2 Linear regression with simple error structures**
    splines, [R] **mkspline**
    tests, *see* estimation, tests after

linearization, *see* linearized variance estimator

linearized variance estimator, [SVY] **variance estimation**, [SVY] **Glossary**

linegap, set subcommand, [R] **set**

linepalette, palette subcommand, [G-2] **palette**

linepatternstyle, [G-4] *linepatternstyle*

lines, [G-4] **concept: lines**
    adding, [G-2] **graph twoway lfit**,
        [G-3] *added_line_options*, *also see* fits, adding
    connecting points, [G-3] *connect_options*,
        [G-4] *connectstyle*
    dashed, [G-4] *linepatternstyle*
    dotted, [G-4] *linepatternstyle*
    grid, [G-3] *axis_label_options*, [G-4] *linestyle*
    look of, [G-3] *fcline_options*, [G-3] *line_options*,
        [G-4] *linestyle*
    long, in do-files and ado-files, [P] **#delimit**,
        [U] **18.11.2 Comments and long lines in ado-files**
    patterns, [G-4] *linepatternstyle*
    suppressing, [G-4] *linestyle*
    thickness, [G-4] *linewidthstyle*

linesize, set subcommand, [R] **log**, [R] **set**

linestyle, [G-4] *linestyle*
    added, [G-4] *addedlinestyle*

linewidthstyle, [G-4] *linewidthstyle*

link function, [R] **glm**, [XT] **xtgee**, [XT] **Glossary**

link, net subcommand, [R] **net**

linkage, [MV] **Glossary**

linktest command, [R] **linktest**

list,
    char subcommand, [P] **char**
    cluster subcommand, [MV] **cluster utility**
    creturn subcommand, [P] **creturn**
    duplicates subcommand, [D] **duplicates**
    ereturn subcommand, [P] **ereturn**, [P] **return**,
        [R] **saved results**

list, *continued*
estat subcommand, [MV] **discrim estat**,
[MV] **discrim knn postestimation**,
[MV] **discrim lda postestimation**, [MV] **discrim logistic postestimation**, [MV] **discrim qda postestimation**
label subcommand, [D] **label**
macro subcommand, [P] **macro**
matrix subcommand, [P] **matrix utility**
notes subcommand, [D] **notes**
odbc subcommand, [D] **odbc**
program subcommand, [P] **program**
return subcommand, [P] **return**, [R] **saved results**
scalar subcommand, [P] **scalar**
snapshot subcommand, [D] **snapshot**
sreturn subcommand, [P] **return**, [R] **saved results**
ssd subcommand, [SEM] **ssd**
sts subcommand, [ST] **sts list**
sysdir subcommand, [P] **sysdir**
timer subcommand, [P] **timer**
list macro extended function, [P] **macro lists**
list manipulation, [P] **macro lists**
list subscripts, *see* subscripts
list command, [D] **list**
listing
data, [D] **edit**, [D] **list**
estimation results, [P] **ereturn**, [P] **_estimates**
macro expanded functions, [P] **macro lists**
values of a variable, [P] **levelsof**
listserver, [U] **3.4 The Stata listserver**
liststruct() function, [M-5] **liststruct( )**
listwise deletion, [MI] **intro substantive**, [MI] **mi estimate**, [MI] **Glossary**
llc, xtunitroot subcommand, [XT] **xtunitroot**
Lmatrix() function, [M-5] **Lmatrix( )**
LMAX value, [ST] **stcox postestimation**, [ST] **Glossary**
ln() function, [D] **functions**, [M-5] **exp( )**
lnfactorial() function, [D] **functions**, [M-5] **factorial( )**
lngamma() function, [D] **functions**, [M-5] **factorial( )**
lnnormal() function, [D] **functions**, [M-5] **normal( )**
lnnormalden() function, [D] **functions**, [M-5] **normal( )**
lnskew0 command, [R] **lnskew0**
load,
bcal subcommand, [D] **bcal**
odbc subcommand, [D] **odbc**
loading, [MV] **Glossary**
loading data, *see* importing data, *see* inputting data interactively, *see* loading saved data, *see* reading data from disk, *see* using data
loading plot, [MV] **scoreplot**, [MV] **Glossary**
loading saved data, [D] **use**
loadingplot command, [MV] **discrim lda postestimation**, [MV] **factor postestimation**, [MV] **pca postestimation**, [MV] **scoreplot**

loadings, estat subcommand, [MV] **ca postestimation**, [MV] **canon postestimation**, [MV] **discrim lda**, [MV] **discrim lda postestimation**, [MV] **pca postestimation**
local
++ command, [P] **macro**
-- command, [P] **macro**
command, [P] **macro**, [U] **18.3.1 Local macros**, [U] **18.3.9 Advanced local macro manipulation**
local,
ereturn subcommand, [P] **ereturn**, [P] **return**
return subcommand, [P] **return**
sreturn subcommand, [P] **return**
Local, class prefix operator, [P] **class**
local linear, [R] **lpoly**
local polynomial smoothing, [G-2] **graph other**, [G-2] **graph twoway lpoly**, [G-2] **graph twoway lpolyci**, [R] **lpoly**
locally weighted smoothing, [R] **lowess**
location, measures of, [R] **lv**, [R] **summarize**, [R] **table**
location, specifying, [G-4] *clockposstyle*, [G-4] *compassdirstyle*, [G-4] *ringposstyle*
locksplitters, set subcommand, [R] **set**
log
close command, [R] **log**
command, [R] **log**, [R] **view**, [U] **15 Saving and printing output—log files**, [U] **16.1.2 Comments and blank lines in do-files**
off command, [R] **log**
on command, [R] **log**
query command, [R] **log**
using command, [R] **log**
.log filename suffix, [U] **11.6 Filenaming conventions**
log files, *see* log command
printing, [R] **translate**, *also see* log command
log() function, [D] **functions**, [M-5] **exp( )**
log likelihood, [SEM] **methods and formulas**
log linear model, [R] **glm**, [R] **poisson**, [R] **zip**
log or nolog option, [R] **maximize**
log scales, [G-3] *axis_scale_options*
log transformations, [R] **boxcox**, [R] **lnskew0**
log10() function, [D] **functions**, [M-5] **exp( )**
logarithms, [M-5] **exp( )**, [M-5] **matexpsym( )**
logical operators, [M-2] **op_logical**, [U] **13.2.4 Logical operators**
logistic and logit regression, [R] **logistic**, [R] **logit**, [SVY] **svy estimation**
complementary log-log, [R] **cloglog**
conditional, [R] **asclogit**, [R] **clogit**, [R] **rologit**
exact, [R] **exlogistic**
fixed-effects, [R] **asclogit**, [R] **clogit**, [XT] **xtlogit**
generalized estimating equations, [XT] **xtgee**
generalized linear model, [R] **glm**
mixed-effects, [XT] **xtmelogit**
multinomial, [R] **asclogit**, [R] **clogit**, [R] **mlogit**
nested, [R] **nlogit**
ordered, [R] **ologit**
polytomous, [R] **mlogit**

logistic and logit regression, *continued*
> population-averaged, [XT] **xtgee**, [XT] **xtlogit**
> random-effects, [XT] **xtlogit**
> rank-ordered, [R] **rologit**
> skewed, [R] **scobit**
> stereotype, [R] **slogit**
> with grouped data, [R] **glogit**
logistic command, [R] **logistic**, [R] **logistic postestimation**
logistic, discrim subcommand, [MV] **discrim logistic**
logistic discriminant analysis, [MV] **discrim logistic**, [MV] **Glossary**
logistic regression imputation, *see* imputation, logistic regression
logit command, [R] **logit**, [R] **logit postestimation**
logit() function, [M-5] **logit( )**
logit function, [D] **functions**
logit regression, *see* logistic and logit regression
log-linear model, [SVY] **svy estimation**
log-log plot, [ST] **stcox PH-assumption tests**
loglogistic survival regression, [ST] **streg**
lognormal distribution, [R] **ameans**
lognormal survival regression, [ST] **streg**
log-rank,
> power, [ST] **stpower logrank**
> sample size, [ST] **stpower logrank**
> test, [ST] **stpower logrank**, [ST] **sts test**
logrank, stpower subcommand, [ST] **stpower logrank**
logtype, set subcommand, [R] **log**, [R] **set**
LOHI, [M-5] **byteorder( )**
loneway command, [R] **loneway**
long (storage type), [D] **data types**, [I] **data types**, [U] **12.2.2 Numeric storage types**
long lines in ado-files and do-files, [P] **#delimit**, [U] **18.11.2 Comments and long lines in ado-files**
long-memory process, [TS] **arfima**, [TS] **Glossary**
long, reshape subcommand, [D] **reshape**
longitudinal data, [MI] **mi estimate**, [XT] **Glossary**, *see* panel data
longitudinal studies, *see* incidence studies
longitudinal survey data, [SVY] **svy estimation**
LOO, [MV] **Glossary**
look of areas, [G-3] *area_options*, [G-3] *fitarea_options*
lookfor command, [D] **lookfor**
lookup,
> icd9 subcommand, [D] **icd9**
> icd9p subcommand, [D] **icd9**
loop, [M-2] **do**, [M-2] **for**, [M-2] **while**
> continuing, [M-2] **continue**
> exiting, [M-2] **break**
> use of semicolons in, [M-2] **semicolons**
loop, endless, *see* endless loop
looping, [P] **continue**, [P] **foreach**, [P] **forvalues**, [P] **while**

Lorenz curve, [R] **inequality**
loss, [MV] **Glossary**
loss to follow-up, [ST] **Glossary**
Lotus 1-2-3, reading data from, *see* spreadsheets
lower() function, [D] **functions**
lowercase-string function, [D] **functions**, [M-5] **strupper( )**
_lowertriangle() function, [M-5] **lowertriangle( )**
lowertriangle() function, [M-5] **lowertriangle( )**
lower-triangular matrix, *see* triangular matrix
lowess, *see* locally weighted smoothing
lowess command, [R] **lowess**
lowess, graph twoway subcommand, [G-2] **graph twoway lowess**
lowess smoothing, [G-2] **graph other**
lpattern() option, [G-3] *connect_options*, [G-3] *rspike_options*
lpoly command, [R] **lpoly**
L-R plots, [G-2] **graph other**
LRECLs, [D] **infile (fixed format)**
lroc command, [R] **logistic postestimation**
lrtest command, [R] **lrtest**, [SEM] **example 10**
ls command, [D] **dir**
lsens command, [R] **logistic postestimation**
lstat command, *see* estat classification command
lstretch, set subcommand, [R] **set**
lstyle() option, [G-3] *rspike_options*
ltable command, [ST] **ltable**
ltolerance() option, [R] **maximize**
ltrim() function, [D] **functions**
LU decomposition, [M-5] **lud( )**
_lud() function, [M-5] **lud( )**
lud() function, [M-5] **lud( )**
_lud_la() function, [M-5] **lud( )**
_luinv() function, [M-5] **luinv( )**
luinv() function, [M-5] **luinv( )**
_luinv_la() function, [M-5] **luinv( )**
_lusolve() function, [M-5] **lusolve( )**
lusolve() function, [M-5] **lusolve( )**
_lusolve_la() function, [M-5] **lusolve( )**
lv command, [R] **lv**
lval, [M-2] **op_assignment**, [M-6] **Glossary**
lvalue, class, [P] **class**
lvr2plot command, [R] **regress postestimation**
lwidth() option, [G-3] *connect_options*, [G-3] *rspike_options*

M

M, [MI] **mi impute**, [MI] **Glossary**
> size recommendations, [MI] **intro substantive**, [MI] **mi estimate**
m, [MI] **Glossary**
MA, [TS] **arch**, [TS] **arfima**, [TS] **arima**, [TS] **sspace**, [TS] **ucm**
ma, tssmooth subcommand, [TS] **tssmooth ma**

Mac,
keyboard use, [U] **10 Keyboard use**
pause, [P] **sleep**
specifying filenames, [U] **11.6 Filenaming
conventions**
machine precision, [M-5] **epsilon( )**, [M-6] **Glossary**
macro
dir command, [P] **macro**
drop command, [P] **macro**
list command, [P] **macro**
shift command, [P] **macro**
macro substitution, [P] **macro**
class, [P] **class**
macros, [P] **creturn**, [P] **scalar**, [P] **syntax**, [P] **macro**,
[U] **18.3 Macros**, *also see* e( )
macval( ) macro expansion function, [P] **macro**
mad( ), egen function, [D] **egen**
MAD regression, [R] **qreg**
Mahalanobis distance, [MV] **Glossary**
Mahalanobis transformation, [MV] **Glossary**
main effects, [MV] **manova**, [R] **anova**
makecns command, [P] **makecns**
_makesymmetric( ) function, [M-5] **makesymmetric( )**
makesymmetric( ) function, [M-5] **makesymmetric( )**
man command, [R] **help**
manage, window subcommand, [P] **window
programming**
MANCOVA, [MV] **manova**, [MV] **Glossary**
mangle option, [G-2] **graph twoway pcarrow**
manifest variables, [SEM] **Glossary**
manipulation commands, [G-2] **graph manipulation**
Mann–Whitney two-sample statistics, [R] **ranksum**
MANOVA, [MV] **manova**, [MV] **Glossary**
manova command, [MV] **manova**
manova, estat subcommand, [MV] **discrim lda
postestimation**
manovatest command, [MV] **manova postestimation**
Mantel–Cox method, [ST] **strate**
Mantel–Haenszel
method, [ST] **strate**
test, [ST] **epitab**, [ST] **stir**
mapping strings to numbers, [D] **destring**, [D] **encode**,
[D] **label**, *also see* real( ) string function
maps, [M-5] **asarray( )**
MAR, *see* missing at random, *see* missing values
marginal
effects, [R] **margins**, [R] **marginsplot**,
[U] **20.15 Obtaining marginal means,
adjusted predictions, and predictive margins**,
[U] **20.19 Graphing margins, marginal effects,
and contrasts**
homogeneity, test of, [R] **symmetry**
means, [R] **contrast**, [R] **margins**, [R] **margins,
contrast**, [R] **margins, pwcompare**,
[R] **marginsplot**, [R] **pwcompare**,
[U] **20.15 Obtaining marginal means, adjusted
predictions, and predictive margins**
tax rate egen function, [D] **egen**

margins command, [R] **margins**, [R] **margins
postestimation**, [R] **margins, contrast**,
[R] **margins, pwcompare**, [R] **marginsplot**,
[SVY] **svy postestimation**, [U] **20.15 Obtaining
marginal means, adjusted predictions, and
predictive margins**
margins, size of, [G-4] *marginstyle*
marginsplot command, [R] **marginsplot**,
[U] **20.19 Graphing margins, marginal effects,
and contrasts**
marginstyle, [G-3] *region_options*,
[G-3] *textbox_options*, [G-4] *marginstyle*
mark command, [P] **mark**
marker labels, [G-3] *marker_label_options*,
[G-4] *markerlabelstyle*
markerlabelstyle, [G-4] *markerlabelstyle*
markers, [G-3] *marker_options*, *also see* marker labels
color, [G-4] *colorstyle*
resizing, [G-3] *scale_option*
shape of, [G-4] *symbolstyle*
size of, [G-4] *markersizestyle*
markersizestyle, [G-4] *markersizestyle*
markerstyle, [G-4] *markerstyle*
markin command, [P] **mark**
marking observations, [P] **mark**
markout command, [P] **mark**
Markov chain Monte Carlo, *see* MCMC
marksample command, [P] **mark**
Marquardt algorithm, [M-5] **moptimize( )**,
[M-5] **optimize( )**
martingale residual, [ST] **stcox postestimation**,
[ST] **streg postestimation**
mass, [MV] **Glossary**
Mata, [D] **putmata**
commands, [M-3] **intro**
error messages, [M-5] **error( )**, *also see* traceback
log
mata, [M-3] **mata clear**, [M-3] **mata describe**,
[M-3] **mata drop**, [M-3] **mata help**, [M-3] **mata
matsave**, [M-3] **mata memory**, [M-3] **mata
mlib**, [M-3] **mata mosave**, [M-3] **mata rename**,
[M-3] **mata set**, [M-3] **mata stata**, [M-3] **mata
which**, [M-3] **namelists**
mata, clear subcommand, [D] **clear**
.mata filename suffix, [M-1] **source**, [M-3] **mata
mlib**, [M-6] **Glossary** [U] **11.6 Filenaming
conventions**
mata invocation command, [M-3] **mata**
matacache, set subcommand, [M-3] **mata set**, [R] **set**
matafavor, set subcommand, [M-3] **mata set**,
[M-5] **favorspeed( )**, [R] **set**
matalibs, set subcommand, [M-3] **mata set**, [R] **set**
matalnum, set subcommand, [M-3] **mata set**, [R] **set**
matamofirst, set subcommand, [M-3] **mata set**,
[R] **set**
mataoptimize, set subcommand, [M-3] **mata set**,
[R] **set**
matastrict, set subcommand, [M-3] **mata set**,
[M-2] **declarations**, [M-1] **ado**, [R] **set**

matched case–control data, [R] **asclogit**, [R] **clogit**,
        [R] **symmetry**, [ST] **epitab**, [ST] **sttocc**,
        [ST] **Glossary**
matched-pairs tests, [R] **signrank**, [R] **ttest**
matching coefficient, [MV] **Glossary**
matching coefficient similarity measure,
        [MV] *measure_option*
matching configuration, [MV] **Glossary**
matcproc command, [P] **makecns**
matdescribe, [M-3] **mata matsave**
_matexpsym( ) function, [M-5] **matexpsym( )**
matexpsym( ) function, [M-5] **matexpsym( )**
math symbols, [G-4] *text*
mathematical functions and expressions, [D] **functions**,
        [M-4] **mathematical**, [M-4] **matrix**,
        [M-4] **scalar**, [M-4] **solvers**, [M-4] **standard**
        [P] **matrix define**, [U] **13.3 Functions**
Matlab, reading data from, [U] **21.4 Transfer programs**
matlist command, [P] **matlist**
_matlogsym( ) function, [M-5] **matexpsym( )**
matlogsym( ) function, [M-5] **matexpsym( )**
matmissing( ) function, [D] **functions**, [P] **matrix
        define**
matname command, [P] **matrix mkmat**
_matpowersym( ) function, [M-5] **matpowersym( )**
matpowersym( ) function, [M-5] **matpowersym( )**
mat_put_rr command, [P] **matrix get**
matrices, [P] **matrix**, [U] **14 Matrix expressions**
    accessing internal, [P] **matrix get**
    accumulating, [P] **matrix accum**
    appending rows and columns, [P] **matrix define**
    Cholesky decomposition, [P] **matrix define**
    coefficient matrices, [P] **ereturn**
    column names, *see* matrices, row and column names
    constrained estimation, [P] **makecns**
    copying, [P] **matrix define**, [P] **matrix get**,
        [P] **matrix mkmat**
    correlation, [MV] **pca**, [P] **matrix define**
    covariance matrix of estimators, [MV] **pca**,
        [P] **ereturn**, [P] **matrix get**
    cross-product, [P] **matrix accum**
    determinant, [P] **matrix define**
    diagonals, [P] **matrix define**
    displaying, [P] **matlist**, [P] **matrix utility**
    dissimilarity, [MV] **matrix dissimilarity**,
        [MV] **Glossary**, [P] **matrix dissimilarity**
    distances, [MV] **matrix dissimilarity**, [P] **matrix
        dissimilarity**
    dropping, [P] **matrix utility**
    eigenvalues, [P] **matrix eigenvalues**, [P] **matrix
        symeigen**
    eigenvectors, [P] **matrix symeigen**
    elements, [P] **matrix define**
    equation names, *see* matrices, row and column
        names
    estimation results, [P] **ereturn**, [P] **_estimates**
    functions, [D] **functions**, [P] **matrix define**
    identity, [P] **matrix define**

matrices, *continued*
    input, [P] **matrix define**, [U] **14.4 Inputting
        matrices by hand**
    inversion, [P] **matrix define**, [P] **matrix svd**
    Kronecker product, [P] **matrix define**
    labeling rows and columns, *see* matrices, row and
        column names
    linear combinations with data, [P] **matrix score**
    listing, [P] **matlist**, [P] **matrix utility**
    namespace and conflicts, [P] **matrix**, [P] **matrix
        define**
    number of rows and columns, [P] **matrix define**
    operators such as addition, [P] **matrix define**,
        [U] **14.7 Matrix operators**
    orthonormal basis, [P] **matrix svd**
    partitioned, [P] **matrix define**
    performing constrained estimation, [P] **makecns**
    posting estimation results, [P] **ereturn**,
        [P] **_estimates**
    renaming, [P] **matrix utility**
    row and column names, [P] **ereturn**, [P] **matrix
        define**, [P] **matrix mkmat**, [P] **matrix
        rownames**, [U] **14.2 Row and column names**
    rows and columns, [P] **matrix define**
    saving matrix, [P] **matrix mkmat**
    scoring, [P] **matrix score**
    similarity, [MV] **matrix dissimilarity**, [P] **matrix
        dissimilarity**
    store variables as matrix, [P] **matrix mkmat**
    submatrix extraction, [P] **matrix define**
    submatrix substitution, [P] **matrix define**
    subscripting, [P] **matrix define**,
        [U] **14.9 Subscripting**
    sweep operator, [P] **matrix define**
    temporary names, [P] **matrix**
    trace, [P] **matrix define**
    transposing, [P] **matrix define**
    variables, make into matrix, [P] **matrix mkmat**
    zero, [P] **matrix define**
matrix( ) function, [D] **functions**, [P] **matrix define**
matrix, [M-4] **intro**, [M-6] **Glossary**
    functions, [M-4] **manipulation**, [M-4] **matrix**,
        [M-4] **solvers**, [M-4] **standard**
    norm, [M-5] **norm( )**
matrix, [M-2] **declarations**
matrix
    accum command, [P] **matrix accum**
    coleq command, [P] **matrix rownames**
    colnames command, [P] **matrix rownames**
    commands, introduction, [P] **matrix**
    define command, [P] **matrix define**
    dir command, [P] **matrix utility**
    dissimilarity command, [MV] **matrix
        dissimilarity**, [P] **matrix dissimilarity**
    drop command, [P] **matrix utility**
    eigenvalues command, [P] **matrix eigenvalues**
    glsaccum command, [P] **matrix accum**
    graph subcommand, [G-2] **graph matrix**

matrix, *continued*
   input command, [P] **matrix define**
   list command, [P] **matrix utility**
   opaccum command, [P] **matrix accum**
   rename command, [P] **matrix utility**
   roweq command, [P] **matrix rownames**
   rownames command, [P] **matrix rownames**
   score command, [P] **matrix score**
   svd command, [P] **matrix svd**
   symeigen command, [P] **matrix symeigen**
   vecaccum command, [P] **matrix accum**
matrix,
   clear subcommand, [D] **clear**
   confirm subcommand, [P] **confirm**
   ereturn subcommand, [P] **ereturn**, [P] **return**
   return subcommand, [P] **return**
matrix graphs, [G-2] **graph matrix**
matsave, [M-3] **mata matsave**
matsize command, [M-1] **limits**, [P] **creturn**,
   [P] **macro**, [U] **14 Matrix expressions**
matsize, set subcommand, [R] **matsize**, [R] **set**
matuniform() function, [D] **functions**, [P] **matrix
   define**
matuse, [M-3] **mata matsave**
max(),
   built-in function, [D] **functions**, [M-5] **minmax()**
   egen function, [D] **egen**
max_memory, set subcommand, [R] **set**
maxbyte() function, [D] **functions**
maxdb, set subcommand, [R] **db**, [R] **set**
maxdouble() function, [D] **functions**,
   [M-5] **mindouble()**
maxes() option, [G-2] **graph matrix**
maxfloat() function, [D] **functions**
maximization, [M-5] **moptimize()**, [M-5] **optimize()**
maximization technique explained, [R] **maximize**,
   [R] **ml**
maximize, ml subcommand, [R] **ml**
maximize_options, *see* sem option *maximize_options*
maximum
   function, [D] **egen**, [D] **functions**
   length of string, [M-1] **limits**
   likelihood, [SEM] **intro 3**, [SEM] **methods and
     formulas**, [SEM] **Glossary**
     with missing values, [SEM] **example 26**,
       [SEM] **Glossary**
   likelihood estimation, [MV] **factor**, [R] **maximize**,
     [R] **ml**
   likelihood factor method, [MV] **Glossary**
   number of observations, [D] **memory**
   number of variables, [D] **describe**, [D] **memory**
   number of variables and observations,
     [U] **6 Managing memory**
   number of variables in a model, [R] **matsize**
   pseudolikelihood estimation, [SVY] **ml for svy**,
     [SVY] **variance estimation**
   restricted-likelihood, [XT] **xtmixed**
   size of dataset, [U] **6 Managing memory**

maximum, *continued*
   size of matrix, [M-1] **limits**
   value dissimilarity measure, [MV] *measure_option*
maximums and minimums,
   creating dataset of, [D] **collapse**
   functions, [D] **egen**, [D] **functions**,
     [M-5] **minindex()**
   reporting, [R] **lv**, [R] **summarize**, [R] **table**
maxindex() function, [M-5] **minindex()**
maxint() function, [D] **functions**
maxiter, set subcommand, [R] **maximize**, [R] **set**
maxlong() function, [D] **functions**
max_memory, set subcommand, [D] **memory**
maxvar, set subcommand, [D] **memory**, [R] **set**
mband, graph twoway subcommand, [G-2] **graph
   twoway mband**
MCA, [MV] **Glossary**
mca command, [MV] **mca**
mcaplot command, [MV] **mca postestimation**
mcaprojection command, [MV] **mca postestimation**
MCAR, *see* missing completely at random
mcc command, [ST] **epitab**
mcci command, [ST] **epitab**
MCE, *see* Monte Carlo error
McFadden's choice model, [R] **asclogit**
MCMC, [MI] **ml impute**, [MI] **mi impute mvn**,
   [MI] **Glossary**
   convergence, [MI] **mi impute**, [MI] **mi impute
     chained**, [MI] **mi impute mvn**
   parameter trace files, [MI] **mi ptrace**
McNemar's chi-squared test, [R] **clogit**, [ST] **epitab**
mcolor() option, [G-3] *marker_options*
md command, [D] **mkdir**
mdev(), egen function, [D] **egen**
MDS, [MV] **Glossary**
mds command, [MV] **mds**
MDS configuration plot, [MV] **Glossary**
mdsconfig command, [MV] **mds**, [MV] **mds
   postestimation**
mdslong command, [MV] **mdslong**
mdsmat command, [MV] **mdsmat**
mdsshepard command, [MV] **mds postestimation**
mdy() function, [D] **datetime**, [D] **functions**,
   [M-5] **date()**
mdyhms() function, [D] **datetime**, [D] **functions**,
   [M-5] **date()**
mean command, [R] **mean**, [R] **mean postestimation**
mean(), egen function, [D] **egen**
mean() function, [M-5] **mean()**
means,
   across variables, not observations, [D] **egen**
   arithmetic, geometric, and harmonic, [R] **ameans**
   confidence interval and standard error, [R] **ci**
   creating
     dataset of, [D] **collapse**
     variable containing, [D] **egen**

means, *continued*
  displaying, [R] **ameans**, [R] **summarize**, [R] **table**,
    [R] **tabstat**, [R] **tabulate, summarize()**,
    [XT] **xtsum**
  estimating, [R] **mean**
  graphing, [R] **grmeanby**
  marginal, [R] **margins**
  pairwise comparisons of, [R] **pwmean**
  pharmacokinetic data, [R] **pksumm**
  robust, [R] **rreg**
  sample size and power for, [R] **sampsi**
  survey data, [SVY] **svy estimation**
  testing equality of, *see* equality tests of means
means, mvtest subcommand, [MV] **mvtest means**
means() option, *see* sem option means()
meanvariance() function, [M-5] **mean( )**
measure, [MV] **Glossary**
measurement
  component, [SEM] **Glossary**
  error, [R] **alpha**, [R] **vwls**
  error models, *see* models, measurement error
  models, [SEM] **Glossary**, *also see* models,
    measurement
  variables, [SEM] **Glossary**
measures, cluster subcommand, [MV] **cluster
  programming utilities**
measures of
  association, [R] **tabulate twoway**
  central tendency, *see* means, *see* medians
  dispersion, *see* standard deviations, displaying, *see*
    variance, displaying, *see* percentiles, displaying,
    see range of data
  inequality, [R] **inequality**
  location, [R] **lv**, [R] **summarize**
median command, [R] **ranksum**
median(), egen function, [D] **egen**
median regression, [R] **qreg**
median test, [R] **ranksum**
medianlinkage,
  clustermat subcommand, [MV] **cluster linkage**
  cluster subcommand, [MV] **cluster linkage**
median-linkage clustering, [MV] **cluster**,
  [MV] **clustermat**, [MV] **cluster linkage**,
  [MV] **Glossary**
medians,
  creating
    dataset of, [D] **collapse**
    variable containing, [D] **egen**
  displaying, [D] **pctile**, [R] **centile**, [R] **lv**,
    [R] **summarize**, [R] **table**, [R] **tabstat**
  graphing, [R] **grmeanby**
  testing equality of, *see* equality tests of medians
MEFF, *see* misspecification effects
MEFT, *see* misspecification effects
member
  function, [M-2] **class**
  programs, [P] **class**
  variables, [M-2] **class**, [P] **class**

memory, [U] **6 Managing memory**
  clearing, [D] **clear**
  determining and resetting limits, [D] **describe**,
    [D] **memory**
  matsize, *see* matsize, set subcommand
  reducing utilization, [D] **compress**, [D] **encode**,
    [D] **recast**, [P] **discard**
memory command, [D] **memory**, [M-3] **mata memory**,
  [U] **6 Managing memory**
memory, query subcommand, [D] **memory**
memory graphs, describing contents, [G-2] **graph
  describe**
memory requirements, estimating for flongsep, [MI] **mi
  convert**
memory settings, [P] **creturn**
memory utilization, [M-1] **limits**, [M-3] **mata memory**
menu, window subcommand, [P] **window programming**
menus, programming, [P] **dialog programming**,
  [P] **window programming**
merge command, [D] **merge**, [U] **22 Combining
  datasets**
merge, mi subcommand, [MI] **mi merge**
_merge variable, [D] **merge**
merged-explicit options, [G-4] **concept: repeated
  options**
merged-implicit options, [G-4] **concept: repeated
  options**
merging data, [MI] **mi merge**, *see* combining datasets
messages and return codes, *see* error messages and
  return codes
meta-analysis, [R] **meta**
method, [M-2] **class**, [SEM] **Glossary**
method() option, *see* sem option method()
metric scaling, [MV] **Glossary**
mfcolor() option, [G-3] *marker_options*
mfp prefix command, [R] **mfp**, [R] **mfp postestimation**
mfx, estat subcommand, [R] **asclogit postestimation**,
  [R] **asmprobit postestimation**, [R] **asroprobit
  postestimation**
MGARCH, *see* multivariate GARCH
mgarch
  ccc command, [TS] **mgarch ccc**
  dcc command, [TS] **mgarch dcc**
  dvech command, [TS] **mgarch dvech**
  vcc command, [TS] **mgarch vcc**
mhodds command, [ST] **epitab**
mi
  add command, [MI] **mi add**
  append command, [MI] **mi append**
  convert command, [MI] **mi convert**
  copy command, [MI] **mi copy**, [MI] **styles**
  describe command, [MI] **mi describe**
  erase command, [MI] **mi erase**, [MI] **styles**
  estimate command, [MI] **mi estimate**,
    [MI] **mi estimate using**, [MI] **mi estimate
    postestimation**, [MI] **mi test**
  estimate postestimation, [MI] **mi estimate
    postestimation**, [MI] **mi predict**, [MI] **mi test**

mi, *continued*

    expand command, [MI] **mi expand**

    export command, [MI] **mi export**, [MI] **mi export ice**, [MI] **mi export nhanes1**

    extract command, [MI] **mi extract**, [MI] **mi replace0**

    fvset command, [MI] **mi XXXset**

    import command, [MI] **mi import**, [MI] **mi import flong**, [MI] **mi import flongsep**, [MI] **mi import ice**, [MI] **mi import nhanes1**, [MI] **mi import wide**

    impute command, [MI] **mi impute**, [MI] **mi impute chained**, [MI] **mi impute intreg**, [MI] **mi impute logit**, [MI] **mi impute mlogit**, [MI] **mi impute monotone**, [MI] **mi impute mvn**, [MI] **mi impute nbreg**, [MI] **mi impute ologit**, [MI] **mi impute pmm**, [MI] **mi impute poisson**, [MI] **mi impute regress**, [MI] **mi impute truncreg**

    merge command, [MI] **mi merge**

    misstable command, [MI] **mi misstable**

    passive command, [MI] **mi passive**

    predict command, [MI] **mi estimate postestimation**, [MI] **mi predict**

    predictnl command, [MI] **mi estimate postestimation**, [MI] **mi predict**

    ptrace command, [MI] **mi ptrace**

    query command, [MI] **mi describe**

    register command, [MI] **mi set**

    rename command, [MI] **mi rename**

    replace0 command, [MI] **mi replace0**

    reset command, [MI] **mi reset**

    reshape command, [MI] **mi reshape**

    select command, [MI] **mi select**, *also see* mi extract command

    set command, [MI] **mi set**

    st command, [MI] **mi XXXset**

    stjoin command, [MI] **mi stsplit**

    streset command, [MI] **mi XXXset**

    stset command, [MI] **mi XXXset**

    stsplit command, [MI] **mi stsplit**

    svyset command, [MI] **mi XXXset**

    test command, [MI] **mi estimate postestimation**, [MI] **mi test**

    testtransform command, [MI] **mi estimate postestimation**, [MI] **mi test**

    tsset command, [MI] **mi XXXset**

    unregister command, [MI] **mi set**

    unset command, [MI] **mi set**

    update command, [MI] **mi update**, [MI] **noupdate option**

    varying command, [MI] **mi varying**

    xeq command, [MI] **mi xeq**

    xtset command, [MI] **mi XXXset**

mi command, [MI] **intro**, [MI] **styles**, [MI] **workflow**

mi data, [MI] **Glossary**

mi() function, [D] **functions**

Microsoft

    Access, reading data from, [D] **odbc**, [U] **21.4 Transfer programs**

    Excel, reading data from, [D] **import excel**, [D] **odbc**

    SpreadsheetML, [D] **xmlsave**

    Windows, *see* Windows

middle suboption, [G-4] *alignmentstyle*

midsummaries, [R] **lv**

mild outliers, [R] **lv**

Mills' ratio, [R] **heckman**, [R] **heckman postestimation**

MIMIC models, *see* models, multiple indicators and multiple causes

min(),

    built-in function, [D] **functions**, [M-5] **minmax( )**

    egen function, [D] **egen**

min_memory, set subcommand, [R] **set**

minbyte() function, [D] **functions**

mindices, estat subcommand, [SEM] **estat mindices**

mindouble() function, [D] **functions**, [M-5] **mindouble( )**

minfloat() function, [D] **functions**

minimal

    detectable difference, hazard ratio, [ST] **stpower**

    effect size, hazard ratio, [ST] **stpower**

minimization, [M-5] **moptimize( )**, [M-5] **optimize( )**

minimum

    absolute deviations, [R] **qreg**

    squared deviations, [R] **areg**, [R] **cnsreg**, [R] **nl**, [R] **regress**, [R] **regress postestimation**

minimum entropy rotation, [MV] **rotate**, [MV] **rotatemat**, [MV] **Glossary**

minimums, [M-5] **minindex( )**

minimums and maximums, *see* maximums and minimums

minindex() function, [M-5] **minindex( )**

minint() function, [D] **functions**

Minkowski dissimilarity measure, [MV] *measure_option*

minlong() function, [D] **functions**

minmax() function, [M-5] **minmax( )**

min_memory, set subcommand, [D] **memory**

minutes() function, [D] **datetime**, [D] **functions**, [M-5] **date( )**

misclassification rate, [MV] **Glossary**

missing at random, [MI] **intro substantive**, [MI] **mi impute**, [MI] **Glossary**

missing completely at random, [MI] **intro substantive**, [MI] **Glossary**

missing data, [MI] **intro substantive**

    arbitrary pattern, [MI] **intro substantive**, [MI] **mi impute**, [MI] **mi impute chained**, [MI] **mi impute mvn**, [MI] **Glossary**

    monotone pattern, [MI] **intro substantive**, [MI] **mi impute**, [MI] **mi impute chained**, [MI] **mi impute monotone**, [MI] **mi impute mvn**, [MI] **Glossary**

missing() function, [D] **functions**, [M-5] **missing( )**

missing not at random, [MI] **intro substantive**, [MI] **Glossary**

missing values, [M-5] **missing( )**, [M-5] **missingof( )**, [M-5] **editmissing( )**, [M-5] **_fillmissing( )**, [D] **missing values**, [I] **missing values**, [R] **misstable**, [SEM] **example 26**, [U] **12.2.1 Missing values**, [U] **13 Functions and expressions**, [U] **17.5 Where does Stata look for ado-files?**

  counting, [D] **codebook**, [D] **inspect**

  encoding and decoding, [D] **mvencode**

  extended, [D] **mvencode**

  hard and soft, [MI] **Glossary**

  ineligible, [MI] **Glossary**

  pattern of, [MI] **mi misstable**

  replacing, [D] **merge**

missingness, pattern, *see* pattern of missingness

missingof() function, [M-5] **missingof( )**

misspecification effects, [SVY] **estat**, [SVY] **Glossary**

misstable

  nested command, [R] **misstable**

  patterns command, [R] **misstable**

  summarize command, [R] **misstable**

  tree command, [R] **misstable**

misstable for mi data, [MI] **mi misstable**

misstable, mi subcommand, [MI] **mi misstable**

mixed designs, [MV] **manova**, [R] **anova**

mixed model, [XT] **xtmelogit**, [XT] **xtmepoisson**, [XT] **xtmixed**, [XT] **Glossary**

mkdir command, [D] **mkdir**

_mkdir() function, [M-5] **chdir( )**

mkdir() function, [M-5] **chdir( )**

mkmat command, [P] **matrix mkmat**

mkspline command, [R] **mkspline**

ML, *see* maximum likelihood

ml, *also see* sem option method()

  check command, [R] **ml**

  clear command, [R] **ml**

  count command, [R] **ml**

  display command, [R] **ml**

  footnote command, [R] **ml**

  graph command, [R] **ml**

  init command, [R] **ml**

  maximize command, [R] **ml**

  model command, [R] **ml**

  plot command, [R] **ml**

  query command, [R] **ml**

  report command, [R] **ml**

  score command, [R] **ml**

  search command, [R] **ml**

  trace command, [R] **ml**

ml command, [SVY] **ml for svy**

mlabangle() option, [G-3] *marker_label_options*

mlabcolor() option, [G-3] *marker_label_options*

mlabel() option, [G-3] *marker_label_options*

mlabgap() option, [G-3] *marker_label_options*

mlabposition() option, [G-3] *marker_label_options*

mlabsize() option, [G-3] *marker_label_options*

mlabstyle() option, [G-3] *marker_label_options*

mlabtextstyle() option, [G-3] *marker_label_options*

mlabvposition() option, [G-3] *marker_label_options*

mlcolor() option, [G-3] *marker_options*

mleval command, [R] **ml**

mlib, [M-3] **mata mlib**

.mlib library files, [M-1] **how**, [M-3] **mata describe**, [M-3] **mata mlib**, [M-3] **mata set**, [M-3] **mata which**, [M-6] **Glossary**, [U] **11.6 Filenaming conventions**

mlmatbysum command, [R] **ml**

mlmatsum command, [R] **ml**

MLMV, *see* maximum likelihood with missing values

mlmv, *see* sem option method()

mlogit command, [R] **mlogit**, [R] **mlogit postestimation**

mlong

  data style, [MI] **styles**, [MI] **Glossary**

  technical description, [MI] **technical**

mlpattern() option, [G-3] *marker_options*

mlstyle() option, [G-3] *marker_options*

mlsum command, [R] **ml**

mlvecsum command, [R] **ml**

mlwidth() option, [G-3] *marker_options*

mm() function, [D] **datetime**, [D] **functions**, [M-5] **date( )**

.mmat files, [M-3] **mata matsave**, [U] **11.6 Filenaming conventions**

mmC() function, [D] **datetime**, [D] **functions**, [M-5] **date( )**

MNAR, *see* missing not at random

MNP, *see* outcomes, multinomial

.mo file, [M-1] **how**, [M-3] **mata mosave**, [M-3] **mata which**, [M-6] **Glossary**, [U] **11.6 Filenaming conventions**

mod() function, [D] **functions**, [M-5] **mod( )**

mode(), egen function, [D] **egen**

model identification, [SEM] **intro 3**

model-implied covariances and correlations, [SEM] **example 11**

model, ml subcommand, [R] **ml**

model simplification tests, [SEM] **example 8**, [SEM] **example 10**

model specification test, *see* specification test

models,

  CFA, *see* models, confirmatory factor analysis

  confirmatory factor analysis, [SEM] **intro 4**, [SEM] **example 15**, [SEM] **Glossary**

  correlated uniqueness, [SEM] **intro 4**, [SEM] **example 17**, [SEM] **Glossary**

  correlation, [SEM] **intro 4**, [SEM] **example 16**, [SEM] **Glossary**

  latent growth, [SEM] **intro 4**, [SEM] **example 18**, [SEM] **Glossary**

models, *continued*
    linear regression, [SEM] **intro 4**, [SEM] **example 6**
    maximum number of variables in, [R] **matsize**
    measurement, [SEM] **intro 4**, [SEM] **example 1**,
        [SEM] **example 3**, [SEM] **example 20**
    measurement error, [SEM] **intro 4**,
        [SEM] **example 1**
    MIMIC, *see* models, multiple indicators and multiple
        causes
    multiple indicators and multiple causes,
        [SEM] **intro 4**, [SEM] **example 10**,
        [SEM] **Glossary**
    multivariate regression, [SEM] **example 12**, *also see*
        seemingly unrelated regression
    nonrecursive, [SEM] **example 7**
    reliability, [SEM] **intro 4**, [SEM] **example 24**
    replaying, [SEM] **intro 6**
    seemingly unrelated regression, *see* seemingly
        unrelated regression
    sensitivity, [R] **regress postestimation**, [R] **rreg**
    structural, [SEM] **intro 4**, [SEM] **example 7**,
        [SEM] **example 9**, [SEM] **Glossary**
modern scaling, [MV] **Glossary**
modification, file, [D] **filefilter**
modification indices, [SEM] **estat mindices**,
    [SEM] **example 5**, [SEM] **methods and
    formulas**, [SEM] **Glossary**
modifying data, [D] **generate**, *also see* editing data
modulus function, [D] **functions**
modulus() function, [M-5] **mod( )**
modulus transformations, [R] **boxcox**
mofd() function, [D] **datetime**, [D] **functions**,
    [M-5] **date( )**
moments (of a distribution), [SEM] **Glossary**
monadic operator, [M-2] **syntax**, [M-6] **Glossary**
monotone imputation, *see* imputation, monotone
monotone missing values, [R] **misstable**
monotone-missing pattern, [MI] **mi impute monotone**,
    [MI] **Glossary**
monotonicity, *see* pattern of missingness
Monte Carlo error, [MI] **mi estimate**, [MI] **mi estimate
    using**, [MI] **Glossary**
Monte Carlo simulations, [P] **postfile**, [R] **permute**,
    [R] **simulate**
month() function, [D] **datetime**, [D] **functions**,
    [M-5] **date( )**, [U] **24.5 Extracting components
    of dates and times**
monthly() function, [D] **datetime**, [D] **datetime
    translation**, [D] **functions**, [M-5] **date( )**
Moore–Penrose inverse, [M-5] **pinv( )**
_moptimize() function, [M-5] **moptimize( )**
moptimize() function, [M-5] **moptimize( )**
moptimize_ado_cleanup() function,
    [M-5] **moptimize( )**
_moptimize_evaluate() function,
    [M-5] **moptimize( )**
moptimize_evaluate() function, [M-5] **moptimize( )**
moptimize_init() function, [M-5] **moptimize( )**

moptimize_init_*() functions, [M-5] **moptimize( )**
moptimize_query() function, [M-5] **moptimize( )**
moptimize_result_*() functions,
    [M-5] **moptimize( )**
moptimize_util_*() functions, [M-5] **moptimize( )**
more command and parameter, [P] **macro**, [P] **more**,
    [R] **more**, [U] **7 –more– conditions**
more condition, [R] **query**, [U] **7 –more– conditions**,
    [U] **16.1.6 Preventing –more– conditions**
more() function, [M-5] **more( )**
more, set subcommand, [R] **more**, [R] **set**
mortality table, *see* life tables
mosave, [M-3] **mata mosave**
moving average
    model, *see* MA
    process, [TS] **Glossary**
    smoother, [TS] **tssmooth**, [TS] **tssmooth ma**
mprobit command, [R] **mprobit**, [R] **mprobit
    postestimation**
mreldif() function, [D] **functions**, [M-5] **reldif( )**,
    [P] **matrix define**
mreldifre() function, [M-5] **reldif( )**
mreldifsym() function, [M-5] **reldif( )**
msize() option, [G-3] *marker_options*,
    [G-3] *rcap_options*
msofhours() function, [D] **datetime**, [D] **functions**,
    [M-5] **date( )**
msofminutes() function, [D] **datetime**, [D] **functions**,
    [M-5] **date( )**
msofseconds() function, [D] **datetime**, [D] **functions**,
    [M-5] **date( )**
mspline, graph twoway subcommand, [G-2] **graph
    twoway mspline**
mstyle() option, [G-3] *marker_options*
msymbol() option, [G-3] *marker_options*
MTMM, *see* multitrait–multimethod data and matrices
mtr(), egen function, [D] **egen**
multiarm trial, [ST] **stpower**, [ST] **Glossary**
multidimensional scaling, [MV] **mds**, [MV] **mds
    postestimation**, [MV] **mdslong**, [MV] **mdsmat**,
    [MV] **Glossary**
multilevel data, [MI] **mi estimate**
multilevel models, [R] **gllamm**, [XT] **xtmelogit**,
    [XT] **xtmepoisson**, [XT] **xtmixed**
multinomial
    logistic regression, [SVY] **svy estimation**
        imputation, *see* imputation, multinomial logistic
        regression
    outcome model, *see* outcomes, multinomial
    probit regression, [SVY] **svy estimation**
multiple comparisons, [R] **contrast**, [R] **margins**,
    [R] **pwcompare**, [R] **pwmean**, [R] **anova
    postestimation**, [R] **correlate**, [R] **mvreg**,
    [R] **oneway**, [R] **regress postestimation**,
    [R] **roccomp**, [R] **spearman**, [R] **test**, [R] **testnl**,
    [R] **tetrachoric**

multiple comparisons, *continued*
> Bonferroni's method, [R] **contrast**, [R] **margins**,
> [R] **pwcompare**, [R] **pwmean**, [R] **anova**
> **postestimation**, [R] **correlate**, [R] **oneway**,
> [R] **regress postestimation**, [R] **roccomp**,
> [R] **spearman**, [R] **test**, [R] **testnl**,
> [R] **tetrachoric**
> Duncan's method, [R] **pwcompare**, [R] **pwmean**
> Dunnett's method, [R] **pwcompare**, [R] **pwmean**
> Holm's method, [R] **anova postestimation**,
> [R] **regress postestimation**, [R] **test**, [R] **testnl**
> multiple-range method, *see* Dunnett's method
> subentry
> Scheffé's method, [R] **contrast**, [R] **margins**,
> [R] **pwcompare**, [R] **pwmean**, [R] **oneway**
> Šidák's method, [R] **contrast**, [R] **margins**,
> [R] **pwcompare**, [R] **pwmean**, [R] **anova**
> **postestimation**, [R] **correlate**, [R] **oneway**,
> [R] **regress postestimation**, [R] **roccomp**,
> [R] **spearman**, [R] **test**, [R] **testnl**,
> [R] **tetrachoric**
> Studentized-range method, *see* Tukey's method
> subentry
> Student–Newman–Keuls' method, [R] **pwcompare**,
> [R] **pwmean**
> Tukey's method, [R] **pwcompare**, [R] **pwmean**
multiple correlation, [SEM] **Glossary**
multiple correspondence analysis, [MV] **Glossary**
multiple imputation, [MI] **intro substantive**, [MI] **intro**,
> [MI] **styles**, [MI] **workflow**, [U] **26.22 Multiple**
> **imputation**
> analysis step, [MI] **intro substantive**, [MI] **mi**
> **estimate**, [MI] **mi estimate using**, [MI] **mi**
> **estimate postestimation**, [MI] **mi test**
> estimation, [MI] **estimation**
> imputation step, [MI] **intro substantive**, [MI] **mi**
> **impute**
> inference, [MI] **intro substantive**
> pooling step, [MI] **intro substantive**, [MI] **mi**
> **estimate**, [MI] **mi estimate using**
> prediction, [MI] **mi predict**
> theory, [MI] **intro substantive**
multiple languages, [D] **label language**
multiple-range multiple-comparison adjustment, *see*
> multiple comparisons, Dunnett's method
multiple-record st data, [ST] **stfill**, [ST] **stvary**,
> [ST] **Glossary**
multiple regression, *see* linear regression
multiplication, [M-2] **op_arith**, [M-2] **op_colon**
multiplication operator, *see* arithmetic operators
multiplicative heteroskedasticity, [TS] **arch**
multistage clustered sampling, [SVY] **survey**,
> [SVY] **svydescribe**, [SVY] **svyset**
multitrait–multimethod data and matrices,
> [SEM] **intro 4**, [SEM] **example 17**
multivariable fractional polynomial regression, [R] **mfp**
multivariate analysis, [MV] **canon**, [MV] **hotelling**,
> [MV] **mvtest**, [U] **26.23 Multivariate and**
> **cluster analysis**

multivariate analysis, *continued*
> bivariate probit, [R] **biprobit**
> of covariance, [MV] **manova**
> of variance, [MV] **manova**
> regression, [R] **mvreg**
> three-stage least squares, [R] **reg3**
> Zellner's seemingly unrelated, [R] **nlsur**, [R] **sureg**
multivariate Behrens–Fisher problem, [MV] **mvtest**
> **means**
multivariate GARCH, [TS] **mgarch**, [TS] **Glossary**
> model,
> > constant conditional correlation, [TS] **mgarch ccc**
> > diagonal vech, [TS] **mgarch dvech**
> > dynamic conditional correlation, [TS] **mgarch**
> > **dcc**
> > varying conditional correlation, [TS] **mgarch vcc**
> postestimation,
> > after ccc model, [TS] **mgarch ccc postestimation**
> > after dcc model, [TS] **mgarch dcc**
> > **postestimation**
> > after dvech model, [TS] **mgarch dvech**
> > **postestimation**
> > after vcc model, [TS] **mgarch vcc postestimation**
multivariate imputation, *see* imputation, multivariate
multivariate kurtosis, [MV] **mvtest normality**
multivariate logistic variable imputation, *see* imputation,
> multivariate
multivariate normal, [MV] **mvtest normality**
multivariate normal imputation, *see* imputation,
> multivariate normal
multivariate normal simulator, [M-5] **ghk( )**,
> [M-5] **ghkfast( )**
multivariate regression, [MV] **Glossary**,
> [SEM] **Glossary**, *see* models, multivariate
> regression
multivariate regression imputation, *see* imputation,
> multivariate
multivariate skewness, [MV] **mvtest normality**
multivariate time-series estimators,
> dynamic-factor models, [TS] **dfactor**
> MGARCH models, *see* multivariate GARCH
> state-space models, [TS] **sspace**
> structural vector autoregressive models, [TS] **var**
> **svar**
> vector autoregressive models, [TS] **var**,
> [TS] **varbasic**
> vector error-correction models, [TS] **vec**
mvdecode command, [D] **mvencode**
mvencode command, [D] **mvencode**
MVN imputation, *see* imputation, multivariate normal
mvreg command, [R] **mvreg**, [R] **mvreg postestimation**
mvreg, estat subcommand, [MV] **procrustes**
> **postestimation**
mvtest, [MV] **mvtest**
> correlations command, [MV] **mvtest**
> **correlations**
> covariances command, [MV] **mvtest covariances**

mvtest, *continued*
    means command, [MV] **mvtest means**
    normality command, [MV] **mvtest normality**

N

_n and _N built in variables, [U] **13.4 System variables**
    (_variables), [U] **13.7 Explicit subscripting**
name() option, [G-3] *name_option*
nameexternal() function, [M-5] **findexternal( )**
namelists, [M-3] **namelists**
names, [U] **11.3 Naming conventions**
    conflicts, [P] **matrix**, [P] **matrix define**, [P] **scalar**
    matrix row and columns, [P] **ereturn**, [P] **matrix define**, [P] **matrix rownames**
names, confirm subcommand, [P] **confirm**
namespace and conflicts, matrices and scalars,
    [P] **matrix**, [P] **matrix define**
naming convention, [M-1] **naming**
naming groups of variables, [D] **rename group**
naming variables, [D] **rename**
NaN, [M-6] **Glossary**
NARCH, [TS] **arch**
NARCHK, [TS] **arch**
natural log function, [D] **functions**
natural splines, [R] **mkspline**
nbetaden() function, [D] **functions**, [M-5] **normal( )**
nbinomial() function, [D] **functions**, [M-5] **normal( )**
nbinomialp() function, [D] **functions**,
    [M-5] **normal( )**
nbinomialtail() function, [D] **functions**,
    [M-5] **normal( )**
nbreg command, [R] **nbreg**, [R] **nbreg postestimation**
nchi2() function, [D] **functions**, [M-5] **normal( )**
n-class command, [P] **program**, [P] **return**
ndots() option, [G-2] **graph twoway dot**
nearest neighbor, [MI] **mi impute pmm**, [MV] **discrim knn**, [MV] **Glossary**
needle plot, [R] **spikeplot**
_negate() function, [M-5] **_negate( )**
negation operator, *see* arithmetic operators
negative binomial
    distribution,
        cumulative, [D] **functions**
        inverse cumulative, [D] **functions**
        inverse reverse cumulative, [D] **functions**
    negative binomial regression, [R] **nbreg**, [SVY] **svy estimation**
        fixed-effects, [XT] **xtnbreg**
        generalized linear models, [R] **glm**
    model, [XT] **Glossary**
    population-averaged, [XT] **xtgee**, [XT] **xtnbreg**
    random-effects, [XT] **xtnbreg**
    truncated, [R] **tnbreg**
    zero-inflated, [R] **zinb**
        reverse cumulative, [D] **functions**
    probability mass function, [D] **functions**

Nelder–Mead method, [M-5] **moptimize( )**,
    [M-5] **optimize( )**
Nelson–Aalen cumulative hazard, [ST] **sts**, [ST] **sts generate**, [ST] **sts graph**, [ST] **sts list**
nested
    case–control data, [ST] **sttocc**
    designs, [MV] **manova**, [R] **anova**
    effects, [MV] **manova**, [R] **anova**
    logit, [R] **nlogit**
    model statistics, [R] **nestreg**
    regression, [R] **nestreg**
nested, misstable subcommand, [R] **misstable**
nestreg prefix command, [R] **nestreg**
net
    cd command, [R] **net**
    describe command, [R] **net**
    from command, [R] **net**
    get command, [R] **net**
    install command, [R] **net**
    link command, [R] **net**
    query command, [R] **net**
    search command, [R] **net search**
    set ado command, [R] **net**
    set other command, [R] **net**
    sj command, [R] **net**
    stb command, [R] **net**
net, view subcommand, [R] **view**
NetCourseNow, [U] **3.7.2 NetCourses**
NetCourses, [U] **3.7.2 NetCourses**
net_d, view subcommand, [R] **view**
.new built-in class function, [P] **class**
new() class function, [M-2] **class**
new lines, data without, [D] **infile (fixed format)**
new, ssc subcommand, [R] **ssc**
newey command, [TS] **newey**, [TS] **newey postestimation**
Newey–West
    covariance matrix, [TS] **Glossary**
    postestimation, [TS] **newey postestimation**
    regression, [TS] **newey**
    standard errors, [P] **matrix accum**, [R] **glm**
_newline(#), display directive, [P] **display**
news command, [R] **news**
news, view subcommand, [R] **view**
newsletter, [U] **3 Resources for learning and using Stata**
Newton–Raphson algorithm, [M-5] **moptimize( )**,
    [M-5] **optimize( )**, [R] **ml**
Neyman allocation, [SVY] **estat**
nFden() function, [D] **functions**, [M-5] **normal( )**
nFtail() function, [D] **functions**, [M-5] **normal( )**
NHANES data, [MI] **mi export nhanes1**, [MI] **mi import nhanes1**
nibeta() function, [D] **functions**, [M-5] **normal( )**
niceness, set subcommand, [D] **memory**, [R] **set**
nl command, [R] **nl**, [R] **nl postestimation**
nl, tssmooth subcommand, [TS] **tssmooth nl**

nlcom command, [R] **nlcom**, [SEM] **intro 6**,
  [SEM] **estat stdize**, [SEM] **nlcom**, [SVY] **svy**
  postestimation
nlogit command, [R] **nlogit**, [R] **nlogit**
  postestimation
nlogitgen command, [R] **nlogit**
nlogittree command, [R] **nlogit**
nlsur command, [R] **nlsur**, [R] **nlsur postestimation**
nm1 option, *see* sem option nm1
noanchor option, *see* sem option noanchor
nobreak command, [P] **break**
nocapslatent option, *see* sem option nocapslatent
nocnsreport option, *see* sem option nocnsreport
noconstant option, *see* sem option noconstant
nodescribe option, *see* sem option nodescribe
nodraw option, [G-3] *nodraw_option*
nofootnote option, *see* sem option nofootnote
noheader option, *see* sem option noheader
noisily prefix, [P] **quietly**
noivstart option, *see* sem option noivstart
nolabel option, *see* sem option nolabel
nolog or log option, [R] **maximize**
nomeans option, *see* sem option nomeans
noncentral
  beta density, [D] **functions**
  beta distribution, [D] **functions**
  chi-squared distribution function, [D] **functions**
  F density, [D] **functions**
  F distribution, [D] **functions**
nonconformities, quality control, [R] **qc**
nonconstant variance, [SVY] **variance estimation**,
  see robust, Huber/White/sandwich estimator of
  variance
noncursive models, *see* nonrecursive models
nonlinear
  combinations of estimators, [R] **nlcom**
  combinations, predictions, and tests, [SVY] **svy**
    postestimation
  estimation, [TS] **arch**
  hypothesis test after estimation, [R] **lrtest**,
    [R] **margins**, [R] **margins, contrast**,
    [R] **margins, pwcompare**, [R] **nlcom**,
    [R] **predictnl**, [R] **testnl**
  least squares, [R] **nl**, [SVY] **svy estimation**
  prediction, *see* multiple imputation, prediction
  regression, [R] **boxcox**, [R] **nl**, [R] **nlsur**
  smoothing, [TS] **tssmooth nl**
  tests, *see* estimation, tests after
nonmetric scaling, [MV] **Glossary**
nonmissing() function, [M-5] **missing( )**
nonnormed fit index, *see* Tucker–Lewis index
nonparametric analysis,
  hypothesis tests,
    association, [R] **spearman**
    cusum, [R] **cusum**
    equality of distributions, [R] **ksmirnov**,
      [R] **kwallis**, [R] **ranksum**, [R] **signrank**
    medians, [R] **ranksum**

nonparametric analysis, hypothesis tests, *continued*
    proportions, [R] **bitest**, [R] **prtest**
    random order, [R] **runtest**
    trend, [R] **nptrend**
  percentiles, [R] **centile**
  quantile regression, [R] **qreg**
  ROC analysis, [R] **roc**
    estimation, [R] **rocreg**
    graphs, [R] **rocregplot**
    test equality of areas, [R] **roccomp**
    without covariates, [R] **roctab**
  smoothing, [R] **kdensity**, [R] **lowess**, [R] **lpoly**,
    [R] **smooth**
  tables, [R] **tabulate twoway**
nonparametric methods, [MV] **Glossary**
nonparametric test, equality of survivor functions,
  [ST] **sts test**
nonrecursive (structural) model (system),
  [SEM] **Glossary**
nonrecursive models, [SEM] **estat stable**
  stability of, [SEM] **estat stable**, [SEM] **example 7**
nonrtolerance option, [R] **maximize**
nonstationary time series, [TS] **dfgls**, [TS] **dfuller**,
  [TS] **pperron**, [TS] **vec intro**, [TS] **vec**
nopreserve option, [P] **nopreserve option**
norm, [M-6] **Glossary**
norm() function, [M-5] **norm( )**
normal() function, [D] **functions**, [M-5] **normal( )**
normal distribution and normality,
  bivariate, [D] **functions**
  cdf, [D] **functions**
  density, [D] **functions**
  examining distributions for, [R] **diagnostic plots**,
    [R] **lv**
  generating multivariate data, [D] **corr2data**,
    [D] **drawnorm**
  inverse, [D] **functions**
  probability and quantile plots, [R] **diagnostic plots**
  test for, [R] **sktest**, [R] **swilk**
  transformations to achieve, [R] **boxcox**, [R] **ladder**,
    [R] **lnskew0**
normal probability plots, [G-2] **graph other**
normalden() function, [D] **functions**, [M-5] **normal( )**
normality, mvtest subcommand, [MV] **mvtest**
  normality
normality test
  after VAR or SVAR, [TS] **varnorm**
  after VEC, [TS] **vecnorm**
normality,
  conditional, [SEM] **intro 3**
  joint, [SEM] **intro 3**
normalization, [MV] **Glossary**
normalization constraints, [SEM] **Glossary**, *see*
  constraints, normalization
normalized residuals, [SEM] **estat residuals**,
  [SEM] **methods and formulas**, [SEM] **Glossary**
normally distributed random numbers, [D] **functions**
Not Elsewhere Classified, *see* Stata Blog

not equal operator, [U] **13.2.3 Relational operators**
not operator, [U] **13.2.4 Logical operators**
notable option, *see* sem option notable
note() option, [G-3] *title_options*
notes
    command, [D] **notes**
    drop command, [D] **notes**
    list command, [D] **notes**
    renumber command, [D] **notes**
    replace command, [D] **notes**
    search command, [D] **notes**
notes,
    cluster subcommand, [MV] **cluster notes**
    estimates subcommand, [R] **estimates notes**
notes,
    cluster analysis, [MV] **cluster notes**
    creating, [D] **notes**, [D] **varmanage**
    editing, [D] **notes**, [D] **varmanage**
notes on estimation results, [R] **estimates notes**
notifyuser, set subcommand, [R] **set**
noupdate option, [MI] **noupdate option**
novarabbrev command, [P] **varabbrev**
noxconditional option, *see* sem option
    noxconditional
NPARCH, [TS] **arch**
npnchi2() function, [D] **functions**, [M-5] **normal( )**
nproc, estat subcommand, [R] **rocreg postestimation**
nptrend command, [R] **nptrend**
NR algorithm, [R] **ml**
nrtolerance() option, [R] **maximize**
NULL, [M-2] **pointers**, [M-6] **Glossary**
nullmat() function, [D] **functions**, [P] **matrix define**
number, confirm subcommand, [P] **confirm**
number to string conversion, *see* string functions,
    expressions, and operators, *see* string functions
numbered
    styles, [G-4] *linestyle*, [G-4] *markerlabelstyle*,
    [G-4] *markerstyle*, [G-4] *pstyle*
numbers, [U] **12.2 Numbers**
    formatting, [D] **format**
    mapping to strings, [D] **encode**, [D] **label**
numeric, [M-2] **declarations**, [M-6] **Glossary**
numeric display formats, [I] **format**
numeric list, [P] **numlist**, [P] **syntax**,
    [U] **11.1.8 numlist**
numeric value labels, [D] **labelbook**
numerical precision, [U] **13.11 Precision and problems
    therein**
numlabel command, [D] **labelbook**
numlist command, [P] **numlist**, [U] **11.1.8 numlist**
N-way analysis of variance, [R] **anova**
N-way multivariate analysis of variance, [MV] **manova**

O

object, [P] **class**
object code, [M-1] **how**, [M-6] **Glossary**

object-oriented programming, [M-2] **class**,
    [M-6] **Glossary**, [P] **class**
objects, size of, [G-4] *relativesize*
.objkey built-in class function, [P] **class**
.objtype built-in class function, [P] **class**
oblimax rotation, [MV] **rotate**, [MV] **rotatemat**,
    [MV] **Glossary**
oblimin rotation, [MV] **rotate**, [MV] **rotatemat**,
    [MV] **Glossary**
oblique rotation, [MV] **factor postestimation**,
    [MV] **rotate**, [MV] **rotatemat**, [MV] **Glossary**
oblique transformation, *see* oblique rotation
obs parameter, [D] **describe**, [D] **obs**
obs, set subcommand, [D] **obs**, [R] **set**
observations,
    built-in counter variable, [U] **11.3 Naming
        conventions**
    complete and incomplete, [MI] **Glossary**
    creating dataset of, [D] **collapse**
    dropping, [D] **drop**
    dropping duplicate, [D] **duplicates**
    duplicating, [D] **expand**
    duplicating, clustered, [D] **expandcl**
    identifying duplicate, [D] **duplicates**
    increasing number of, [D] **obs**
    marking, [P] **mark**
    maximum number of, [D] **memory**, [U] **6 Managing
        memory**
    ordering, [D] **gsort**, [D] **sort**
    transposing with variables, [D] **xpose**
observed information matrix, [SEM] **Glossary**, *see* OIM
observed variables, [SEM] **intro 3**, [SEM] **Glossary**
Ochiai coefficient similarity measure,
    [MV] *measure_option*
odbc
    describe command, [D] **odbc**
    exec() command, [D] **odbc**
    insert command, [D] **odbc**
    list command, [D] **odbc**
    load command, [D] **odbc**
    query command, [D] **odbc**
    sqlfile() command, [D] **odbc**
ODBC data source, reading data from, [D] **odbc**,
    [U] **21.4 Transfer programs**, [U] **21.5 ODBC
    sources**
odbcmgr, set subcommand, [D] **odbc**, [R] **set**
odds, [ST] **Glossary**
odds ratio, [R] **asclogit**, [R] **binreg**, [R] **clogit**,
    [R] **cloglog**, [R] *eform_option*, [R] **exlogistic
    postestimation**, [R] **glm**, [R] **glogit**, [R] **logistic**,
    [R] **logit**, [R] **mlogit**, [R] **scobit**, [ST] **epitab**,
    [SVY] **svy estimation**, [ST] **Glossary**,
    [XT] **xtcloglog**, [XT] **xtgee**, [XT] **xtlogit**,
    [XT] **xtmelogit**
    differences, [R] **lincom**, [R] **nlcom**, [SVY] **svy
    postestimation**
_OEx, [SEM] **sem option covstructure( )**

off,
 cmdlog subcommand, [R] **log**
 log subcommand, [R] **log**
 timer subcommand, [P] **timer**
offset between axes and data, setting,
 [G-3] *region_options*
offset variable, [ST] **Glossary**
ograph, irf subcommand, [TS] **irf ograph**
OIM, [R] **ml**, [R] *vce_option*, [XT] *vce_options*, *see*
 observed information matrix
oim, *see* sem option vce( )
OLDPLACE directory, [P] **sysdir**, [U] **17.5 Where does
 Stata look for ado-files?**
OLE Automation, [P] **automation**
ologit command, [R] **ologit**, [R] **ologit postestimation**
OLS regression, *see* linear regression
omitted variables test, [R] **regress postestimation**
on,
 cmdlog subcommand, [R] **log**
 log subcommand, [R] **log**
 timer subcommand, [P] **timer**
one-level model, [XT] **Glossary**
one-way analysis of variance, [R] **kwallis**, [R] **loneway**,
 [R] **oneway**
oneway command, [R] **oneway**
online help, [U] **7 –more– conditions**, [M-1] **help**,
 [M-3] **mata help**, [R] **help**, [R] **hsearch**,
 [R] **search**, [U] **4 Stata's help and search
 facilities**
opaccum, matrix subcommand, [P] **matrix accum**
open, file subcommand, [P] **file**
OpenOffice dates, [D] **datetime**
operating system command, [D] **cd**, [D] **copy**, [D] **dir**,
 [D] **erase**, [D] **mkdir**, [D] **rmdir**, [D] **shell**,
 [D] **type**
operator,
 difference, [U] **11.4.4 Time-series varlists**
 lag, [U] **11.4.4 Time-series varlists**
 lead, [U] **11.4.4 Time-series varlists**
 seasonal lag, [U] **11.4.4 Time-series varlists**
operators, [M-2] **op_arith**, [M-2] **op_assignment**,
 [M-2] **op_colon**, [M-2] **op_conditional**,
 [M-2] **op_increment**, [M-2] **op_join**,
 [M-2] **op_kronecker**, [M-2] **op_logical**,
 [M-2] **op_range**, [M-2] **op_transpose**,
 [M-6] **Glossary**, [P] **matrix define**,
 [U] **13.2 Operators**
 order of evaluation, [U] **13.2.5 Order of evaluation,
 all operators**
OPG, [R] **ml**, [R] *vce_option*, [SEM] **Glossary**,
 [XT] *vce_options*
oprobit command, [R] **oprobit**, [R] **oprobit
 postestimation**
optimization, [M-3] **mata set**, [M-5] **moptimize( )**,
 [M-5] **optimize( )**, [M-6] **Glossary**
_optimize( ) function, [M-5] **optimize( )**
optimize( ) function, [M-5] **optimize( )**
_optimize_evaluate( ) function, [M-5] **optimize( )**

optimize_evaluate( ) function, [M-5] **optimize( )**
optimize_init( ) function, [M-5] **optimize( )**
optimize_init_*( ) functions, [M-5] **optimize( )**
optimize_query( ) function, [M-5] **optimize( )**
optimize_result_*( ) functions, [M-5] **optimize( )**
options, [U] **11 Language syntax**
 in a programming context, [P] **syntax**, [P] **unab**
 repeated, [G-4] **concept: repeated options**
or operator, [U] **13.2.4 Logical operators**
ORACLE, reading data from, [D] **odbc**,
 [U] **21.4 Transfer programs**
order command, [D] **order**
order( ) function, [M-5] **sort( )**
order statistics, [D] **egen**, [R] **lv**
ordered
 logistic regression, [SVY] **svy estimation**
 imputation, *see* imputation, ordered logistic
 regression
 logit, [R] **ologit**
 probit, [R] **oprobit**
 probit regression, [SVY] **svy estimation**
ordering
 observations, [D] **gsort**, [D] **sort**
 variables, [D] **order**, [D] **sort**
ordinal outcome model, *see* outcomes, ordinal
ordinary least squares, *see* linear regression
ordination, [MV] **Glossary**
orgtype, [M-2] **declarations**, [M-6] **Glossary**
orgtype( ) function, [M-5] **eltype( )**
orientationstyle, [G-4] *orientationstyle*
original data, [MI] **Glossary**
orthog command, [R] **orthog**
orthogonal matrix, [M-6] **Glossary**
orthogonal polynomials, [R] **contrast**, [R] **margins,
 contrast**, [R] **orthog**
orthogonal rotation, [MV] **factor postestimation**,
 [MV] **rotate**, [MV] **rotatemat**, [MV] **Glossary**
orthogonal transformation, *see* orthogonal rotation
orthogonalized impulse–response function, [TS] **irf**,
 [TS] **var intro**, [TS] **vec intro**, [TS] **vec**,
 [TS] **Glossary**
orthonormal basis, [P] **matrix svd**
orthpoly command, [R] **orthog**
other graph commands, [G-2] **graph other**
.out filename suffix, [D] **outsheet**,
 [U] **11.6 Filenaming conventions**
outcomes,
 binary,
 complementary log-log, [R] **cloglog**
 glm for binomial family, [R] **binreg**, [R] **glm**
 grouped data, [R] **glogit**
 logistic, [R] **exlogistic**, [R] **logistic**, [R] **logit**,
 [R] **scobit**
 probit, [R] **biprobit**, [R] **heckprob**, [R] **hetprob**,
 [R] **ivprobit**, [R] **probit**
 ROC analysis, [R] **rocfit**, [R] **rocreg**

outcomes, *continued*
  categorical,
    logistic, [R] **asclogit**, [R] **clogit**, [R] **mlogit**,
      [R] **nlogit**, [R] **slogit**
    probit, [R] **asmprobit**, [R] **mprobit**
  count,
    negative binomial, [R] **nbreg**, [R] **tnbreg**,
      [R] **zinb**
    Poisson, [R] **expoisson**, [R] **poisson**,
      [R] **tpoisson**, [R] **zip**
  multinomial, *see* categorical subentry, *see* ordinal
    subentry, *see* rank subentry
  ordinal,
    logistic, [R] **ologit**
    probit, [R] **oprobit**
  polytomous, *see* categorical subentry, *see* ordinal
    subentry, *see* rank subentry
  rank,
    logistic, [R] **rologit**
    probit, [R] **asroprobit**
outer fence, [R] **lv**
outer product, [D] **cross**
outer product of the gradient, *see* OPG
outfile command, [D] **outfile**
outliers, [R] **lv**, [R] **qreg**, [R] **regress postestimation**,
  [R] **rreg**
outlines, suppressing, [G-4] *linestyle*
outlining regions, [G-3] *region_options*
out-of-sample predictions, [R] **predict**, [R] **predictnl**,
  [U] **20.10.3 Making out-of-sample predictions**
output,
  coefficient table,
    automatically widen, [R] **set**
    display settings, [R] **set showbaselevels**
    format settings, [R] **set cformat**
  controlling the scrolling of, [R] **more**
  displaying, [P] **display**, [P] **smcl**
  formatting numbers, [D] **format**
  printing, [R] **translate**, [U] **15 Saving and printing
    output—log files**
  recording, [R] **log**
  suppressing, [P] **quietly**
output, set subcommand, [P] **quietly**, [R] **set**
output settings, [P] **creturn**
outsheet command, [D] **outsheet**
outside values, [R] **lv**
over() option, [G-2] **graph bar**, [G-2] **graph box**,
  [G-2] **graph dot**
overdispersion, *see* imputation, overdispersed count data
overid, estat subcommand, [R] **gmm postestimation**,
  [R] **ivregress postestimation**
overidentifying restrictions, [XT] **Glossary**
  tests of, [R] **gmm postestimation**, [R] **ivregress
    postestimation**, [XT] **xtabond**, [XT] **xtabond
    postestimation**, [XT] **xtdpd**, [XT] **xtdpdsys**,
    [XT] **xtdpdsys postestimation**
overloading, class program names, [P] **class**

ovtest, estat subcommand, [R] **regress
  postestimation**

P

P charts, [G-2] **graph other**
p-value, [SEM] **Glossary**
pac command, [TS] **corrgram**
pagesize, set subcommand, [R] **more**, [R] **set**
paging of screen output, controlling, [P] **more**,
  [R] **more**
paired-coordinate plots, [G-2] **graph twoway pcarrow**,
  [G-2] **graph twoway pccapsym**, [G-2] **graph
  twoway pcscatter**, [G-2] **graph twoway pcspike**
pairwise combinations, [D] **cross**, [D] **joinby**
pairwise comparisons, [MV] **intro**, [R] **margins,
  pwcompare**, [R] **marginsplot**, [R] **pwcompare**,
  [R] **pwmean**, [U] **20.17 Obtaining pairwise
  comparisons**
pairwise correlation, [R] **correlate**
pairwise, estat subcommand, [MV] **mds
  postestimation**
palette command, [G-2] **palette**
panel data, [M-5] **panelsetup( )**, [MI] **mi estimate**,
  [U] **26.17 Panel-data models**, [XT] **Glossary**
panel-corrected standard error (PCSE), [XT] **xtpcse**,
  [XT] **Glossary**
panels, variable identifying, [XT] **xtset**
panelsetup() function, [M-5] **panelsetup( )**
panelstats() function, [M-5] **panelsetup( )**
panelsubmatrix() function, [M-5] **panelsetup( )**
panelsubview() function, [M-5] **panelsetup( )**
Paradox, reading data from, [U] **21.4 Transfer
  programs**
parameter constraints, [SEM] **estat ginvariant**,
  [SEM] **Glossary**
parameter trace files, [MI] **mi impute mvn**, [MI] **mi
  ptrace**
parameterized curves, [D] **range**
parameters, [SEM] **Glossary**
  combinations of, [SEM] **lincom**, [SEM] **nlcom**
  system, *see* system parameters
parametric methods, [MV] **Glossary**
parametric survival models, [ST] **streg**, [SVY] **svy
  estimation**
PARCH, [TS] **arch**
parsedistance, cluster subcommand, [MV] **cluster
  programming utilities**
parsimax rotation, [MV] **rotate**, [MV] **rotatemat**,
  [MV] **Glossary**
parsing, [P] **gettoken**, [P] **numlist**, [P] **tokenize**,
  [M-5] **tokenget( )**, [M-5] **tokens( )**, [P] **syntax**,
  [U] **18.4 Program arguments**
partial
  autocorrelation function, [TS] **corrgram**,
    [TS] **Glossary**
  correlation, [R] **pcorr**
  DFBETA, [ST] **Glossary**
  effects, [R] **margins**, [R] **marginsplot**

partial, *continued*
   likelihood displacement value, [ST] **Glossary**
   LMAX value, [ST] **Glossary**
   regression leverage plot, [R] **regress postestimation**
   regression plot, [R] **regress postestimation**
   residual plot, [R] **regress postestimation**
partially specified target rotation, [MV] **rotate**,
   [MV] **rotatemat**, [MV] **Glossary**
partition cluster-analysis methods, [MV] **cluster kmeans
   and kmedians**, [MV] **Glossary**
partition clustering, *see* partition cluster-analysis
   methods
partitioned matrices, [P] **matrix define**
partitioning memory, [U] **6 Managing memory**
Parzen kernel density function, [R] **kdensity**
passive imputation, *see* imputation, passive
passive, mi subcommand, [MI] **mi passive**
passive variables, *see* variables, passive
past history, [ST] **stset**, [ST] **Glossary**
path, [SEM] **Glossary**
   adding, [SEM] **intro 5**
   coefficients, [SEM] **Glossary**
     constraining, [SEM] **intro 3**
   diagrams, [SEM] **intro 2**, [SEM] **Glossary**
   notation, [SEM] **intro 2**, [SEM] **sem path notation**,
     [SEM] **Glossary**
pathasciisuffix() function, [M-5] **pathjoin( )**
pathbasename() function, [M-5] **pathjoin( )**
pathisabs() function, [M-5] **pathjoin( )**
pathisurl() function, [M-5] **pathjoin( )**
pathjoin() function, [M-5] **pathjoin( )**
pathlist() function, [M-5] **pathjoin( )**
pathrmsuffix() function, [M-5] **pathjoin( )**
paths, [U] **11.6 Filenaming conventions**
paths and directories, [P] **creturn**
pathsearchlist() function, [M-5] **pathjoin( )**
pathsplit() function, [M-5] **pathjoin( )**
pathstatasuffix() function, [M-5] **pathjoin( )**
pathsubsysdir() function, [M-5] **pathjoin( )**
pathsuffix() function, [M-5] **pathjoin( )**
pattern matching, [M-5] **strmatch( )**
pattern of missing values, [R] **misstable**
pattern of missingness, [MI] **intro substantive**, [MI] **mi
   impute**, [MI] **mi misstable**, [MI] **Glossary**
patterns, misstable subcommand, [R] **misstable**
patterns of data, [D] **egen**
pause command, [P] **pause**
pausing until key is pressed, [P] **more**, [R] **more**
pc(), egen function, [D] **egen**
PCA, [MV] **Glossary**
pca command, [MV] **pca**
pcamat command, [MV] **pca**
pcarrow, graph twoway subcommand, [G-2] **graph
   twoway pcarrow**
pcarrowi, graph twoway subcommand, [G-2] **graph
   twoway pcarrowi**

pcbarrow, graph twoway subcommand, [G-2] **graph
   twoway pcarrow**
pccapsym, graph twoway subcommand, [G-2] **graph
   twoway pccapsym**
pchart command, [R] **qc**
pchi command, [R] **diagnostic plots**
pci, graph twoway subcommand, [G-2] **graph twoway
   pci**
pclose, [SEM] **example 4**
p-conformability, [M-6] **Glossary**
pcorr command, [R] **pcorr**
pcscatter, graph twoway subcommand, [G-2] **graph
   twoway pcscatter**
PCSE, *see* panel-corrected standard error
pcspike, graph twoway subcommand, [G-2] **graph
   twoway pcspike**
pctile(), egen function, [D] **egen**
_pctile command, [D] **pctile**
pctile command, [D] **pctile**
PDF files, [G-2] **graph export**, [R] **translate**
Pearson coefficient similarity measure,
   [MV] *measure_option*
Pearson goodness-of-fit test, [R] **logistic postestimation**,
   [R] **poisson postestimation**
Pearson product-moment correlation coefficient,
   [R] **correlate**
Pearson residual, [R] **binreg postestimation**, [R] **glm
   postestimation**, [R] **logistic postestimation**,
   [R] **logit postestimation**
penalized log-likelihood function, [ST] **stcox**,
   [ST] **Glossary**
percentiles,
   create
     dataset of, [D] **collapse**
     displaying, [R] **centile**, [R] **lv**, [R] **summarize**,
       [R] **table**, [R] **tabstat**
       variable containing, [D] **codebook**, [D] **egen**,
         [D] **pctile**
perfect prediction, *see* imputation, perfect prediction
pergram command, [TS] **pergram**
_perhapsequilc() function, [M-5] **_equilrc( )**
_perhapsequilr() function, [M-5] **_equilrc( )**
_perhapsequilrc() function, [M-5] **_equilrc( )**
periodogram, [G-2] **graph other**, [TS] **pergram**,
   [TS] **psdensity**, [TS] **Glossary**
permname macro extended function, [P] **macro**
permutation matrix and vector, [M-1] **permutation**,
   [M-5] **invorder( )**, [M-6] **Glossary**
permutation tests, [R] **permute**
permutations, [M-5] **cvpermute( )**
permute prefix command, [R] **permute**
person time, [ST] **stptime**
personal command, [P] **sysdir**
PERSONAL directory, [P] **sysdir**, [U] **17.5 Where does
   Stata look for ado-files?**
pformat, set subcommand, [R] **set**, [R] **set cformat**
pharmaceutical statistics, [R] **pk**, [R] **pksumm**
pharmacokinetic data, *see* pk (pharmacokinetic data)

pharmacokinetic plots, [G-2] **graph other**

phase function, [TS] **Glossary**

Phillips–Perron test, [TS] **pperron**

phtest, estat subcommand, [ST] **stcox PH-assumption tests**

_pi built-in variable, [U] **11.3 Naming conventions**

pi( ) function, [M-5] **sin( )**

pi, value of, [U] **11.3 Naming conventions**, [U] **13.4 System variables (_variables)**

pie charts, [G-2] **graph pie**

pie, graph subcommand, [G-2] **graph pie**

piece macro extended function, [P] **macro**

piecewise
    cubic functions, [R] **mkspline**
    linear functions, [R] **mkspline**

Pillai's trace statistic, [MV] **canon**, [MV] **manova**, [MV] **mvtest means**, [MV] **Glossary**

pinnable, set subcommand, [R] **set**

_pinv( ) function, [M-5] **pinv( )**

pinv( ) function, [M-5] **pinv( )**

pk (pharmacokinetic data), [R] **pk**, [R] **pkcollapse**, [R] **pkcross**, [R] **pkequiv**, [R] **pkexamine**, [R] **pkshape**, [R] **pksumm**, [U] **26.24 Pharmacokinetic data**,

pkcollapse command, [R] **pkcollapse**

pkcross command, [R] **pkcross**

pkequiv command, [R] **pkequiv**

pkexamine command, [R] **pkexamine**

.pkg filename suffix, [R] **net**

pkshape command, [R] **pkshape**

pksumm command, [R] **pksumm**

Plackett–Luce model, [R] **rologit**

platforms for which Stata is available, [U] **5.1 Platforms**

play, graph subcommand, [G-2] **graph play**

play( ) option, [G-3] *play_option*

playsnd, set subcommand, [R] **set**

plegend( ) option, [G-3] *legend_options*

plot, definition, [G-4] *pstyle*

plot, ml subcommand, [R] **ml**

plot region, [G-3] *region_options*
    suppressing border around, [G-3] *region_options*

plotregion( ) option, [G-3] *region_options*

plotregionstyle, [G-4] *plotregionstyle*

plottypes
    base, [G-3] *advanced_options*
    derived, [G-3] *advanced_options*

plugin, loading, [P] **plugin**

plugin option, [P] **plugin**, [P] **program**

plural( ) function, [D] **functions**

PLUS directory, [P] **sysdir**, [U] **17.5 Where does Stata look for ado-files?**

PMM imputation, *see* imputation, predictive mean matching

PNG, [G-3] *png_options*

pnorm command, [R] **diagnostic plots**

point estimate, [SVY] **Glossary**

pointers, [M-2] **ftof**, [M-2] **pointers**, [M-5] **findexternal( )**, [M-6] **Glossary**

points, connecting, [G-3] *connect_options*, [G-4] *connectstyle*

poisson( ) function, [D] **functions**

Poisson
    distribution,
        cdf, [D] **functions**
        confidence intervals, [R] **ci**
        cumulative, [D] **functions**
        inverse cumulative, [D] **functions**
        inverse reverse cumulative, [D] **functions**
        regression, *see* Poisson regression
        reverse cumulative, [D] **functions**
    probability mass function, [D] **functions**

poisson command, [R] **nbreg**, [R] **poisson**, [R] **poisson postestimation**

poisson( ) function, [M-5] **normal( )**

Poisson regression, [R] **nbreg**, [R] **poisson**, [ST] **stcox**, [SVY] **svy estimation**
    fixed-effects, [XT] **xtpoisson**
    generalized linear model, [R] **glm**
    mixed-effects, [XT] **xtmepoisson**
    model, [XT] **Glossary**
    population-averaged, [XT] **xtgee**, [XT] **xtpoisson**
    random-effects, [XT] **xtpoisson**
    truncated, [R] **tpoisson**
    zero-inflated, [R] **zip**

poissonp( ) function, [D] **functions**, [M-5] **normal( )**

poissontail( ) function, [D] **functions**, [M-5] **normal( )**

polar coordinates, [D] **range**

polyadd( ) function, [M-5] **polyeval( )**

polyderiv( ) function, [M-5] **polyeval( )**

polydiv( ) function, [M-5] **polyeval( )**

polyeval( ) function, [M-5] **polyeval( )**

polyinteg( ) function, [M-5] **polyeval( )**

polymorphism, [P] **class**

polymult( ) function, [M-5] **polyeval( )**

polynomials, [M-5] **polyeval( )**
    fractional, [R] **fracpoly**, [R] **mfp**
    orthogonal, [R] **orthog**
    smoothing, *see* local polynomial smoothing

polyroots( ) function, [M-5] **polyeval( )**

polysolve( ) function, [M-5] **polyeval( )**

polytomous logistic regression, [SVY] **svy estimation**

polytomous outcome model, *see* outcomes, polytomous

polytrim( ) function, [M-5] **polyeval( )**

pooled estimates, [ST] **epitab**

pooled estimator, [XT] **Glossary**

pooling step, [MI] **intro substantive**, [MI] **mi estimate**, [MI] **mi estimate using**, [MI] **mi predict**

population attributable risk, [ST] **epitab**

population error, [SEM] **estat gof**, [SEM] **example 4**

population pyramid, [G-2] **graph twoway bar**

population standard deviation, *see* subpopulation, standard deviations of

population-averaged model, [XT] **xtcloglog**, [XT] **xtgee**,
[XT] **xtlogit**, [XT] **xtnbreg**, [XT] **xtpoisson**,
[XT] **xtprobit**, [XT] **xtreg**, [XT] **Glossary**
populations,
diagnostic plots, [R] **diagnostic plots**
examining, [R] **histogram**, [R] **lv**, [R] **stem**,
[R] **summarize**, [R] **table**
standard, [R] **dstdize**
testing equality of, *see* distributions, testing equality
of
testing for normality, [R] **sktest**, [R] **swilk**
portmanteau statistic, [TS] **Glossary**
portmanteau test, [TS] **corrgram**, [TS] **wntestq**
post command, [P] **postfile**
post, ereturn subcommand, [P] **ereturn**,
[P] **makecns**, [P] **return**
postclose command, [P] **postfile**
posterior probabilities, [MV] **Glossary**
postestimation, [MI] **mi estimate postestimation**,
[SVY] **estat**, [SVY] **svy postestimation**
predicted values, [SEM] **intro 6**, [SEM] **example 14**,
[SEM] **predict**
postestimation command, [I] **postestimation
commands**, [P] **estat programming**,
[R] **contrast**, [R] **estat**, [R] **estimates**,
[R] **hausman**, [R] **lincom**, [R] **linktest**,
[R] **lrtest**, [R] **margins**, [R] **margins,
contrast**, [R] **margins, pwcompare**,
[R] **marginsplot**, [R] **nlcom**, [R] **predict**,
[R] **predictnl**, [R] **pwcompare**, [R] **suest**,
[R] **test**, [R] **testnl**, [SEM] **intro 6**, [SEM] **sem
postestimation**, [ST] **stcurve**, [TS] **fcast
compute**, [TS] **fcast graph**, [TS] **irf**,
[TS] **psdensity**, [TS] **vargranger**, [TS] **varlmar**,
[TS] **varnorm**, [TS] **varsoc**, [TS] **varstable**,
[TS] **varwle**, [TS] **veclmar**, [TS] **vecnorm**,
[TS] **vecstable**
postfile command, [P] **postfile**
PostScript, [G-3] *eps_options*, [G-3] *ps_options*,
[G-2] **graph export**, [G-4] *text*
poststratification, [SVY] **poststratification**,
[SVY] **Glossary**
postutil
clear command, [P] **postfile**
dir command, [P] **postfile**
poverty indices, [R] **inequality**
power, [M-2] **op_arith**, [M-2] **op_colon**,
[M-5] **matpowersym( )**, [ST] **Glossary**
Cox proportional hazards regression, [ST] **stpower**,
[ST] **stpower cox**
exponential survival, [ST] **stpower**, [ST] **stpower
exponential**
exponential test, [ST] **stpower**, [ST] **stpower
exponential**
log-rank, [ST] **stpower**, [ST] **stpower logrank**
power of a test, [R] **sampsi**
power, raise to, function, *see* arithmetic operators
power transformations, [R] **boxcox**, [R] **lnskew0**
P–P plot, [R] **diagnostic plots**

pperron command, [TS] **pperron**
pragma, [M-2] **pragma**, [M-6] **Glossary**
pragma unset, [M-2] **pragma**
pragma unused, [M-2] **pragma**
prais command, [TS] **prais**, [TS] **prais
postestimation**, [TS] **Glossary**
Prais–Winsten regression, [XT] **xtpcse**, *also see* prais
command
precision, [U] **13.11 Precision and problems therein**
predetermined variable, [XT] **Glossary**
_predict command, [P] **_predict**
predict command, [P] **ereturn**, [P] **_estimates**,
[I] **postestimation commands**, [MV] **factor
postestimation**, [MV] **pca postestimation**,
[R] **predict**, [R] **regress postestimation**,
[SEM] **intro 6**, [SEM] **example 14**,
[SEM] **predict**, [SVY] **svy postestimation**,
[U] **20.10 Obtaining predicted values**
predict, estat subcommand, [R] **exlogistic
postestimation**
predict, mi subcommand, [MI] **mi predict**
predicted values, *see* postestimation, predicted values
prediction, *see* multiple imputation, prediction
prediction, standard error of, [R] **glm**, [R] **predict**,
[R] **regress postestimation**
predictions, [MI] **mi predict**, [P] **_predict**, [R] **predict**,
[R] **predictnl**, [SVY] **svy postestimation**, *also
see* postestimation, predicted values
predictive margins, [SVY] **Glossary**,
[U] **20.15 Obtaining marginal means, adjusted
predictions, and predictive margins**
predictive mean matching imputation, *see* imputation,
predictive mean matching
predictnl command, [R] **predictnl**, [SVY] **svy
postestimation**
predictnl, mi subcommand, [MI] **mi predict**
prefix command, [R] **bootstrap**, [R] **fracpoly**,
[R] **jackknife**, [R] **mfp**, [R] **nestreg**,
[R] **permute**, [R] **simulate**, [R] **stepwise**, [R] **xi**,
[U] **11.1.10 Prefix commands**
Pregibon delta beta influence statistic, *see* delta beta
influence statistic
preprocessor commands, [R] **#review**
preserve command, [P] **preserve**
preserving user's data, [P] **preserve**
prevalence studies, *see* case–control data, *see* cross-
sectional studies
prevented fraction, [ST] **epitab**, [ST] **Glossary**
prewhiten, [XT] **Glossary**
primary sampling unit, [SVY] **svydescribe**,
[SVY] **svyset**, [SVY] **Glossary**
priming values, [TS] **Glossary**
principal
component analysis, [MV] **pca**, [MV] **Glossary**
factors analysis, [MV] **factor**
print,
graph set subcommand, [G-2] **graph set**
graph subcommand, [G-2] **graph print**
print command, [R] **translate**

printcolor, set subcommand, [G-2] **set printcolor**, [R] **set**
printf() function, [M-5] **printf( )**
printing graphs, [G-2] **graph print**, [G-3] *pr_options*
    exporting options, [G-2] **graph set**
    settings, [G-2] **graph set**
printing, logs (output), [R] **translate**, [U] **15 Saving and printing output—log files**
prior probabilities, [MV] **Glossary**
private, [M-2] **class**
probability weight, *see* sampling weight
probit command, [R] **probit**, [R] **probit postestimation**
probit regression, [R] **probit**, [SVY] **svy estimation**
    alternative-specific multinomial probit, [R] **asmprobit**
    alternative-specific rank-ordered, [R] **asroprobit**
    bivariate, [R] **biprobit**
    generalized estimating equations, [XT] **xtgee**
    generalized linear model, [R] **glm**
    heteroskedastic, [R] **hetprob**
    multinomial, [R] **mprobit**
    ordered, [R] **oprobit**
    population-averaged, [XT] **xtgee**, [XT] **xtprobit**
    random-effects, [XT] **xtprobit**
    two-equation, [R] **biprobit**
    with endogenous regressors, [R] **ivprobit**, [SVY] **svy estimation**
    with grouped data, [R] **glogit**
    with panel data, [XT] **xtgee**, [XT] **xtprobit**
    with sample selection, [R] **heckprob**, [SVY] **svy estimation**
procedure codes, [D] **icd9**
processors, set subcommand, [R] **set**
procoverlay command, [MV] **procrustes postestimation**
procrustes command, [MV] **procrustes**
Procrustes rotation, [MV] **procrustes**, [MV] **Glossary**
Procrustes transformation, *see* Procrustes rotation
product, [M-2] **op_arith**, [M-2] **op_colon**, [M-2] **op_kronecker**, [M-5] **cross( )**, [M-5] **crossdev( )**, [M-5] **quadcross( )**
product-moment correlation, [R] **correlate**
    between ranks, [R] **spearman**
production
    frontier model, [R] **frontier**, [XT] **xtfrontier**
    function, [XT] **Glossary**
profile plots, [R] **marginsplot**
profiles, estat subcommand, [MV] **ca postestimation**
program
    define command, [P] **plugin**, [P] **program**, [P] **program properties**
    dir command, [P] **program**
    drop command, [P] **program**
    list command, [P] **program**
program properties, [P] **program properties**

programmer's commands and utilities, [MI] **mi select**, [MI] **styles**, [MI] **technical**
programming, [P] **syntax**
    cluster analysis, [MV] **cluster programming utilities**
    cluster subcommands, [MV] **cluster programming subroutines**
    cluster utilities, [MV] **cluster programming subroutines**
    dialog, [P] **dialog programming**
    estat, [P] **estat programming**
    functions, [M-4] **programming**
    Mac, [P] **window programming**
    menus, [P] **window programming**
    rotations, [MV] **rotate**
    use, [M-1] **ado**
    Windows, [P] **window programming**
programs,
    adding comments to, [P] **comments**
    debugging, [P] **trace**
    dropping, [P] **discard**
    looping, [P] **continue**
    user-written, *see* ado-files
programs, clear subcommand, [D] **clear**
projection matrix, diagonal elements of, [R] **logistic postestimation**, [R] **logit postestimation**, [R] **probit postestimation**, [R] **regress postestimation**, [R] **rreg**
projection plot, [G-2] **graph twoway contour**, [G-2] **graph twoway contourline**
promax rotation, [MV] **rotate**, [MV] **rotatemat**, [MV] **Glossary**
proper() function, [D] **functions**
proper imputation method, [MI] **intro substantive**
proper values, [M-5] **eigensystem( )**
properties, [P] **program properties**
properties macro extended function, [P] **macro**
proportion command, [R] **proportion**, [R] **proportion postestimation**
proportional
    hazards models, [ST] **Glossary**, [SVY] **svy estimation**, *see* survival analysis, *also see* Cox proportional hazards model
    odds model, [R] **ologit**, [R] **slogit**
    sampling, [D] **sample**, [R] **bootstrap**
proportions,
    confidence intervals for, [R] **ci**
    estimating, [R] **proportion**
    sample size and power for, [R] **sampsi**
    survey data, [SVY] **svy estimation**, [SVY] **svy: tabulate oneway**, [SVY] **svy: tabulate twoway**
    testing equality of, [R] **bitest**, [R] **prtest**
prospective study, [ST] **epitab**, [ST] **Glossary**
protected, [M-2] **class**
proximity, [MV] **Glossary**
prtest command, [R] **prtest**
prtesti command, [R] **prtest**

psdensity command, [TS] **psdensity**

pseudo *R*-squared, [R] **maximize**

pseudofunctions, [D] **datetime**, [D] **functions**

pseudoinverse, [M-5] **pinv( )**

pseudolikelihood, [SVY] **Glossary**

pseudosigmas, [R] **lv**

psi function, [D] **functions**

pstyle, [G-4] *pstyle*

pstyle( ) option, [G-2] **graph twoway scatter**,
[G-3] *rspike_options*, [G-4] *pstyle*

PSU, *see* primary sampling unit

.ptrace files, [MI] **mi impute mvn**, [MI] **mi ptrace**

ptrace, mi subcommand, [MI] **mi ptrace**

public, [M-2] **class**

push, window subcommand, [P] **window programming**

putmata command, [D] **putmata**

pwcompare command, [R] **pwcompare**,
[R] **pwcompare postestimation**, [SVY] **svy
postestimation**, [U] **20.17 Obtaining pairwise
comparisons**

pwcorr command, [R] **correlate**

pwd command, [D] **cd**

pwd( ) function, [M-5] **chdir( )**

pweight, *see* sampling weight

[pweight=*exp*] modifier, [U] **11.1.6 weight**,
[U] **20.22.3 Sampling weights**

pwmean command, [R] **pwmean**, [R] **pwmean
postestimation**

pyramid, population, [G-2] **graph twoway bar**

Q

Q statistic, [TS] **wntestq**

qc charts, *see* quality control charts

qchi command, [R] **diagnostic plots**

QDA, [MV] **discrim qda**, [MV] **Glossary**

qda, discrim subcommand, [MV] **discrim qda**

qfit, graph twoway subcommand, [G-2] **graph
twoway qfit**

qfitci, graph twoway subcommand, [G-2] **graph
twoway qfitci**

qladder command, [R] **ladder**

QML, *see* quasimaximum likelihood

qnorm command, [R] **diagnostic plots**

qofd( ) function, [D] **datetime**, [D] **functions**,
[M-5] **date( )**

Q–Q plot, [R] **diagnostic plots**

qqplot command, [R] **diagnostic plots**

QR decomposition, [M-5] **qrd( )**

qrd( ) function, [M-5] **qrd( )**

qrdp( ) function, [M-5] **qrd( )**

_qreg command, [R] **qreg**

qreg command, [R] **qreg**, [R] **qreg postestimation**

_qrinv( ) function, [M-5] **qrinv( )**

qrinv( ) function, [M-5] **qrinv( )**

_qrsolve( ) function, [M-5] **qrsolve( )**

qrsolve( ) function, [M-5] **qrsolve( )**

qtolerance( ) option, [R] **maximize**

quad precision, [M-5] **runningsum( )**, [M-5] **sum( )**,
[M-5] **mean( )**, [M-5] **quadcross( )**

quadchk command, [XT] **quadchk**

quadcolsum( ) function, [M-5] **sum( )**

quadcorrelation( ) function, [M-5] **mean( )**

quadcross( ) function, [M-5] **quadcross( )**

quadcrossdev( ) function, [M-5] **quadcross( )**

quadmeanvariance( ) function, [M-5] **mean( )**

quadrant( ) function, [M-5] **sign( )**

quadratic discriminant analysis, [MV] **discrim qda**,
[MV] **Glossary**

quadratic terms, [SVY] **svy postestimation**

quadrature, [XT] **quadchk**, [XT] **Glossary**

quadrowsum( ) function, [M-5] **sum( )**

_quadrunningsum( ) function, [M-5] **runningsum( )**

quadrunningsum( ) function, [M-5] **runningsum( )**

quadsum( ) function, [M-5] **sum( )**

quadvariance( ) function, [M-5] **mean( )**

qualitative dependent variables, [R] **asclogit**,
[R] **asmprobit**, [R] **asroprobit**, [R] **binreg**,
[R] **biprobit**, [R] **brier**, [R] **clogit**, [R] **cloglog**,
[R] **cusum**, [R] **exlogistic**, [R] **glm**,
[R] **glogit**, [R] **heckprob**, [R] **hetprob**,
[R] **ivprobit**, [R] **logistic**, [R] **logit**, [R] **mlogit**,
[R] **mprobit**, [R] **nlogit**, [R] **ologit**, [R] **oprobit**,
[R] **probit**, [R] **rocfit**, [R] **rocreg**, [R] **rologit**,
[R] **scobit**, [R] **slogit**, [SVY] **svy estimation**,
[U] **26.6 Binary-outcome qualitative dependent-
variable models**, [U] **26.9 Multiple-outcome
qualitative dependent-variable models**,
[U] **26.17.4 Qualitative dependent-variable
models with panel data**, [XT] **xtcloglog**,
[XT] **xtgee**, [XT] **xtlogit**, [XT] **xtmelogit**,
[XT] **xtprobit**

quality control charts, [G-2] **graph other**, [R] **qc**,
[R] **serrbar**

quantile command, [R] **diagnostic plots**

quantile–normal plots, [R] **diagnostic plots**

quantile plots, [G-2] **graph other**, [R] **diagnostic plots**

quantile–quantile plots, [G-2] **graph other**,
[R] **diagnostic plots**

quantile regression, [R] **qreg**

quantiles, [D] **pctile**, *also see* percentiles, *see*
percentiles, displaying

quantiles, estat subcommand, [MV] **mds
postestimation**

quarter( ) function, [D] **datetime**, [D] **functions**,
[M-5] **date( )**

quarterly( ) function, [D] **datetime**, [D] **datetime
translation**, [D] **functions**, [M-5] **date( )**

quartimax rotation, [MV] **rotate**, [MV] **rotatemat**,
[MV] **Glossary**

quartimin rotation, [MV] **rotate**, [MV] **rotatemat**,
[MV] **Glossary**

quasimaximum likelihood, [SEM] **Glossary**

Quattro Pro, reading data from, [U] **21.4 Transfer
programs**

query, [M-3] **mata set**

query,
  cluster subcommand, [MV] **cluster programming utilities**
  estimates subcommand, [R] **estimates store**
  file subcommand, [P] **file**
  graph subcommand, [G-2] **graph query**
  icd9 subcommand, [D] **icd9**
  icd9p subcommand, [D] **icd9**
  log subcommand, [R] **log**
  mi subcommand, [MI] **mi describe**
  ml subcommand, [R] **ml**
  net subcommand, [R] **net**
  odbc subcommand, [D] **odbc**
  translator subcommand, [R] **translate**
  transmap subcommand, [R] **translate**
  update subcommand, [R] **update**
  webuse subcommand, [D] **webuse**
query command, [R] **query**
query graphics, [G-2] **set graphics**, [G-2] **set printcolor**, [G-2] **set scheme**
query memory command, [D] **memory**
querybreakintr() function, [M-5] **setbreakintr( )**
quick reference, [D] **data types**, [D] **missing values**, [I] **data types**, [I] **estimation commands**, [I] **file extensions**, [I] **format**, [I] **immediate commands**, [I] **import and export data**, [I] **missing values**, [I] **postestimation commands**, [I] **prefix commands**,
quietly prefix, [P] **quietly**
quit Mata, [M-3] **end**
quitting Stata, *see* exit command
quotes
  to delimit strings, [U] **18.3.5 Double quotes**
  to expand macros, [P] **macro**, [U] **18.3.1 Local macros**

R

r() saved results, [R] **saved results**, [U] **18.8 Accessing results calculated by other programs**, [U] **18.10.1 Saving results in r()**
r() scalars, macros, matrices, functions, [D] **functions**, [P] **discard**, [P] **return**
r(functions) macro extended function, [P] **macro**
r(macros) macro extended function, [P] **macro**
r(matrices) macro extended function, [P] **macro**
r(scalars) macro extended function, [P] **macro**
R charts, [G-2] **graph other**
R dates, [D] **datetime**
r-class command, [U] **18.8 Accessing results calculated by other programs**
r1title() option, [G-3] *title_options*
r2title() option, [G-3] *title_options*
radians, [D] **functions**
raise to a power function, [U] **13.2.1 Arithmetic operators**
Ramsey test, [R] **regress postestimation**

random
  number function, [D] **functions**, [D] **generate**, [M-5] **runiform( )**
  numbers, normally distributed, [D] **functions**, [D] **generate**
  order, test for, [R] **runtest**
  sample, [D] **sample**, [R] **bootstrap**, [U] **21.3 If you run out of memory**
  variates, [M-5] **runiform( )**
  walk, [TS] **Glossary**
random-coefficients
  linear regression, [XT] **xtrc**
  model, [U] **26.17.6 Random-coefficients model with panel data**, [XT] **Glossary**
random-effects model, [R] **anova**, [R] **loneway**, [XT] **xtabond**, [XT] **xtcloglog**, [XT] **xtdpd**, [XT] **xtdpdsys**, [XT] **xtgee**, [XT] **xthtaylor**, [XT] **xtintreg**, [XT] **xtivreg**, [XT] **xtlogit**, [XT] **xtnbreg**, [XT] **xtpoisson**, [XT] **xtprobit**, [XT] **xtreg**, [XT] **xtregar**, [XT] **xttobit**, [XT] **Glossary**
random-number seed, [MI] **mi impute**
random-number generators, [P] **version**
range
  chart, [R] **qc**
  of data, [D] **codebook**, [D] **inspect**, [R] **lv**, [R] **stem**, [R] **summarize**, [R] **table**, [R] **tabstat**, [XT] **xtsum**
  operators, [M-2] **op_range**
  plots, [G-3] *rcap_options*
  spikes, [G-3] *rspike_options*
  subscripts, *see* subscripts
  vector, [M-5] **range( )**
range command, [D] **range**
range() function, [M-5] **range( )**
rangen() function, [M-5] **range( )**
rank, [M-5] **rank( )**, [M-6] **Glossary**
rank correlation, [R] **spearman**
rank(), egen function, [D] **egen**
rank() function, [M-5] **rank( )**
rank-order statistics, [D] **egen**, [R] **signrank**, [R] **spearman**
ranking data, [R] **rologit**
rank ordered logistic regression, *see* outcomes, rank
ranks of observations, [D] **egen**
ranksum command, [R] **ranksum**
Rao's canonical-factor method, [MV] **factor**
rarea, graph twoway subcommand, [G-2] **graph twoway rarea**
rate ratio, [ST] **epitab**, [ST] **stci**, [ST] **stir**, [ST] **stptime**, [ST] **stsum**, *see* incidence-rate ratio
ratio command, [R] **ratio**, [R] **ratio postestimation**
ratios, estimating, [R] **ratio**
ratios, survey data, [SVY] **svy estimation**, [SVY] **svy: tabulate twoway**
raw data, [U] **12 Data**
.raw filename suffix, [U] **11.6 Filenaming conventions**
raw residuals, [SEM] **methods and formulas**

rbar, graph twoway subcommand, [G-2] **graph twoway rbar**

rbeta() function, [D] **functions**, [M-5] **runiform( )**

rbinomial() function, [D] **functions**, [M-5] **runiform( )**

rc (return codes), *see* error messages and return codes

_rc built-in variable, [P] **capture**, [U] **13.4 System variables (_variables)**

rcap, graph twoway subcommand, [G-2] **graph twoway rcap**

rcapsym, graph twoway subcommand, [G-2] **graph twoway rcapsym**

rchart command, [R] **qc**

rchi2() function, [D] **functions**, [M-5] **runiform( )**

r-class command, [P] **program**, [P] **return**

r-conformability, [M-5] **normal( )**, [M-6] **Glossary**

rconnected, graph twoway subcommand, [G-2] **graph twoway rconnected**

rdiscrete() function, [M-5] **runiform( )**

Re() function, [M-5] **Re( )**

read, file subcommand, [P] **file**

reading console input in programs, *see* console, obtaining input from

reading data from disk, [U] **21 Inputting and importing data**, [U] **21.4 Transfer programs**, *see* importing data

real, [M-2] **declarations**, [M-6] **Glossary**

real() function, [D] **functions**

real number to string conversion, [D] **destring**, [D] **encode**, [D] **functions**

real part, [M-5] **Re( )**

realization, [M-6] **Glossary**

recase() function, [D] **functions**

recast command, [D] **recast**

recast() option, [G-3] *advanced_options*, [G-3] *rcap_options*, [G-3] *rspike_options*

receiver operating characteristic (ROC) analysis, *see* ROC analysis

reciprocal averaging, [MV] **ca**

recode command, [D] **recode**

recode() function, [D] **functions**, [U] **25.1.2 Converting continuous variables to categorical variables**

recoding data, [D] **recode**

recoding data autocode() function, [D] **functions**

reconstructed correlations, [MV] **factor postestimation**

record I/O versus stream I/O, [U] **21 Inputting and importing data**

recording sessions, [U] **15 Saving and printing output—log files**

recovariance, estat subcommand, [XT] **xtmelogit postestimation**, [XT] **xtmepoisson postestimation**, [XT] **xtmixed postestimation**

recruitment period, *see* accrual period

rectangularize dataset, [D] **fillin**

recursive (structural) model (system), [SEM] **Glossary**

recursive estimation, [TS] **rolling**

recursive regression analysis, [TS] **Glossary**

redisplay graph, [G-2] **graph display**

reexpression, [R] **boxcox**, [R] **ladder**, [R] **lnskew0**

.ref built-in class function, [P] **class**

references, class, [P] **class**

reflection, [MV] **Glossary**

.ref_n built-in class function, [P] **class**

reg3 command, [R] **reg3**, [R] **reg3 postestimation**

regexm() function, [D] **functions**

regexr() function, [D] **functions**

regexs() function, [D] **functions**

regions
    look of, [G-4] *areastyle*
    outlining, [G-3] *region_options*
    shading, [G-3] *region_options*

register, mi subcommand, [MI] **mi set**

registered variables, *see* variables, registered

regress command, [R] **regress**, [R] **regress postestimation**, [R] **regress postestimation time series**

regression, [SEM] **Glossary**, *also see* models
    competing risks, [ST] **stcrreg**
    creating orthogonal polynomials for, [R] **orthog**
    diagnostic plots, [G-2] **graph other**
    diagnostics, [R] **predict**, [R] **ladder**, [R] **logistic postestimation**, [R] **regress postestimation**, [R] **regress postestimation time series**
    dummy variables, with, [R] **anova**, [R] **areg**, [R] **xi**
    fixed-effects, [R] **areg**
    fractional polynomial, [R] **fracpoly**, [R] **mfp**
    function, estimating, [R] **lpoly**
    graphing, [R] **logistic**, [R] **regress postestimation**
    grouped data, [R] **intreg**
    increasing number of variables allowed, [R] **matsize**
    instrumental variables, [R] **gmm**, [R] **ivprobit**, [R] **ivregress**, [R] **ivtobit**, [R] **nlsur**
    linear, *see* linear regression
    lines, *see* fits, adding
    scoring, [MV] **factor postestimation**
    system, [R] **mvreg**, [R] **reg3**, [R] **sureg**
    truncated, [R] **truncreg**

regression (in generic sense), *also see* estimation commands
    accessing coefficients and standard errors, [P] **matrix get**, [U] **13.5 Accessing coefficients and standard errors**
    dummy variables, with, [XT] **xtreg**
    fixed-effects, [XT] **xtreg**
    instrumental variables, [XT] **xtabond**, [XT] **xtdpd**, [XT] **xtdpdsys**, [XT] **xthtaylor**, [XT] **xtivreg**
    random-effects, [XT] **xtgee**, [XT] **xtreg**

regular expressions, [D] **functions**

regular variables, *see* variables, regular

relational operators, [U] **13.2.3 Relational operators**

relative difference function, [D] **functions**

relative efficiency, [MI] **mi estimate**, [MI] **mi predict**, [MI] **Glossary**

relative risk, [ST] **epitab**

relative variance increase, [MI] **mi estimate**, [MI] **mi predict**, [MI] **Glossary**
reldif() function, [D] **functions**, [M-5] **reldif( )**
release marker, [P] **version**
releases, compatibility of Stata programs across, [P] **version**
reliability, [MV] **factor**, [R] **alpha**, [R] **eivreg**, [R] **loneway**, [SEM] **Intro 4**, [SEM] **sem model description options**, [SEM] **sem option reliability( )**, [SEM] **Glossary**
reliability option, *see* sem option reliability()
reliability theory, *see* survival analysis
remainder function, [D] **functions**
REML, *see* restricted maximum likelihood
removing
directories, [D] **rmdir**
files, [D] **erase**
r._En, [SEM] **sem option covstructure( )**
rename,
char subcommand, [P] **char**
cluster subcommand, [MV] **cluster utility**
graph subcommand, [G-2] **graph rename**
irf subcommand, [TS] **irf rename**
matrix subcommand, [P] **matrix utility**
rename, mi subcommand, [MI] **mi rename**
rename command, [D] **rename**, [D] **rename group**, [M-3] **mata rename**, [MI] **ml rename**
rename graph, [G-2] **graph rename**
renamevar, cluster subcommand, [MV] **cluster utility**
renaming variables, [D] **rename**, [D] **rename group**, [MI] **mi rename**
renumber, notes subcommand, [D] **notes**
reordering data, [D] **order**, [D] **gsort**, [D] **sort**
reorganizing data, [D] **reshape**, [D] **xpose**
repair, ssd subcommand, [SEM] **ssd**
repeated measures, [MV] **Glossary**
ANOVA, [R] **anova**
MANOVA, [MV] **manova**
repeated options, [G-4] **concept: repeated options**
repeating and editing commands, [R] **#review**, [U] **10 Keyboard use**
repeating commands, [D] **by**, [P] **continue**, [P] **foreach**, [P] **forvalues**
replace command, [D] **generate**, [MI] **mi passive**, [MI] **mi xeq**
replace, notes subcommand, [D] **notes**
replace option, [U] **11.2 Abbreviation rules**
replace0, mi subcommand, [MI] **mi replace0**
replay, estimates subcommand, [R] **estimates replay**
replay() function, [P] **ereturn**, [P] **_estimates**, [D] **functions**
replicate weight variable, [SVY] **survey**, [SVY] **svy bootstrap**, [SVY] **svy brr**, [SVY] **svy jackknife**, [SVY] **svy sdr**, [SVY] **svyset**, [SVY] **Glossary**

replicating
clustered observations, [D] **expandcl**
observations, [D] **expand**
replication method, [SVY] **svy bootstrap**, [SVY] **svy brr**, [SVY] **svy jackknife**, [SVY] **svy sdr**, [SVY] **svyset**, [SVY] **variance estimation**
report,
datasignature subcommand, [D] **datasignature**
duplicates subcommand, [D] **duplicates**
fvset subcommand, [R] **fvset**
ml subcommand, [R] **ml**
reporting options, [SEM] **sem reporting options**
repost, ereturn subcommand, [P] **ereturn**, [P] **return**
_request(*macname*), display directive, [P] **display**
resampling, [SVY] **Glossary**
reserved names, [U] **11.3 Naming conventions**
reserved words, [M-2] **reswords**
reset, mi subcommand, [MI] **mi reset**
RESET test, [R] **regress postestimation**
reset, translator subcommand, [R] **translate**
reset_id, serset subcommand, [P] **serset**
reshape
command, [D] **reshape**, [MI] **mi reshape**
error command, [D] **reshape**
long command, [D] **reshape**
wide command, [D] **reshape**
reshape, mi subcommand, [MI] **mi reshape**
residual-versus-fitted plot, [G-2] **graph other**, [R] **regress postestimation**
residual versus-predictor plot, [G-2] **graph other**, [R] **regress postestimation**
residuals, [R] **logistic**, [R] **predict**, [R] **regress postestimation**, [R] **rreg**, [SEM] **estat gof**, [SEM] **estat residuals**, [SEM] **example 4**, [SEM] **Glossary**
residuals, estat subcommand, [MV] **factor postestimation**, [MV] **pca postestimation**, [SEM] **estat residuals**
resistant smoothers, [R] **smooth**
restore command, [P] **preserve**
restore,
estimates subcommand, [R] **estimates store**
_return subcommand, [P] **_return**
snapshot subcommand, [D] **snapshot**
restoring data, [D] **snapshot**
restricted cubic splines, [R] **mkspline**
restricted maximum likelihood (REML), [XT] **Glossary**
results, clear subcommand, [D] **clear**
results,
clearing, [P] **ereturn**, [P] **_estimates**, [P] **_return**
listing, [P] **ereturn**, [P] **_estimates**, [P] **_return**
returning, [P] **_return**, [P] **return**
saved, [R] **saved results**
saving, [P] **ereturn**, [P] **_estimates**, [P] **postfile**, [P] **_return**, [P] **return**, [R] **estimates save**

_return
 dir command, [P] _return
 drop command, [P] _return
 hold command, [P] _return
 restore command, [P] _return
return, [M-2] return
 add command, [P] return
 clear command, [P] return
 list command, [P] return, [R] saved results
 local command, [P] return
 matrix command, [P] return
 scalar command, [P] return
return codes, [P] rmsg, see error messages and return
  codes
return() function, [D] functions
return value, [P] class
returning results, [P] return
 class programs, [P] class
reventries, set subcommand, [R] set
reverse() function, [D] functions
reversed scales, [G-3] axis_scale_options
#review command, [R] #review, [U] 10 Keyboard
  use, [U] 15 Saving and printing output—log
  files
revkeyboard, set subcommand, [R] set
reorder() function, [M-5] invorder()
rgamma() function, [D] functions, [M-5] runiform()
rhypergeometric() function, [D] functions,
  [M-5] runiform()
ridge prior, [MI] mi impute mvn
right-censoring, [ST] Glossary, see imputation, interval-
  censored data
right eigenvectors, [M-5] eigensystem()
right suboption, [G-4] justificationstyle
right-truncation, see imputation, truncated data
rightmost options, [G-4] concept: repeated options
risk
 difference, [ST] epitab
 factor, [ST] epitab, [ST] Glossary
 pool, [ST] stcox, [ST] stcrreg, [ST] stset,
  [ST] Glossary
 ratio, [R] binreg, [ST] epitab, [ST] Glossary
rline, graph twoway subcommand, [G-2] graph
  twoway rline
rm command, [D] erase
_rmcoll command, [P] _rmcoll
_rmdcoll command, [P] _rmcoll
rmdir command, [D] rmdir
_rmdir() function, [M-5] chdir()
rmdir() function, [M-5] chdir()
rmexternal() function, [M-5] findexternal()
RMSEA, see root mean squared error of approximation
rmsg, [P] creturn, [P] error, [U] 8 Error messages
  and return codes
 set subcommand, [P] rmsg, [R] set
rnbinomial() function, [D] functions,
  [M-5] runiform()
rnormal() function, [D] functions, [M-5] runiform()

robust, [SEM] Glossary
robust, see sem option vce()
_robust command, [P] _robust
robust, Huber/White/sandwich estimator of variance,
  [P] _robust, [R] vce_option, [SVY] variance
  estimation, [XT] vce_options
 alternative-specific
  conditional logit model, [R] asclogit
  multinomial probit regression, [R] asmprobit
  rank-ordered probit regression, [R] asroprobit
 ARCH, [TS] arch
 ARFIMA, [TS] arfima
 ARIMA and ARMAX, [TS] arima
 complementary log-log regression, [R] cloglog
 Cox proportional hazards model, [ST] stcox,
  [ST] stcrreg
 dynamic-factor model, [TS] dfactor
 fixed-effects models,
  linear, [XT] xtreg
  Poisson, [XT] xtpoisson
 GARCH model, see GARCH
 generalized linear models, [R] glm
  for binomial family, [R] binreg
 generalized method of moments, [R] gmm
 heckman selection model, [R] heckman
 instrumental-variables regression, [R] ivregress
 interval regression, [R] intreg
 linear dynamic panel-data estimation, [XT] xtabond,
  [XT] xtdpd, [XT] xtdpdsys
 linear regression, [R] regress
  constrained, [R] cnsreg
  truncated, [R] truncreg
  with dummy-variable set, [R] areg
 logistic regression, [R] logistic, [R] logit, also see
  logit regression subentry
  conditional, [R] clogit
  multinomial, [R] mlogit
  ordered, [R] ologit
  rank-ordered, [R] rologit
  skewed, [R] scobit
  stereotype, [R] slogit
 logit regression, [R] logistic, [R] logit, also see
  logistic regression subentry
  for grouped data, [R] glogit
  nested, [R] nlogit
 maximum likelihood estimation, [R] ml
 multilevel mixed-effects models, [XT] xtmixed
 multinomial
  logistic regression, [R] mlogit
  probit regression, [R] mprobit
 negative binomial regression, [R] nbreg
  truncated, [R] tnbreg
  zero-inflated, [R] zinb
 Newey–West regression, [TS] newey
 nonlinear
  least-squares estimation, [R] nl
  systems of equations, [R] nlsur

robust, Huber/White/sandwich estimator of variance,
continued
  parametric survival models, [ST] **streg**
  Poisson regression, [R] **poisson**
    truncated, [R] **tpoisson**
    zero-inflated, [R] **zip**
  population-averaged models, [XT] **xtgee**
    cloglog, [XT] **xtcloglog**
    logit, [XT] **xtlogit**
    negative binomial, [XT] **xtnbreg**
    Poisson, [XT] **xtpoisson**
    probit, [XT] **xtprobit**
  Prais–Winsten and Cochrane–Orcutt regression,
    [TS] **prais**
  probit regression, [R] **probit**
    bivariate, [R] **biprobit**
    for grouped data, [R] **glogit**
    heteroskedastic, [R] **hetprob**
    multinomial, [R] **mprobit**
    ordered, [R] **oprobit**
    with endogenous regressors, [R] **ivprobit**
    with sample selection, [R] **heckprob**
  random-effects models, linear, [XT] **xtreg**
  state-space model, [TS] **sspace**
  structural equation modeling, [SEM] **intro 7**,
    [SEM] **sem option method( )**
  summary statistics,
    mean, [R] **mean**
    proportion, [R] **proportion**
    ratio, [R] **ratio**
    total, [R] **total**
  tobit model, [R] **tobit**
    with endogenous regressors, [R] **ivtobit**
  treatment effects model, [R] **treatreg**
  truncated
    negative binomial regression, [R] **tnbreg**
    Poisson regression, [R] **tpoisson**
    regression, [R] **truncreg**
  unobserved-components model, [TS] **ucm**
  with endogenous regressors,
    instrumental-variables regression, [R] **ivregress**
    probit regression, [R] **ivprobit**
    tobit regression, [R] **ivtobit**
  zero inflated
    negative binomial regression, [R] **zinb**
    Poisson regression, [R] **zip**
robust, other methods of, [R] **qreg**, [R] **rreg**,
  [R] **smooth**
robust regression, [R] **regress**, [R] **rreg**, *also see* robust,
  Huber/White/sandwich estimator of variance
robust standard errors, [XT] **Glossary**
robust test for equality of variance, [R] **sdtest**
robvar command, [R] **sdtest**
ROC analysis, [G-2] **graph other**, [R] **roc**,
  [U] **26.7 ROC analysis**
  area under ROC curve, [R] **logistic postestimation**

ROC analysis, *continued*
  nonparametric analysis without covariates,
    [R] **roctab**
  parametric analysis without covariates, [R] **rocfit**
  regression models, [R] **rocreg**
  ROC curves after rocfit, [R] **rocfit postestimation**
  ROC curves after rocreg, [R] **rocregplot**
  test equality of ROC areas, [R] **roccomp**
roccomp command, [R] **roc**, [R] **roccomp**
rocfit command, [R] **rocfit**, [R] **rocfit postestimation**
rocgold command, [R] **roc**, [R] **roccomp**
rocplot command, [R] **rocfit postestimation**
rocreg command, [R] **rocreg**, [R] **rocreg**
  postestimation, [R] **rocregplot**
rocregplot command, [R] **rocregplot**
roctab command, [R] **roc**, [R] **roctab**
Rogers and Tanimoto similarity measure,
  [MV] *measure_option*
roh, [R] **loneway**
rolling command, [TS] **rolling**
rolling regression, [TS] **rolling**, [TS] **Glossary**
rologit command, [R] **rologit**, [R] **rologit**
  postestimation
root mean squared error of approximation, [SEM] **estat**
  gof, [SEM] **example 4**, [SEM] **methods and**
  formulas
rootograms, [G-2] **graph other**, [R] **spikeplot**
roots of polynomials, [M-5] **polyeval( )**
rotate command, [MV] **factor postestimation**,
  [MV] **pca postestimation**, [MV] **rotate**
rotate, estat subcommand, [MV] **canon**
  postestimation
rotatecompare, estat subcommand, [MV] **canon**
  postestimation, [MV] **factor postestimation**,
  [MV] **pca postestimation**
rotated factor loadings, [MV] **factor postestimation**
rotated principal components, [MV] **pca postestimation**
rotatemat command, [MV] **rotatemat**
rotation, [MV] **factor postestimation**, [MV] **pca**
  postestimation, [MV] **rotate**, [MV] **rotatemat**,
  [MV] **Glossary**
  Bentler's invariant pattern simplicity, *see* Bentler's
    invariant pattern simplicity rotation
  biquartimax, *see* biquartimax rotation
  biquartimin, *see* biquartimin rotation
  Comrey's tandem 1, *see* Comrey's tandem 1 and 2
    rotations
  Comrey's tandem 2, *see* Comrey's tandem 1 and 2
    rotations
  covarimin, *see* covarimin rotation
  Crawford–Ferguson, *see* Crawford–Ferguson rotation
  equamax, *see* equamax rotation
  factor parsimony, *see* factor parsimony rotation
  minimum entropy, *see* minimum entropy rotation
  oblimax, *see* oblimax rotation
  oblimin, *see* oblimin rotation
  oblique, *see* oblique rotation
  orthogonal, *see* orthogonal rotation

rotation, *continued*
 parsimax, *see* parsimax rotation
 partially specified target, *see* partially specified target rotation
 Procrustes, *see* Procrustes rotation
 promax, *see* promax rotation
 quartimax, *see* quartimax rotation
 quartimin, *see* quartimin rotation
 toward a target, *see* toward a target rotation
 varimax, *see* varimax rotation
round() function, [D] **functions**, [M-5] **trunc( )**
roundoff error, [M-5] **epsilon( )**, [M-5] **edittozero( )**, [M-5] **edittoint( )**, [U] **13.11 Precision and problems therein**
row-join operator, [M-2] **op_join**
row of matrix, selecting, [M-5] **select( )**
row operators for data, [D] **egen**
roweq macro extended function, [P] **macro**
roweq, matrix subcommand, [P] **matrix rownames**
rowfirst(), egen function, [D] **egen**
rowfullnames macro extended function, [P] **macro**
rowlast(), egen function, [D] **egen**
row-major order, [M-6] **Glossary**
rowmax(), egen function, [D] **egen**
rowmax() function, [M-5] **minmax( )**
rowmaxabs() function, [M-5] **minmax( )**
rowmean(), egen function, [D] **egen**
rowmedian(), egen function, [D] **egen**
rowmin(), egen function, [D] **egen**
rowmin() function, [M-5] **minmax( )**
rowminmax() function, [M-5] **minmax( )**
rowmiss(), egen function, [D] **egen**
rowmissing() function, [M-5] **missing( )**
rownames macro extended function, [P] **macro**
rownames, matrix subcommand, [P] **matrix rownames**
rownonmiss(), egen function, [D] **egen**
rownonmissing() function, [M-5] **missing( )**
rownumb() function, [P] **matrix define**, [D] **functions**
rowpctile(), egen function, [D] **egen**
rows() function, [M-5] **rows( )**
rows of matrix
 appending to, [P] **matrix define**
 names, [P] **ereturn**, [P] **matrix define**, [P] **matrix rownames**
 operators, [P] **matrix define**
rowscalefactors() function, [M-5] **_equilrc( )**
rowsd(), egen function, [D] **egen**
rowshape() function, [M-5] **rowshape( )**
rowsof() function, [D] **functions**, [P] **matrix define**
rowsum() function, [M-5] **sum( )**
rowtotal(), egen function, [D] **egen**
rowvector, [M-2] **declarations**, [M-6] **Glossary**
Roy's
 largest root test, [MV] **canon**, [MV] **manova**, [MV] **mvtest means**, [MV] **Glossary**

Roy's, *continued*
 union-intersection test, [MV] **canon**, [MV] **manova**, [MV] **mvtest means**
rpoisson() function, [D] **functions**, [M-5] **runiform( )**
rreg command, [R] **rreg**, [R] **rreg postestimation**
rscatter, graph twoway subcommand, [G-2] **graph twoway rscatter**
rseed() function, [D] **functions**, [M-5] **runiform( )**
rspike, graph twoway subcommand, [G-2] **graph twoway rspike**
R^2, [SEM] **estat eqgof**
rt() function, [D] **functions**, [M-5] **runiform( )**
rtrim() function, [D] **functions**
Rubin's combination rules, [MI] **mi estimate**, [MI] **mi estimate using**, [MI] **mi predict**
run command, [R] **do**, [U] **16 Do-files**
runiform() function, [D] **functions**, [M-5] **runiform( )**, [R] **set seed**
_runningsum() function, [M-5] **runningsum( )**
runningsum() function, [M-5] **runningsum( )**
runtest command, [R] **runtest**
Russell and Rao coefficient similarity measure, [MV] *measure_option*
rvalue, class, [P] **class**
rvfplot command, [R] **regress postestimation**
RVI, *see* relative variance increase
rvpplot command, [R] **regress postestimation**

S

s() function, [D] **functions**
s() saved results, [D] **functions**, [P] **return**, [R] **saved results**, [U] **18.8 Accessing results calculated by other programs**, [U] **18.10.3 Saving results in s()**
s(macros) macro extended function, [P] **macro**
S_ macros, [R] **saved results**
s-class command, [U] **18.8 Accessing results calculated by other programs**
s1color scheme, [G-4] **scheme s1**
s1manual scheme, [G-4] **scheme s1**
s1mono scheme, [G-4] **scheme s1**
s1rcolor scheme, [G-4] **scheme s1**
s2color scheme, [G-4] **scheme s2**
s2gcolor scheme, [G-4] **scheme s2**
s2gmanual scheme, [G-4] **scheme s2**
s2manual scheme, [G-4] **scheme s2**
s2mono scheme, [G-4] **scheme s2**
SAARCH, [TS] **arch**
Sammon mapping criterion, [MV] **Glossary**
sample, [SVY] **Glossary**
sample command, [D] **sample**
sample, random, *see* random sample
sample size, [R] **sampsi**
 Cox proportional hazards regression, [ST] **stpower**, [ST] **stpower cox**
 exponential survival, [ST] **stpower**, [ST] **stpower exponential**

sample size, *continued*
> exponential test, [ST] **stpower**, [ST] **stpower exponential**
> log-rank, [ST] **stpower**, [ST] **stpower logrank**

sampling, [D] **sample**, [R] **bootstrap**, [R] **bsample**, [SVY] **survey**, [SVY] **svydescribe**, [SVY] **svyset**, [SVY] **Glossary**, *also see* cluster sampling
> stage, [SVY] **estat**, [SVY] **Glossary**
> unit, [SVY] **survey**, [SVY] **Glossary**, *also see* primary sampling unit
> weight, [SVY] **poststratification**, [SVY] **survey**, [SVY] **Glossary**, [U] **11.1.6 weight**, [U] **20.22.3 Sampling weights**, *also see* survey data
> with and without replacement, [SVY] **Glossary**

sampsi command, [R] **sampsi**

sandwich/Huber/White estimator of variance, *see* robust, Huber/White/sandwich estimator of variance

sargan, estat subcommand, [XT] **xtabond postestimation**, [XT] **xtdpd postestimation**, [XT] **xtdpdsys postestimation**

Sargan test, [XT] **xtabond postestimation**, [XT] **xtdpd postestimation**, [XT] **xtdpdsys postestimation**

SAS dates, [D] **datetime**

SAS, reading data from, [U] **21.4 Transfer programs**

SAS XPORT format, [D] **import sasxport**

sasxport,
> export subcommand, [D] **import sasxport**
> import subcommand, [D] **import sasxport**

satopts() option, *see* sem option satopts()

saturated model, [SEM] **estat gof**, [SEM] **example 4**, [SEM] **methods and formulas**, [SEM] **Glossary**

save,
> estimates subcommand, [R] **estimates save**
> graph subcommand, [G-2] **graph save**
> label subcommand, [D] **label**
> snapshot subcommand, [D] **snapshot**

save estimation results, [P] **ereturn**, [P] **_estimates**

save command, [D] **save**

saved results, [P] **_return**, [P] **return**, [R] **saved results**, [SEM] **intro 6**, [U] **18.8 Accessing results calculated by other programs**, [U] **18.9 Accessing results calculated by estimation commands**, [U] **18.10 Saving results**
> clearing, [M-5] **st_rclear()**
> hidden or historical, [M-5] **st_global()**, [M-5] **st_matrix()**, [M-5] **st_numscalar()**

saveold command, [D] **save**

saving data, [D] **outfile**, [D] **outsheet**, [D] **save**, [D] **snapshot**, *also see* exporting data

saving() option, [G-3] *saving_option*

saving results, [P] **ereturn**, [P] **_estimates**, [P] **postfile**, [P] **_return**, [P] **return**, [R] **estimates save**

scalar, [M-2] **declarations**, [M-6] **Glossary**, [P] **scalar**
> confirm subcommand, [P] **confirm**
> define command, [P] **scalar**

scalar, *continued*
> dir command, [P] **scalar**
> drop command, [P] **scalar**
> ereturn subcommand, [P] **ereturn**, [P] **return**
> list command, [P] **scalar**
> return subcommand, [P] **return**

scalar() function, [D] **functions**, [P] **scalar**

scalar functions, [M-4] **scalar**

scalars, [P] **scalar**
> namespace and conflicts, [P] **matrix**, [P] **matrix define**

scale() option, [G-3] *scale_option*

scale,
> log, [G-3] *axis_scale_options*
> range of, [G-3] *axis_scale_options*
> reversed, [G-3] *axis_scale_options*

scaling, [MV] **mds**, [MV] **mds postestimation**, [MV] **mdslong**, [MV] **mdsmat**

scatter, graph twoway subcommand, [G-2] **graph twoway scatter**

scatteri, graph twoway subcommand, [G-2] **graph twoway scatteri**

scatterplot matrices, [G-2] **graph matrix**

Scheffé's multiple-comparison adjustment, *see* multiple comparisons, Scheffé's method

scheme() option, [G-3] *scheme_option*

scheme, set subcommand, [G-2] **set scheme**, [R] **set**

schemes, [G-2] **set scheme**, [G-3] *scheme_option*, [G-4] **scheme economist**, [G-4] **scheme s1**, [G-4] **scheme s2**, [G-4] **scheme sj**, [G-3] *play_option*, [G-4] **schemes intro**
> changing, [G-2] **graph display**
> creating your own, [G-4] **schemes intro**
> default, [G-2] **set scheme**

Schoenfeld residual, [ST] **stcox postestimation**, [ST] **streg postestimation**

Schur
> decomposition, [M-5] **schurd()**, [M-6] **Glossary**
> form, [M-6] **Glossary**

_schurd() function, [M-5] **schurd()**

schurd() function, [M-5] **schurd()**

_schurdgroupby() function, [M-5] **schurd()**

schurdgroupby() function, [M-5] **schurd()**

_schurdgroupby_la() function, [M-5] **schurd()**

_schurd_la() function, [M-5] **schurd()**

Schwarz information criterion, *see* Bayesian information criterion

scientific notation, [U] **12.2 Numbers**

s-class command, [P] **program**, [P] **return**, [R] **saved results**

scobit command, [R] **scobit**, [R] **scobit postestimation**

scope, class, [P] **class**

score, [MV] **Glossary**

score, matrix subcommand, [P] **matrix score**

score, ml subcommand, [R] **ml**

score plot, [MV] **scoreplot**, [MV] **Glossary**

score tests, [SEM] **intro 6**, [SEM] **estat ginvariant**,
  [SEM] **estat mindices**, [SEM] **estat scoretests**,
  [SEM] **methods and formulas**, [SEM] **Glossary**
scoreplot command, [MV] **discrim lda**
  postestimation, [MV] **factor postestimation**,
  [MV] **pca postestimation**, [MV] **scoreplot**
scores, [R] **predict**, [SEM] **Glossary**
scores, obtaining, [U] **20.21 Obtaining scores**
scoretests, estat subcommand, [SEM] **estat**
  scoretests
scoring, [MV] **factor postestimation**, [MV] **pca**
  postestimation, [P] **matrix score**
scree plot, [MV] **screeplot**, [MV] **Glossary**
screeplot command, [MV] **discrim lda**
  postestimation, [MV] **factor postestimation**,
  [MV] **mca postestimation**, [MV] **pca**
  postestimation, [MV] **screeplot**
scrollbufsize, set subcommand, [R] **set**
scrolling of output, controlling, [P] **more**, [R] **more**
sd(), egen function, [D] **egen**
sd, estat subcommand, [SVY] **estat**
SDR, *see* successive difference replication
sdr_options, [SVY] *sdr_options*
sdtest command, [R] **sdtest**
sdtesti command, [R] **sdtest**
se, estat subcommand, [R] **exlogistic postestimation**,
  [R] **expoisson postestimation**
_se[], [U] **13.5 Accessing coefficients and standard**
  errors
search
  help, [R] **hsearch**
  Internet, [R] **net search**
search,
  icd9 subcommand, [D] **icd9**
  icd9p subcommand, [D] **icd9**
  ml subcommand, [R] **ml**
  net subcommand, [R] **net**
  view subcommand, [R] **view**
search command, [R] **search**, [U] **4 Stata's help and**
  search facilities
search, notes subcommand, [D] **notes**
search_d, view subcommand, [R] **view**
searchdefault, set subcommand, [R] **search**, [R] **set**
seasonal
  ARIMA, [TS] **arima**
  difference operator, [TS] **Glossary**
  smoothing, [TS] **tssmooth**, [TS] **tssmooth shwinters**
seasonal lag operator, [U] **11.4.4 Time-series varlists**
second-order latent variables, [SEM] **Glossary**
secondary sampling unit, [SVY] **Glossary**
seconds() function, [D] **datetime**, [D] **functions**,
  [M-5] **date()**
seed, set subcommand, [R] **set**, [R] **set seed**
seek, file subcommand, [P] **file**
seemingly unrelated
  estimation, [R] **suest**

seemingly unrelated, *continued*
  regression, [R] **nlsur**, [R] **reg3**, [R] **sureg**,
    [SEM] **intro 4**, [SEM] **example 12**,
    [SEM] **Glossary**
segmentsize, set subcommand, [D] **memory**, [R] **set**
select() function, [M-5] **select()**
select() option, *see* sem option select()
select, mi subcommand, [MI] **mi select**
selection models, [R] **heckman**, [R] **heckprob**,
  [SVY] **svy estimation**
selection-order statistics, [TS] **varsoc**
SEM, [MV] **intro**, [SEM] **methods and formulas**,
  [SEM] **Glossary**, *see* structural equation
  modeling
sem command, [SEM] **GUI**, [SEM] **methods and**
  formulas, [SEM] **sem**, [SEM] **sem model**
  description options, [SEM] **sem path notation**,
  [SEM] **Glossary**
  missing values, [SEM] **example 26**
  with constraints, [SEM] **example 8**
sem option
  allmissing, [SEM] **sem estimation options**
  baseopts(), [SEM] **sem estimation options**
  coeflegend, [SEM] **example 8**,
    [SEM] **example 16**, [SEM] **sem reporting**
    options
  constraints(), [SEM] **sem model description**
    options, [SEM] **sem option constraints()**
  covariance(), [SEM] **sem model description**
    options, [SEM] **sem path notation**
  covstructure(), [SEM] **intro 4**,
    [SEM] **example 17**, [SEM] **sem model**
    description options, [SEM] **sem option**
    covstructure()
  forcecorrelations, [SEM] **sem ssd options**
  forcenoanchor, [SEM] **sem model description**
    options
  forcexconditional, [SEM] **sem option**
    noxconditional
  from(), [SEM] **sem model description options**,
    [SEM] **sem option from()**
  ginvariant(), [SEM] **intro 5**, [SEM] **example 23**,
    [SEM] **sem group options**
  groups(), [SEM] **intro 5**, [SEM] **example 20**,
    [SEM] **example 23**, [SEM] **sem group options**,
    [SEM] **sem option select()**, [SEM] **sem path**
    notation
  latent(), [SEM] **sem syntax options**
  level(), [SEM] **sem reporting options**
  maximize_options, [SEM] **sem estimation options**
  means(), [SEM] **intro 4**, [SEM] **example 18**,
    [SEM] **sem model description options**,
    [SEM] **sem path notation**
  method(), [SEM] **intro 3**, [SEM] **intro 7**,
    [SEM] **intro 8**, [SEM] **example 26**, [SEM] **sem**
    estimation options, [SEM] **sem option**
    method(), [SEM] **Glossary**
  nm1, [SEM] **sem estimation options**
  noanchor, [SEM] **sem model description options**

sem option, *continued*
   nocapslatent, [SEM] **sem syntax options**
   nocnsreport, [SEM] **sem reporting options**
   noconstant, [SEM] **sem model description options**
   nodescribe, [SEM] **sem reporting options**
   nofootnote, [SEM] **sem reporting options**
   noheader, [SEM] **sem reporting options**
   noivstart, [SEM] **sem estimation options**
   nolabel, [SEM] **sem reporting options**
   nomeans, [SEM] **sem model description options**
   notable, [SEM] **sem reporting options**
   noxconditional, [SEM] **sem estimation options**,
     [SEM] **sem option noxconditional**
   reliability(), [SEM] **example 24**, [SEM] **sem model description options**, [SEM] **sem option reliability( )**
   satopts(), [SEM] **sem estimation options**
   select(), [SEM] **sem option select( )**, [SEM] **sem ssd options**
   showginvariant, [SEM] **sem reporting options**
   standardized, [SEM] **sem reporting options**
   variance(), [SEM] **sem model description options**, [SEM] **sem path notation**
   vce(), [SEM] **intro 3**, [SEM] **intro 7**, [SEM] **intro 8**, [SEM] **sem estimation options**, [SEM] **sem option method( )**, [SEM] **Glossary**
   wrap(), [SEM] **sem reporting options**
sem postestimation commands, [SEM] **intro 6**
semicolons, [M-2] **semicolons**
semiparametric imputation method, *see* imputation, predictive mean matching
semiparametric model, [ST] **stcox**, [ST] **stcrreg**, [ST] **stset**, [ST] **Glossary**
semirobust standard errors, [XT] **Glossary**
sensitivity, [R] **logistic postestimation**, *also see* ROC analysis, *see* ROC analysis
   model, [R] **regress postestimation**, [R] **rreg**
separate command, [D] **separate**
separating string variables into parts, [D] **split**
seq(), egen function, [D] **egen**
sequential imputation, [MI] **mi impute**, [MI] **mi impute chained**, [MI] **mi impute monotone**
sequential limit theory, [XT] **Glossary**
sequential regression multivariate imputation, *see* imputation, chained equations
serial correlation, *see* autocorrelation
serial independence, test for, [R] **runtest**
serrbar command, [R] **serrbar**
serset, [P] **serset**
   clear command, [P] **serset**
   create command, [P] **serset**
   create_cspline command, [P] **serset**
   create_xmedians command, [P] **serset**
   dir command, [P] **serset**
   drop command, [P] **serset**
   reset_id command, [P] **serset**
   set command, [P] **serset**

serset, *continued*
   sort command, [P] **serset**
   summarize command, [P] **serset**
   use command, [P] **serset**
sersetread, file subcommand, [P] **serset**
sersetwrite, file subcommand, [P] **serset**
session, recording, [R] **log**, [U] **15 Saving and printing output—log files**
set, [M-3] **mata set**
   adosize command, [P] **sysdir**, [R] **set**
   autotabgraphs command, [R] **set**
   cformat command, [R] **set**, [R] **set cformat**
   charset command, [P] **smcl**
   checksum command, [D] **checksum**, [R] **set**
   command, [R] **query**, [R] **set**
   conren command, [R] **set**
   copycolor command, [R] **set**
   dockable command, [R] **set**
   dockingguides command, [R] **set**
   doublebuffer command, [R] **set**
   dp command, [D] **format**, [R] **set**
   emptycells command, [R] **set**, [R] **set emptycells**
   eolchar command, [R] **set**
   fastscroll command, [R] **set**
   floatresults command, [R] **set**
   floatwindows command, [R] **set**
   graphics command, [G-2] **set graphics**, [R] **set**
   httpproxy command, [R] **netio**, [R] **set**
   httpproxyauth command, [R] **netio**, [R] **set**
   httpproxyhost command, [R] **netio**, [R] **set**
   httpproxyport command, [R] **netio**, [R] **set**
   httpproxypw command, [R] **netio**, [R] **set**
   httpproxyuser command, [R] **netio**, [R] **set**
   include_bitmap command, [R] **set**
   level command, [R] **level**, [R] **set**
   linegap command, [R] **set**
   linesize command, [R] **log**, [R] **set**
   locksplitters command, [R] **set**
   logtype command, [R] **log**, [R] **set**
   lstretch command, [R] **set**
   matacache command, [R] **set**
   matafavor command, [R] **set**
   matalibs command, [R] **set**
   matalnum command, [R] **set**
   matamofirst command, [R] **set**
   mataoptimize command, [R] **set**
   matastrict command, [R] **set**
   matsize command, [R] **matsize**, [R] **set**
   max_memory command, [R] **set**
   maxdb command, [R] **db**, [R] **set**
   maxiter command, [R] **maximize**, [R] **set**
   max_memory command, [D] **memory**
   maxvar command, [D] **memory**, [R] **set**
   min_memory command, [D] **memory**, [R] **set**
   more command, [P] **more**, [R] **more**, [R] **set**
   niceness command, [D] **memory**, [R] **set**
   notifyuser command, [R] **set**

set, *continued*
    obs command, [D] **obs**, [R] **set**
    odbcmgr command, [D] **odbc**, [R] **set**
    output command, [P] **quietly**, [R] **set**
    pagesize command, [R] **more**, [R] **set**
    pformat command, [R] **set**, [R] **set cformat**
    pinnable command, [R] **set**
    playsnd command, [R] **set**
    printcolor command, [G-2] **set printcolor**, [R] **set**
    processors command, [R] **set**
    reventries command, [R] **set**
    revkeyboard command, [R] **set**
    rmsg command, [P] **rmsg**, [R] **set**
    scheme command, [G-2] **set scheme**, [G-4] **schemes intro**, [R] **set**
    scrollbufsize command, [R] **set**
    searchdefault command, [R] **search**, [R] **set**
    seed command, [R] **set**, [R] **set seed**
    segmentsize command, [D] **memory**, [R] **set**
    sformat command, [R] **set**, [R] **set cformat**
    showbaselevels command, [R] **set**, [R] **set showbaselevels**
    showemptycells command, [R] **set**, [R] **set showbaselevels**
    showomitted command, [R] **set**, [R] **set showbaselevels**
    smoothfonts command, [R] **set**
    timeout1 command, [R] **netio**, [R] **set**
    timeout2 command, [R] **netio**, [R] **set**
    trace command, [P] **trace**, [R] **set**
    tracedepth command, [P] **trace**, [R] **set**
    traceexpand command, [P] **trace**, [R] **set**
    tracehilite command, [P] **trace**, [R] **set**
    traceindent command, [P] **trace**, [R] **set**
    tracenumber command, [P] **trace**, [R] **set**
    tracesep command, [P] **trace**, [R] **set**
    type command, [D] **generate**, [R] **set**
    update_interval command, [R] **set**, [R] **update**
    update_prompt command, [R] **set**, [R] **update**
    update_query command, [R] **set**, [R] **update**
    varabbrev command, [R] **set**
    varkeyboard command, [R] **set**
set,
    cluster subcommand, [MV] **cluster programming utilities**
    datasignature subcommand, [D] **datasignature**
    file subcommand, [P] **file**
    irf subcommand, [TS] **irf set**
    mi subcommand, [MI] **mi set**
    serset subcommand, [P] **serset**
    ssd subcommand, [SEM] **ssd**
    sysdir subcommand, [P] **sysdir**
    translator subcommand, [R] **translate**
    webuse subcommand, [D] **webuse**
set ado, net subcommand, [R] **net**
set other, net subcommand, [R] **net**
setbreakintr() function, [M-5] **setbreakintr( )**

set_defaults command, [R] **set_defaults**
setmore() function, [M-5] **more( )**
setmoreonexit() function, [M-5] **more( )**
setting M, [MI] **mi add**, [MI] **mi set**
setting mi data, [MI] **mi set**
settings,
    display, [R] **set showbaselevels**
    efficiency, [P] **creturn**
    format, [R] **set cformat**
    graphics, [P] **creturn**
    interface, [P] **creturn**
    memory, [P] **creturn**
    network, [P] **creturn**
    output, [P] **creturn**
    program debugging, [P] **creturn**
    trace, [P] **creturn**
sformat, set subcommand, [R] **set**, [R] **set cformat**
sfrancia command, [R] **swilk**
shadestyle, [G-4] ***shadestyle***
shading region, [G-3] ***region_options***
shape parameter, [ST] **Glossary**
Shapiro–Francia test for normality, [R] **swilk**
Shapiro–Wilk test for normality, [R] **swilk**
shared object, [P] **class**, [P] **plugin**
shell command, [D] **shell**
Shepard
    diagram, [MV] **mds postestimation**, [MV] **Glossary**
    plot, [MV] **mds postestimation**
shewhart command, [R] **qc**
shift, macro subcommand, [P] **macro**
showbaselevels, set subcommand, [R] **set**, [R] **set showbaselevels**
showemptycells, set subcommand, [R] **set**, [R] **set showbaselevels**
showginvariant option, *see* sem option showginvariant
shownrtolerance option, [R] **maximize**
showomitted, set subcommand, [R] **set**, [R] **set showbaselevels**
showstep option, [R] **maximize**
showtolerance option, [R] **maximize**
shwinters, tssmooth subcommand, [TS] **tssmooth shwinters**
Šidák's multiple-comparison adjustment, *see* multiple comparisons, Šidák's method
SIF, *see* Stata internal form
sign() function, [D] **functions**, [M-5] **dsign( )**, [M-5] **sign( )**
signature of data, [D] **checksum**, [D] **datasignature**, [P] **_datasignature**, [P] **signestimationsample**
signestimationsample command, [P] **signestimationsample**
significance levels, [R] **level**, [R] **query**, [U] **20.7 Specifying the width of confidence intervals**
signing digitally data, *see* datasignature command
signrank command, [R] **signrank**
signtest command, [R] **signrank**

signum function, [D] **functions**
similarity, [MV] **Glossary**
    matrices, [MV] **matrix dissimilarity**, [P] **matrix dissimilarity**
    measures, [MV] **cluster**, [MV] **cluster programming utilities**, [MV] **matrix dissimilarity**, [MV] *measure_option*, [P] **matrix dissimilarity**
        Anderberg coefficient, [MV] *measure_option*
        angular, [MV] *measure_option*
        correlation, [MV] *measure_option*
        Dice coefficient, [MV] *measure_option*
        Gower coefficient, [MV] *measure_option*
        Hamann coefficient, [MV] *measure_option*
        Jaccard coefficient, [MV] *measure_option*
        Kulczynski coefficient, [MV] *measure_option*
        matching coefficient, [MV] *measure_option*
        Ochiai coefficient, [MV] *measure_option*
        Pearson coefficient, [MV] *measure_option*
        Rogers and Tanimoto coefficient, [MV] *measure_option*
        Russell and Rao coefficient, [MV] *measure_option*
        Sneath and Sokal coefficient, [MV] *measure_option*
        Yule coefficient, [MV] *measure_option*
simple random sample, [SVY] **Glossary**
simulate prefix command, [R] **simulate**
simulations, Monte Carlo, [P] **postfile**, [R] **simulate**, [R] **permute**
simultaneous
    quantile regression, [R] **qreg**
    systems, [R] **reg3**
sin() function, [D] **functions**, [M-5] **sin( )**
sine function, [D] **functions**
single-imputation methods, [MI] **intro substantive**
single-linkage clustering, [MV] **cluster**, [MV] **clustermat**, [MV] **cluster linkage**, [MV] **Glossary**
single-precision floating point number, [U] **12.2.2 Numeric storage types**
single-record st data, *see* st data
singlelinkage,
    clustermat subcommand, [MV] **cluster linkage**
    cluster subcommand, [MV] **cluster linkage**
singleton-group data, [ST] **stcox**, [ST] **Glossary**
singleton strata, [SVY] **estat**, [SVY] **variance estimation**
singular value decomposition, [M-5] **svd( )**, [M-5] **fullsvd( )**, [MV] **Glossary**, [P] **matrix svd**
sinh() function, [D] **functions**, [M-5] **sin( )**
SITE directory, [P] **sysdir**, [U] **17.5 Where does Stata look for ado-files?**
size, estat subcommand, [SVY] **estat**
size of
    all text and markers, [G-3] *scale_option*
    graph, [G-3] *region_options*
        changing, [G-2] **graph display**

size of, *continued*
    markers, [G-3] *marker_options*
    objects, [G-4] *relativesize*
    text, [G-3] *textbox_options*
sizeof() function, [M-5] **sizeof( )**
SJ, *see* Stata Journal and Stata Technical Bulletin
sj, net subcommand, [R] **net**
sj scheme, [G-4] **scheme sj**
skew(), egen function, [D] **egen**
skewed logistic regression, [R] **scobit**, [SVY] **svy estimation**
skewness, [MV] **mvtest normality**, [R] **summarize**, [TS] **varnorm**, [R] **lnskew0**, [R] **lv**, [R] **pksumm**, [R] **sktest**, [R] **tabstat**
_skip(#), display directive, [P] **display**
sktest command, [R] **sktest**
sleep command, [P] **sleep**
slogit command, [R] **slogit**, [R] **slogit postestimation**
S_ macros, [P] **creturn**, [P] **macro**
Small Stata, [U] **5 Flavors of Stata**
smallestdouble() function, [D] **functions**, [M-5] **mindouble( )**
smc, estat subcommand, [MV] **factor postestimation**, [MV] **pca postestimation**
SMCL, *see* Stata Markup and Control Language
    tags, [G-4] *text*
.smcl filename suffix, [U] **11.6 Filenaming conventions**
smooth command, [R] **smooth**
smoothers, [TS] **tssmooth**, [TS] **Glossary**
    double exponential, [TS] **tssmooth dexponential**
    exponential, [TS] **tssmooth exponential**
    Holt–Winters,
        nonseasonal, [TS] **tssmooth hwinters**
        seasonal, [TS] **tssmooth shwinters**
    moving average, [TS] **tssmooth ma**
    nonlinear, [TS] **tssmooth nl**
smoothfonts, set subcommand, [R] **set**
smoothing, [G-2] **graph twoway lpoly**, [R] **lpoly**, [R] **smooth**
    graphs, [G-2] **graph other**, [R] **kdensity**, [R] **lowess**
SMR, [ST] **epitab**, [ST] **stptime**, [ST] **Glossary**
snapshot
    erase command, [D] **snapshot**
    label command, [D] **snapshot**
    list command, [D] **snapshot**
    restore command, [D] **snapshot**
    save command, [D] **snapshot**
snapshot data, [D] **snapshot**, [ST] **snapspan**, [ST] **stset**, [ST] **Glossary**
snapspan command, [ST] **snapspan**
Sneath and Sokel coefficient similarity measure, [MV] *measure_option*
soft missing value, [MI] **mi impute**, [MI] **Glossary**
solve AX=B, [M-4] **solvers**, [M-5] **cholsolve( )**, [M-5] **lusolve( )**, [M-5] **qrsolve( )**, [M-5] **svsolve( )**, [M-5] **solvelower( )**, [M-5] **solve_tol( )**

_solvelower() function, [M-5] **solvelower( )**
solvelower() function, [M-5] **solvelower( )**
solve_tol() function, [M-5] **solve_tol( )**
_solvetolerance, [M-5] **solve_tol( )**
_solveupper() function, [M-5] **solvelower( )**
solveupper() function, [M-5] **solvelower( )**
sort command, [D] **sort**
_sort() function, [M-5] **sort( )**
sort() function, [M-5] **sort( )**
sort option, [G-3] *connect_options*
sort order, [D] **describe**, [P] **byable**, [P] **macro**,
    [P] **sortpreserve**, [U] **13.2.3 Relational
    operators**
sort, serset subcommand, [P] **serset**
sortedby macro extended function, [P] **macro**
sortpreserve option, [P] **sortpreserve**
soundex() function, [D] **functions**, [M-5] **soundex( )**
soundex_nara() function, [D] **functions**,
    [M-5] **soundex( )**
source code, [M-1] **how**, [M-1] **source**, [M-6] **Glossary**
    view, [P] **viewsource**
Spearman–Brown prophecy formula, [R] **alpha**
spearman command, [R] **spearman**
Spearman's rho, [R] **spearman**
specialized graphs, [G-2] **graph other**
specification test, [R] **gmm postestimation**,
    [R] **hausman**, [R] **ivregress postestimation**,
    [R] **linktest**, [R] **lnskew0**, [R] **regress
    postestimation**, [R] **suest**, [XT] **xtreg
    postestimation**
specificity, [MV] **factor**, [R] **logistic postestimation**,
    also see ROC analysis, *see* ROC analysis
spectral
    analysis, [TS] **Glossary**
    density, [TS] **psdensity**, [TS] **Glossary**
    distribution, [TS] **cumsp**, [TS] **pergram**,
        [TS] **psdensity**, [TS] **Glossary**
    distribution plots, cumulative, [G-2] **graph other**
spectrum, [TS] **psdensity**, [TS] **Glossary**
spell data, [ST] **Glossary**
spherical covariance, [MV] **mvtest covariances**
sphericity, [MV] **Glossary**
Spiegelhalter's Z statistic, [R] **brier**
spike, graph twoway subcommand, [G-2] **graph
    twoway spike**
spike plot, [R] **spikeplot**
spikeplot command, [R] **spikeplot**
spline3() function, [M-5] **spline3( )**
spline3eval() function, [M-5] **spline3( )**
splines
    linear, [R] **mkspline**
    restricted cubic, [R] **mkspline**
split command, [D] **split**
split-plot designs, [MV] **manova**, [R] **anova**
splitting time-span records, [ST] **stsplit**
S-Plus, reading data from, [U] **21.4 Transfer programs**
spread, [R] **lv**

spreadsheets, transferring
    from Stata, [D] **edit**, [D] **export**, [D] **import excel**,
        [D] **odbc**, [D] **outfile**, [D] **outsheet**, [D] **xmlsave**,
        [U] **21.4 Transfer programs**
    into Stata, [D] **edit**, [D] **import**, [D] **import
        excel**, [D] **infile (fixed format)**, [D] **infile (free
        format)**, [D] **insheet**, [D] **odbc**, [D] **xmlsave**,
        [U] **21 Inputting and importing data**,
        [U] **21.4 Transfer programs**
sprintf() function, [M-5] **printf( )**
SPSS dates, [D] **datetime**
SPSS, reading data from, [U] **21.4 Transfer programs**
SQL, [D] **odbc**
sqlfile(), odbc subcommand, [D] **odbc**
sqreg command, [R] **qreg**, [R] **qreg postestimation**
sqrt() function, [D] **functions**, [M-5] **sqrt( )**
square
    matrix, [M-6] **Glossary**
    root, [D] **functions**, [M-5] **sqrt( )**, [M-5] **cholesky( )**
squared multiple correlations, [MV] **factor
    postestimation**, [SEM] **methods and formulas**
sreturn
    clear command, [P] **return**
    list command, [P] **return**, [R] **saved results**
    local command, [P] **return**
SRMI, *see* imputation, chained equations
SRMR, *see* standardized, root mean squared residual
SRS, *see* simple random sample
ss() function, [D] **datetime**, [D] **functions**,
    [M-5] **date( )**
ssc
    copy command, [R] **ssc**
    describe command, [R] **ssc**
    hot command, [R] **ssc**
    install command, [R] **ssc**
    new command, [R] **ssc**
    type command, [R] **ssc**
    uninstall command, [R] **ssc**
SSC archive, [R] **ssc**
ssC() function, [D] **datetime**, [D] **functions**,
    [M-5] **date( )**
SSCP matrix, [MV] **Glossary**
SSD, *see* summary statistics data
ssd
    addgroup command, [SEM] **ssd**
    build command, [SEM] **ssd**
    describe command, [SEM] **ssd**
    init command, [SEM] **ssd**
    list command, [SEM] **ssd**
    repair command, [SEM] **ssd**
    set command, [SEM] **ssd**
    status command, [SEM] **ssd**
    unaddgroup command, [SEM] **ssd**
sspace command, [TS] **sspace**
SSU, *see* secondary sampling unit
_st_addobs() function, [M-5] **st_addobs( )**
st_addobs() function, [M-5] **st_addobs( )**
_st_addvar() function, [M-5] **st_addvar( )**

st_addvar() function, [M-5] **st_addvar( )**

st command, [ST] **stset**

st commands for mi data, [MI] **mi stsplit**, [MI] **mi XXXset**

st data, [ST] **st**, [ST] **Glossary**

_st_data() function, [M-5] **st_data( )**

st_data() function, [M-5] **st_data( )**

st_dir() function, [M-5] **st_dir( )**

st_dropobsif() function, [M-5] **st_dropvar( )**

st_dropobsin() function, [M-5] **st_dropvar( )**

st_dropvar() function, [M-5] **st_dropvar( )**

st_eclear() function, [M-5] **st_rclear( )**

st_global() function, [M-5] **st_global( )**

st_global_hcat() function, [M-5] **st_global( )**

st_isfmt() function, [M-5] **st_isfmt( )**

st_islmname() function, [M-5] **st_isname( )**

st_isname() function, [M-5] **st_isname( )**

st_isnumfmt() function, [M-5] **st_isfmt( )**

st_isnumvar() function, [M-5] **st_vartype( )**

st_isstrfmt() function, [M-5] **st_isfmt( )**

st_isstrvar() function, [M-5] **st_vartype( )**

st_keepobsif() function, [M-5] **st_dropvar( )**

st_keepobsin() function, [M-5] **st_dropvar( )**

st_keepvar() function, [M-5] **st_dropvar( )**

st_local() function, [M-5] **st_local( )**

_st_macroexpand() function, [M-5] **st_macroexpand( )**

st_macroexpand() function, [M-5] **st_macroexpand( )**

st_matrix() function, [M-5] **st_matrix( )**

st_matrix_hcat() function, [M-5] **st_matrix( )**

st_matrixcolstripe() function, [M-5] **st_matrix( )**

st_matrixrowstripe() function, [M-5] **st_matrix( )**

st, mi subcommand, [MI] **mi XXXset**

st_nobs() function, [M-5] **st_nvar( )**

st_numscalar() function, [M-5] **st_numscalar( )**

st_numscalar_hcat() function, [M-5] **st_numscalar( )**

st_nvar() function, [M-5] **st_nvar( )**

st_rclear() function, [M-5] **st_rclear( )**

st_replacematrix() function, [M-5] **st_matrix( )**

st_sclear() function, [M-5] **st_rclear( )**

_st_sdata() function, [M-5] **st_data( )**

st_sdata() function, [M-5] **st_data( )**

st_select() function, [M-5] **select( )**

_st_sstore() function, [M-5] **st_store( )**

st_sstore() function, [M-5] **st_store( )**

_st_store() function, [M-5] **st_store( )**

st_store() function, [M-5] **st_store( )**

st_strscalar() function, [M-5] **st_numscalar( )**

st_subview() function, [M-5] **st_subview( )**

st_tempfilename() function, [M-5] **st_tempname( )**

st_tempname() function, [M-5] **st_tempname( )**

_st_tsrevar() function, [M-5] **st_tsrevar( )**

st_tsrevar() function, [M-5] **st_tsrevar( )**

st_updata() function, [M-5] **st_updata( )**

st_varformat() function, [M-5] **st_varformat( )**

_st_varindex() function, [M-5] **st_varindex( )**

st_varindex() function, [M-5] **st_varindex( )**

st_varlabel() function, [M-5] **st_varformat( )**

st_varname() function, [M-5] **st_varname( )**

st_varrename() function, [M-5] **st_varrename( )**

st_vartype() function, [M-5] **st_vartype( )**

st_varvaluelabel() function, [M-5] **st_varformat( )**

st_view() function, [M-5] **st_view( )**

st_viewobs() function, [M-5] **st_viewvars( )**

st_viewvars() function, [M-5] **st_viewvars( )**

st_vldrop() function, [M-5] **st_vlexists( )**

st_vlexists() function, [M-5] **st_vlexists( )**

st_vlload() function, [M-5] **st_vlexists( )**

st_vlmap() function, [M-5] **st_vlexists( )**

st_vlmodify() function, [M-5] **st_vlexists( )**

st_vlsearch() function, [M-5] **st_vlexists( )**

stability, [TS] **var intro**, [TS] **var**, [TS] **var svar**, [TS] **vecstable**

    after VAR or SVAR, [TS] **varstable**

    after VEC, [TS] **vec intro**, [TS] **vec**

    of nonrecursive models, *see* nonrecursive models, stability of

stable, estat subcommand, [SEM] **estat stable**

stack command, [D] **stack**

stacked variables, [MV] **Glossary**

stacking data, [D] **stack**

stacking variables, [MV] **Glossary**

stairstep, connecting points with, [G-4] **connectstyle**

standard

    deviations, *also see* subpopulation, standard deviations of

        creating dataset of, [D] **collapse**

        creating variable containing, [D] **egen**

        displaying, [R] **lv**, [R] **summarize**, [R] **table**, [R] **tabstat**, [R] **tabulate, summarize()**, [XT] **xtsum**

        testing equality of, [R] **sdtest**

    error bar charts, [G-2] **graph other**

    errors, [SVY] **svy bootstrap**, [SVY] **svy brr**, [SVY] **svy jackknife**, [SVY] **svy sdr**, [SVY] **variance estimation**, *also see* sem option vce()

        accessing, [P] **matrix get**, [U] **13.5 Accessing coefficients and standard errors**

        bootstrap, *see* bootstrap standard errors

        for general predictions, [R] **predictnl**

        forecast, [R] **predict**, [R] **regress postestimation**

        jackknife, *see* jackknife standard errors

        mean, [R] **ci**, [R] **mean**

        panel-corrected, *see* panel-corrected standard error (PCSE)

        prediction, [R] **glm**, [R] **predict**, [R] **regress postestimation**

        residuals, [R] **predict**, [R] **regress postestimation**

        robust, *see* robust standard errors, *see* robust, Huber/White/sandwich estimator of variance

        semirobust, *see* semirobust standard errors

standard, *continued*
  strata, *see* direct standardization
  weights, *see* direct standardization
standardized
  coefficients, [SEM] **example 3**, [SEM] **example 6**,
    [SEM] **Glossary**, *also see* standardized
    parameters
  covariance, [SEM] **Glossary**
  covariance residual, [SEM] **methods and formulas**
  data, [MV] **Glossary**
  margins, [R] **margins**
  mean residual, [SEM] **methods and formulas**
  means, [R] **mean**
  mortality ratio, *see* SMR
  option, [SEM] **example 11**
  parameters, [SEM] **estat stdize**, [SEM] **methods and**
    formulas
  proportions, [R] **proportion**
  rates, [R] **dstdize**, [ST] **epitab**
  ratios, [R] **ratio**
  residuals, [R] **binreg postestimation**, [R] **glm**
    postestimation, [R] **logistic postestimation**,
    [R] **logit postestimation**, [R] **predict**, [R] **regress**
    postestimation, [SEM] **estat residuals**,
    [SEM] **methods and formulas**, [SEM] **Glossary**
  root mean squared residual, [SEM] **estat**
    ggof, [SEM] **estat gof**, [SEM] **example 4**,
    [SEM] **example 21**, [SEM] **methods and**
    formulas
standardized option, *see* sem option standardized
standardized, variables, [D] **egen**
start() option, [G-2] **graph twoway histogram**
starting values, [SEM] **intro 3**, [SEM] **sem**
    option from( ), [SEM] **sem path notation**,
    [SEM] **Glossary**
Stata
  Blog, [U] **3.2.2 The Stata Blog—Not Elsewhere**
    Classified
  c()-class results, [M-5] **st_global( )**
  characteristics, [M-5] **st_global( )**, [M-5] **st_dir( )**
  conference, [U] **3.7.1 Conferences and users group**
    meetings
  description, [U] **2 A brief description of Stata**
  documentation, [U] **1 Read this—it will help**
  e()-class results, [M-5] **st_global( )**, [M-5] **st_dir( )**,
    [M-5] **st_rclear( )**
  error message, [M-5] **error( )**
  example datasets, [U] **1.2.2 Example datasets**
  execute command, [M-3] **mata stata**, [M-5] **stata( )**
  exiting, *see* exit command
  for Mac, *see* Mac
  for Unix, *see* Unix
  for Windows, *see* Windows
  limits, [U] **5 Flavors of Stata**
  listserver, [U] **3.4 The Stata listserver**
  logo, [G-2] **graph print**, [G-3] *pr_options*
  macros, [M-5] **st_global( )**, [M-5] **st_local( )**,
    [M-5] **st_dir( )**

Stata, *continued*
  matrices, [M-5] **st_matrix( )**, [M-5] **st_dir( )**
  NetCourseNow, [U] **3.7.2 NetCourses**
  NetCourses, [U] **3.7.2 NetCourses**
  on Facebook, [U] **3.2.3 Stata on Twitter and**
    Facebook
  on Twitter, [U] **3.2.3 Stata on Twitter and**
    Facebook
  op.varname, *see* Stata, time-series–operated variable
  platforms, [U] **5.1 Platforms**
  Press, [U] **3.3 Stata Press**
  r()-class results, [M-5] **st_global( )**, [M-5] **st_dir( )**,
    [M-5] **st_rclear( )**
  s()-class results, [M-5] **st_global( )**, [M-5] **st_dir( )**,
    [M-5] **st_rclear( )**
  scalars, [M-5] **st_numscalar( )**, [M-5] **st_dir( )**
  Small, *see* Small Stata
  Stata/IC, *see* Stata/IC
  Stata/MP, *see* Stata/MP
  Stata/SE, *see* Stata/SE
  supplementary material, [U] **3 Resources for**
    learning and using Stata
  support, [U] **3 Resources for learning and using**
    Stata
  temporary
    filenames, [M-5] **st_tempname( )**
    names, [M-5] **st_tempname( )**
  time-series–operated variable, [M-5] **st_tsrevar( )**,
    [M-6] **Glossary**
  training, [U] **3.7 Conferences and training**
  updates, *see* updates to Stata
  users group meeting, [U] **3.7.1 Conferences and**
    users group meetings
  value labels, [M-5] **st_varformat( )**,
    [M-5] **st_vlexists( )**
  variable
    formats, [M-5] **st_varformat( )**
    labels, [M-5] **st_varformat( )**
  website, [U] **3.2.1 The Stata website**
    (www.stata.com)
Stata,
  data file format, technical description, [P] **file**
    formats .dta
  exiting, *see* exit command
  pause, [P] **sleep**
stata, [M-3] **mata stata**
STATA directory, [P] **sysdir**
_stata() function, [M-5] **stata( )**
stata() function, [M-5] **stata( )**
Stata internal form, [D] **datetime**, [D] **datetime display**
    formats, [D] **datetime translation**
Stata Journal, [G-4] **scheme sj**
Stata Journal and *Stata Technical Bulletin*, [U] **3.5 The**
    Stata Journal
  installation of, [R] **net**, [R] **sj**, [U] **17.6 How do I**
    install an addition?
  keyword search of, [R] **search**, [U] **4 Stata's help**
    and search facilities

Stata Markup and Control Language, [M-5] **display( )**,
     [M-5] **printf( )**, [M-5] **errprintf( )**, [P] **smcl**
Stata News, [U] **3 Resources for learning and using
     Stata**
Stata Technical Bulletin Reprints, [U] **3.5 The Stata
     Journal**
Stata/IC, [U] **5 Flavors of Stata**
Stata/MP, [U] **5 Flavors of Stata**
Stata/SE, [U] **5 Flavors of Stata**
stata.key file, [R] **search**
Statalist, [U] **3.4 The Stata listserver**
statasetversion( ) function, [M-5] **stataversion( )**
stataversion( ) function, [M-5] **stataversion( )**
state-space model, [TS] **sspace**, [TS] **sspace
     postestimation**, [TS] **Glossary**, *also see* ARIMA
     model, *also see* dynamic factor model
static, [M-2] **class**
stationary time series, [TS] **dfgls**, [TS] **dfuller**,
     [TS] **pperron**, [TS] **var intro**, [TS] **var**, [TS] **vec
     intro**, [TS] **vec**
statistical
     density functions, [M-5] **normal( )**
     distribution functions, [M-5] **normal( )**
Statistical Software Components (SSC) archive, [R] **ssc**
stats, estimates subcommand, [R] **estimates stats**
statsby prefix command, [D] **statsby**
Stat/Transfer, [U] **21.4 Transfer programs**
status, ssd subcommand, [SEM] **ssd**
STB, *see Stata Journal and Stata Technical Bulletin*
stb, net subcommand, [R] **net**
stbase command, [ST] **stbase**
.stbcal file suffix, [D] **bcal**, [D] **datetime business
     calendars**, [D] **datetime business calendars
     creation**, [U] **11.6 Filenaming conventions**
stbcal-files, [D] **bcal**, [D] **datetime business calendars**,
     [D] **datetime business calendars creation**
stci command, [ST] **stci**
stcox command, [ST] **stcox**
stcox, fractional polynomials, [R] **fracpoly**, [R] **mfp**
stcoxkm command, [ST] **stcox PH-assumption tests**
stcrreg command, [ST] **stcrreg**
st_ct, [ST] **st_is**
stcurve command, [ST] **stcurve**
std( ), egen function, [D] **egen**
stdescribe command, [ST] **stdescribe**
stdize, estat subcommand, [SEM] **estat stdize**
steady-state equilibrium, [TS] **Glossary**
steepest descent (ascent), [M-5] **moptimize( )**,
     [M-5] **optimize( )**
stem command, [R] **stem**
stem-and-leaf displays, [R] **stem**
stepwise estimation, [R] **stepwise**
stepwise prefix command, [R] **stepwise**
.ster files, [MI] **mi estimate**, [MI] **mi estimate
     using**, [MI] **mi predict**, [U] **11.6 Filenaming
     conventions**

stereotype logistic regression, [R] **slogit**, [SVY] **svy
     estimation**
stfill command, [ST] **stfill**
stgen command, [ST] **stgen**
.sthlp filename suffix, [U] **11.6 Filenaming
     conventions**
.sthlp files, [U] **4 Stata's help and search facilities**,
     [U] **18.11.6 Writing online help**
stir command, [ST] **stir**
st_is 2, [ST] **st_is**
stjoin command, [ST] **stsplit**
stjoin for mi data, [MI] **mi stsplit**
stjoin, mi subcommand, [MI] **mi stsplit**
stmc command, [ST] **strate**
stmh command, [ST] **strate**
stochastic frontier models, [R] **frontier**,
     [U] **26.13 Stochastic frontier models**,
     [XT] **xtfrontier**
stochastic trend, [TS] **Glossary**
stop,
     clustermat subcommand, [MV] **cluster stop**
     cluster subcommand, [MV] **cluster stop**
stopbox, window subcommand, [P] **window
     programming**
stopping command execution, [U] **10 Keyboard use**
stopping rules, [MV] **Glossary**
     adding, [MV] **cluster programming subroutines**
     Caliński and Harabasz index, [MV] **cluster**,
          [MV] **cluster stop**
     Duda and Hart index, [MV] **cluster**, [MV] **cluster
          stop**
     stepsize, [MV] **cluster programming subroutines**
storage types, [U] **11.4 varlists**, [D] **codebook**,
     [D] **compress**, [D] **describe**, [D] **encode**,
     [D] **format**, [D] **generate**, [D] **recast**,
     [D] **varmanage**, [U] **12.2.2 Numeric storage
     types**, [U] **12.4.4 String storage types**
store, estimates subcommand, [R] **estimates store**
storing and restoring estimation results, [R] **estimates
     store**
stphplot command, [ST] **stcox PH-assumption tests**
stpower
     cox command, [ST] **stpower**, [ST] **stpower cox**
     exponential command, [ST] **stpower**,
          [ST] **stpower exponential**
     logrank command, [ST] **stpower**, [ST] **stpower
          logrank**
stptime command, [ST] **stptime**
.stptrace filename suffix, [U] **11.6 Filenaming
     conventions**
str#, [D] **data types**, [I] **data types**, [U] **12.4.4 String
     storage types**
strata, estat subcommand, [SVY] **estat**
strata with one sampling unit, [SVY] **variance
     estimation**
strata,
     graphs, [R] **dotplot**

strata, *continued*
  models, [R] **asclogit**, [R] **asmprobit**, [R] **asroprobit**,
    [R] **clogit**, [R] **exlogistic**, [R] **expoisson**,
    [R] **rocreg**, [R] **rologit**
  resampling, [R] **bootstrap**, [R] **bsample**, [R] **bstat**,
    [R] **permute**
  standardization, [R] **dstdize**
  summary statistics, [R] **mean**, [R] **proportion**,
    [R] **ratio**, [R] **total**
strate command, [ST] **strate**
stratification, [ST] **epitab**, [ST] **stcox**, [ST] **streg**, *see*
    stratified sampling
stratified model,
  model, [ST] **Glossary**
  sampling, [SVY] **survey**, [SVY] **svydescribe**,
    [SVY] **svyset**, [SVY] **Glossary**
  tables, [ST] **epitab**
  test, [ST] **Glossary**
stratum collapse, [SVY] **svydescribe**
strdup() function, [M-5] **strdup( )**
stream I/O versus record I/O, [U] **21 Inputting and
    importing data**
streg command, [ST] **streg**
streset command, [MI] **mi XXXset**, [ST] **stset**
streset, mi subcommand, [MI] **mi XXXset**
stress, [MV] **mds postestimation**, [MV] **Glossary**
stress, estat subcommand, [MV] **mds
    postestimation**
strict stationarity, [TS] **Glossary**
string
  duplication, [M-5] **strdup( )**
  functions, [M-4] **string**
  concatenation, [M-5] **invtokens( )**
  display formats, [I] **format**
  functions, expressions, and operators, [D] **functions**,
    [U] **12.4 Strings**, [U] **23 Working with strings**
  pattern matching, [M-5] **strmatch( )**
  to real, convert, [M-5] **strtoreal( )**
  variables, [D] **data types**, [D] **infile (free
    format)**, [I] **data types**, [U] **12.4 Strings**,
    [U] **23 Working with strings**
    converting to numbers, [D] **functions**
    encoding, [D] **encode**
    exporting, [D] **export**
    formatting, [D] **format**
    importing, [D] **import**
    inputting, [D] **edit**, [D] **input**, [U] **21 Inputting
      and importing data**
    making from value labels, [D] **encode**
    mapping to numbers, [D] **destring**, [D] **encode**,
      [D] **label**, *also see* real( ) function
    sort order, [U] **13.2.3 Relational operators**
    splitting into parts, [D] **split**
    parsing, [P] **gettoken**, [P] **tokenize**
string, [M-2] **declarations**, [M-6] **Glossary**
string() function, [D] **functions**
stritrim() function, [M-5] **strtrim( )**

strlen() function, [D] **functions**, [M-5] **strlen( )**
strlower() function, [D] **functions**, [M-5] **strupper( )**
strltrim() function, [D] **functions**, [M-5] **strtrim( )**
strmatch() function, [D] **functions**, [M-5] **strmatch( )**
strofreal() function, [D] **functions**,
    [M-5] **strofreal( )**
strongly balanced, [XT] **Glossary**
strpos() function, [D] **functions**, [M-5] **strpos( )**
strproper() function, [D] **functions**,
    [M-5] **strupper( )**
strreverse() function, [D] **functions**,
    [M-5] **strreverse( )**
strrtrim() function, [D] **functions**, [M-5] **strtrim( )**
strtoname() function, [D] **functions**,
    [M-5] **strtoname( )**
_strtoreal() function, [M-5] **strtoreal( )**
strtoreal() function, [M-5] **strtoreal( )**
strtrim() function, [D] **functions**, [M-5] **strtrim( )**
struct, [M-2] **struct**
structural
  equation modeling, [MV] **intro**, [SEM] **Glossary**,
    [SVY] **svy estimation**, [U] **26.3 Structural
    equation modeling (SEM)**, *also see* constraints
  models, *see* models, structural
  time-series model, [TS] **psdensity**, [TS] **sspace**,
    [TS] **ucm**, [TS] **Glossary**
  vector autoregression, *see* SVAR
structure, [MV] **Glossary**
structure, estat subcommand, [MV] **discrim lda
    postestimation**, [MV] **factor postestimation**
structured, [SEM] **Glossary**
structured (correlation or covariance), [SEM] **Glossary**
structures, [M-2] **struct**, [M-5] **liststruct( )**,
    [M-6] **Glossary**
strupper() function, [D] **functions**, [M-5] **strupper( )**
sts
  command, [ST] **sts**, [ST] **sts generate**, [ST] **sts
    graph**, [ST] **sts list**, [ST] **sts test**
  generate command, [ST] **sts**, [ST] **sts generate**
  graph command, [ST] **sts**, [ST] **sts graph**
  list command, [ST] **sts**, [ST] **sts list**
  test command, [ST] **sts**, [ST] **sts test**
stset command, [MI] **mi XXXset**, [ST] **stset**
stset, mi subcommand, [MI] **mi XXXset**
st_show, [ST] **st_is**
stsplit command, [MI] **mi stsplit**, [ST] **stsplit**
stsplit, mi subcommand, [MI] **mi stsplit**
stsum command, [ST] **stsum**
sttocc command, [ST] **sttocc**
sttoct command, [ST] **sttoct**
Studentized residuals, [R] **predict**, [R] **regress
    postestimation**
Studentized-range multiple-comparison adjustment, *see*
    multiple comparisons, Tukey's method
Student–Newman–Keuls' multiple-comparison
    adjustment, *see* multiple comparisons, Student–
    Newman–Keuls' method

Student's *t* distribution
  cdf, [D] **functions**
  confidence interval for mean, [R] **ci**, [R] **mean**
  testing equality of means, [R] **ttest**
stvary command, [ST] **stvary**
style
  added line, [G-4] *addedlinestyle*
  area, [G-4] *areastyle*
  axis, [G-4] *tickstyle*
  by-graphs, [G-4] *bystyle*
  clock position, [G-4] *clockposstyle*
  color, [G-4] *colorstyle*
  compass direction, [G-4] *compassdirstyle*
  connect points, [G-4] *connectstyle*
  grid lines, [G-4] *gridstyle*
  legends, [G-4] *legendstyle*
  lines, [G-4] *linepatternstyle*, [G-4] *linestyle*,
    [G-4] *linewidthstyle*
  lists, [G-4] *stylelists*
  margins, [G-4] *marginstyle*
  marker labels, [G-4] *markerlabelstyle*,
    [G-4] *markersizestyle*, [G-4] *markerstyle*
  markers, [G-4] *symbolstyle*
  plot, [G-4] *pstyle*
  text, [G-4] *textsizestyle*, [G-4] *textstyle*
  text display angle, [G-4] *anglestyle*
  text justification, [G-4] *justificationstyle*
  textboxes, [G-4] *orientationstyle*,
    [G-4] *textboxstyle*
  vertical alignment of text, [G-4] *alignmentstyle*
style, [MI] **mi convert**, [MI] **styles**, [MI] **Glossary**
style,
  flong, *see* flong
  flongsep, *see* flongsep
  mlong, *see* mlong
  wide, *see* wide
stylelist, [G-4] *stylelists*
subclass, [M-2] **class**
subdirectories, [U] **11.6 Filenaming conventions**
subhazard, [ST] **Glossary**
subhazard ratio, [ST] **Glossary**
subinertia, estat subcommand, [MV] **mca**
  postestimation
subinstr macro extended function, [P] **macro**
subinstr() function, [D] **functions**, [M-5] **subinstr( )**
subinword() function, [D] **functions**,
  [M-5] **subinstr( )**
_sublowertriangle() function,
  [M-5] **sublowertriangle( )**
sublowertriangle() function,
  [M-5] **sublowertriangle( )**
subpopulation
  differences, [SVY] **survey**, [SVY] **svy**
    postestimation
  estimation, [SVY] **subpopulation estimation**,
    [SVY] **svy estimation**, [SVY] **Glossary**
  means, [SVY] **svy estimation**

subpopulation, *continued*
  proportions, [SVY] **svy estimation**,
    [SVY] **svy: tabulate oneway**,
    [SVY] **svy: tabulate twoway**
  ratios, [SVY] **svy estimation**, [SVY] **svy: tabulate**
    oneway, [SVY] **svy: tabulate twoway**
  standard deviations of, [SVY] **estat**
  totals, [SVY] **svy estimation**, [SVY] **svy: tabulate**
    oneway, [SVY] **svy: tabulate twoway**
subroutines, adding, [MV] **cluster programming**
  utilities
subscripting matrices, [P] **matrix define**
subscripts, [G-4] *text*, [M-2] **subscripts**,
  [M-6] **Glossary**
subscripts in expressions, [U] **13.7 Explicit**
  subscripting
substantive constraints, [SEM] **Glossary**, *see* constraints
_substr() function, [M-5] **_substr( )**
substr() function, [D] **functions**, [M-5] **substr( )**
substring function, [D] **functions**
subtitle() option, [G-3] *title_options*
subtraction operator, *see* arithmetic operators
successive difference replication, [SVY] *sdr_options*,
  [SVY] **svy sdr**, [SVY] **variance estimation**,
  [SVY] **Glossary**
suest command, [R] **suest**, [SVY] **svy postestimation**
.sum filename suffix, [U] **11.6 Filenaming conventions**
sum() function, [D] **functions**, [M-5] **sum( )**
sum of vector, [M-5] **runningsum( )**
summarize,
  estat subcommand, [MV] **ca postestimation**,
    [MV] **discrim estat**, [MV] **discrim**
    knn postestimation, [MV] **discrim lda**
    postestimation, [MV] **discrim logistic**
    postestimation, [MV] **discrim qda**
    postestimation, [MV] **factor postestimation**,
    [MV] **mca postestimation**, [MV] **mds**
    postestimation, [MV] **pca postestimation**,
    [MV] **procrustes postestimation**, [R] **estat**,
    [SEM] **estat summarize**
  misstable subcommand, [R] **misstable**
  serset subcommand, [P] **serset**
summarize command, [D] **format**, [R] **summarize**,
  [R] **tabulate, summarize()**
summarizing data, [D] **codebook**, [D] **inspect**,
  [R] **summarize**, [R] **tabstat**, [SVY] **svy: tabulate**
  twoway, [XT] **xtsum**, [R] **lv**, [R] **table**,
  [R] **tabulate oneway**, [R] **tabulate twoway**,
  [R] **tabulate, summarize()**
summary statistics, *see* descriptive statistics
summary statistics data, [SEM] **intro 10**,
  [SEM] **example 2**, [SEM] **example 19**,
  [SEM] **example 25**, [SEM] **sem option**
  select( ), [SEM] **sem ssd options**, [SEM] **ssd**,
  [SEM] **Glossary**
summary variables, generating, [MV] **cluster generate**
summative (Likert) scales, [R] **alpha**
sums, over observations, [R] **summarize**

sums,
    creating dataset containing, [D] **collapse**
    over observations, [D] **egen**, [D] **functions**
    over variables, [D] **egen**
sunflower command, [R] **sunflower**
sunflower plots, [R] **sunflower**
Super, class prefix operator, [P] **class**
super-varying variables, [MI] **mi varying**,
    [MI] **Glossary**
.superclass built-in class function, [P] **class**
superscripts, [G-4] *text*
supplementary rows or columns, [MV] **Glossary**
supplementary variables, [MV] **Glossary**
support of Stata, [U] **3 Resources for learning and
    using Stata**
suppressing graphs, [G-3] *nodraw_option*
suppressing terminal output, [P] **quietly**
SUR, [TS] **dfactor**
sureg command, [R] **sureg**, [R] **sureg postestimation**,
    also see seemingly unrelated regression
survey
    data, [MI] **intro substantive**, [MI] **mi
        estimate**, [SEM] **intro 9**, [SVY] **survey**,
        [SVY] **svydescribe**, [SVY] **svyset**,
        [SVY] **Glossary**, [U] **26.21 Survey data**
    design, [SVY] **Glossary**
    postestimation, [SVY] **svy postestimation**
    prefix command, [SVY] **svy**
    sampling, [ST] **stcox**, [ST] **streg**, [ST] **sts**,
        [SVY] **survey**, [SVY] **svydescribe**,
        [SVY] **svyset**, *also see* cluster sampling
survival analysis, [G-2] **graph other**, [R] **cloglog**,
    [R] **exlogistic**, [R] **expoisson**, [R] **gllamm**,
    [R] **glm**, [R] **intreg**, [R] **logistic**, [R] **logit**,
    [R] **poisson**, [R] **tobit**, [ST] **cttost**,
    [ST] **st**, [ST] **stcox**, [ST] **stcox PH-
    assumption tests**, [ST] **stdescribe**,
    [ST] **streg**, [ST] **stset**, [ST] **sttoct**, [SVY] **svy
    estimation**, [XT] **xtnbreg**, [XT] **xtpoisson**,
    [U] **26.18 Survival-time (failure-time) models**
survival clinical trial, [ST] **stpower**
survival data, [MI] **mi estimate**, [MI] **mi predict**
survival models, [SVY] **svy estimation**
survival-time data, *see* survival analysis
survivor function, [G-2] **graph other**, [ST] **sts**,
    [ST] **sts generate**, [ST] **sts list**, [ST] **sts test**,
    [ST] **Glossary**
    graph of, [ST] **stcurve**, [ST] **sts graph**
SVAR, [TS] **var intro**, [TS] **var svar**, [TS] **Glossary**
    postestimation, [R] **regress postestimation time
        series**, [TS] **fcast compute**, [TS] **fcast graph**,
        [TS] **irf**, [TS] **irf create**, [TS] **vargranger**,
        [TS] **varlmar**, [TS] **varnorm**, [TS] **varsoc**,
        [TS] **varstable**, [TS] **varwle**, [TS] **var svar
        postestimation**
svar command, [TS] **var svar**, [TS] **var svar
    postestimation**
SVD, [MV] **Glossary**, *see* singular value decomposition
_svd() function, [M-5] **svd( )**

svd() function, [M-5] **svd( )**
svd, matrix subcommand, [P] **matrix svd**
_svd_la() function, [M-5] **svd( )**, [M-5] **fullsvd( )**
_svdsv() function, [M-5] **svd( )**
svdsv() function, [M-5] **svd( )**
svmat command, [P] **matrix mkmat**
_svsolve() function, [M-5] **svsolve( )**
svsolve() function, [M-5] **svsolve( )**
svy: biprobit command, [SVY] **svy estimation**
svy: clogit command, [SVY] **svy estimation**
svy: cloglog command, [SVY] **svy estimation**
svy: cnsreg command, [SVY] **svy estimation**
svy: glm command, [SVY] **svy estimation**
svy: gnbreg command, [SVY] **svy estimation**
svy: heckman command, [SVY] **svy estimation**
svy: heckprob command, [SVY] **svy estimation**
svy: hetprob command, [SVY] **svy estimation**
svy: intreg command, [SVY] **svy estimation**
svy: ivprobit command, [SVY] **svy estimation**
svy: ivregress command, [SVY] **svy estimation**
svy: ivtobit command, [SVY] **svy estimation**
svy: logistic command, [SVY] **svy estimation**,
    [SVY] **svy postestimation**
svy: logit command, [SVY] **svy estimation**
svy: mean command, [SVY] **estat**,
    [SVY] **poststratification**, [SVY] **subpopulation
    estimation**, [SVY] **survey**, [SVY] **svy**,
    [SVY] **svy estimation**, [SVY] **svy
    postestimation**, [SVY] **svydescribe**,
    [SVY] **svyset**
svy: mlogit command, [SVY] **svy estimation**
svy: mprobit command, [SVY] **svy estimation**
svy: nbreg command, [SVY] **svy estimation**
svy: nl command, [SVY] **svy estimation**
svy: ologit command, [SVY] **svy estimation**,
    [SVY] **svy postestimation**
svy: oprobit command, [SVY] **svy estimation**
svy: poisson command, [SVY] **svy estimation**
svy: probit command, [SVY] **svy estimation**
svy: proportion command, [SVY] **svy estimation**
svy: ratio command, [SVY] **direct standardization**,
    [SVY] **svy brr**, [SVY] **svy estimation**,
    [SVY] **svy: tabulate twoway**
svy: regress command, [SVY] **survey**, [SVY] **svy**,
    [SVY] **svy estimation**, [SVY] **svy jackknife**,
    [SVY] **svy postestimation**
svy: scobit command, [SVY] **svy estimation**
svy: sem command, [SVY] **svy estimation**
svy: slogit command, [SVY] **svy estimation**
svy: stcox command, [SVY] **svy estimation**
svy: streg command, [SVY] **svy estimation**
svy: tabulate command, [SVY] **svy: tabulate
    oneway**, [SVY] **svy: tabulate twoway**
svy: tnbreg command, [SVY] **svy estimation**
svy: tobit command, [SVY] **svy estimation**
svy: total command, [SVY] **svy brr**, [SVY] **svy
    estimation**
svy: tpoisson command, [SVY] **svy estimation**

svy: treatreg command, [SVY] **svy estimation**
svy: truncreg command, [SVY] **svy estimation**
svy: zinb command, [SVY] **svy estimation**
svy: zip command, [SVY] **svy estimation**
svy bootstrap prefix command, [SVY] **svy bootstrap**
svy brr prefix command, [SVY] **svy brr**
svydescribe command, [SVY] **survey**,
        [SVY] **svydescribe**
svy jackknife prefix command, [SVY] **svy jackknife**
svymarkout command, [P] **mark**, [SVY] **svymarkout**
svy prefix command, [SVY] **svy**
svy sdr prefix command, [SVY] **svy sdr**
svyset command, [MI] **mi XXXset**, [SVY] **survey**,
        [SVY] **svyset**
svyset, estat subcommand, [SVY] **estat**
svyset, mi subcommand, [MI] **mi XXXset**
swap() function, [M-5] **swap( )**
sweep() function, [D] **functions**, [P] **matrix define**
swilk command, [R] **swilk**
switching styles, [MI] **mi convert**
Sybase, reading data from, [U] **21.4 Transfer programs**
symbolic forms, [R] **anova**
symbolpalette, palette subcommand, [G-2] **palette**
symbols, [G-4] *text*, *also see* markers
symbolstyle, [G-4] *symbolstyle*
symeigen, matrix subcommand, [P] **matrix symeigen**
_symeigen_la() function, [M-5] **eigensystem( )**
_symeigensystem() function, [M-5] **eigensystem( )**
symeigensystem() function, [M-5] **eigensystem( )**
_symeigensystem_select() functions,
        [M-5] **eigensystemselect( )**
symeigensystemselect*() functions,
        [M-5] **eigensystemselect( )**
_symeigenvalues() function, [M-5] **eigensystem( )**
symeigenvalues() function, [M-5] **eigensystem( )**
symmetric matrices, [M-5] **issymmetric( )**,
        [M-5] **makesymmetric( )**, [M-6] **Glossary**
symmetriconly, [M-6] **Glossary**
symmetry command, [R] **symmetry**
symmetry plots, [G-2] **graph other**, [R] **diagnostic
        plots**
symmetry, test of, [R] **symmetry**
symmi command, [R] **symmetry**
symplot command, [R] **diagnostic plots**
syntax, [M-2] **syntax**
syntax command, [P] **syntax**
syntax diagrams explained, [R] **intro**
syntax of Stata's language, [P] **syntax**,
        [U] **11 Language syntax**
sysdir
    list command, [P] **sysdir**
    macro extended function, [P] **macro**
    set command, [P] **sysdir**
sysmiss, *see* missing values
Systat, reading data from, [U] **21.4 Transfer programs**

system
    estimators, [R] **gmm**, [R] **ivregress**, [R] **nlsur**,
        [R] **reg3**, [R] **sureg**, [U] **26.14 Regression
        with systems of equations**, *also see* generalized
        method of moments (GMM)
    limits, [P] **creturn**
    parameters, [P] **creturn**, [R] **query**, [R] **set**,
        [R] **set_defaults**
    values, [P] **creturn**
    variables, [U] **13.4 System variables (_variables)**
system1 command, [R] **gmm**
sysuse
    command, [D] **sysuse**
    dir command, [D] **sysuse**
szroeter, estat subcommand, [R] **regress
        postestimation**
Szroeter's test for heteroskedasticity, [R] **regress
        postestimation**

T

t distribution
    cdf, [D] **functions**
    confidence interval for mean, [R] **ci**, [R] **mean**
    testing equality of means, [R] **ttest**
%t formats, [D] **format**
%t values and formats, [D] **datetime**
t1title() option, [G-3] *title_options*
t2title() option, [G-3] *title_options*
tab characters, show, [D] **type**
tab expansion of variable names, [U] **10.6 Tab
        expansion of variable names**
tab1 command, [R] **tabulate oneway**
tab2 command, [R] **tabulate twoway**
tabdisp command, [P] **tabdisp**
tabi command, [R] **tabulate twoway**
table command, [R] **table**
table, estat subcommand, [MV] **ca postestimation**
table, estimates subcommand, [R] **estimates table**
tables, [TS] **irf ctable**, [TS] **irf table**
    actuarial, *see* life tables
    coefficient,
        display in exponentiated form, [R] *eform_option*
        display settings, [R] **estimation options**, [R] **set
            showbaselevels**
        format settings, [R] **set cformat**
        maximum likelihood display options, [R] **ml**
        system parameter settings, [R] **set**
    contingency, [R] **table**, [R] **tabulate twoway**,
        [SVY] **svy: tabulate twoway**
    epidemiological, *see* epidemiological tables
    estimation results, [R] **estimates table**
    failure, *see* failure tables
    formatting numbers in, [D] **format**
    fourfold, *see* fourfold tables

tables, *continued*
frequency, [R] **tabulate oneway**, [R] **tabulate twoway**, [R] **table**, [R] **tabstat**, [R] **tabulate, summarize()** [SVY] **svy: tabulate oneway**, [SVY] **svy: tabulate twoway**, [XT] **xttab**
hazard, *see* hazard tables
life, *see* life tables
missing values, [R] **misstable**
N-way, [P] **tabdisp**
of statistics, [P] **tabdisp**
printing, [U] **15 Saving and printing output—log files**
summary statistics, [R] **table**, [R] **tabstat**, [R] **tabulate, summarize()**
tabodds command, [ST] **epitab**
tabstat command, [R] **tabstat**
tabulate
one-way, [SVY] **svy: tabulate oneway**
two-way, [SVY] **svy: tabulate twoway**
tabulate command, [R] **tabulate oneway**, [R] **tabulate twoway**
summarize(), [R] **tabulate, summarize()**
tag, duplicates subcommand, [D] **duplicates**
tag(), egen function, [D] **egen**
tan() function, [D] **functions**, [M-5] **sin( )**
tangent function, [D] **functions**
tanh() function, [D] **functions**, [M-5] **sin( )**
TARCH, [TS] **arch**
target rotation, [MV] **procrustes**, [MV] **rotate**, [MV] **rotatemat**, [MV] **Glossary**
tau, [R] **spearman**
taxonomy, [MV] **Glossary**
Taylor linearization, *see* linearized variance estimator
tC() pseudofunction, [D] **datetime**, [D] **functions**
tc() pseudofunction, [D] **datetime**, [D] **functions**
td() pseudofunction, [D] **datetime**, [D] **functions**
tden() function, [D] **functions**, [M-5] **normal( )**
TDT test, [R] **symmetry**
technical support, [U] **3.9 Technical support**
technique, [SEM] **Glossary**
technique() option, [R] **maximize**
teffects, estat subcommand, [SEM] **estat teffects**
tempfile command, [P] **macro**
tempfile macro extended function, [P] **macro**
tempname, class, [P] **class**
tempname command, [P] **macro**, [P] **matrix**, [P] **scalar**
tempname macro extended function, [P] **macro**
temporary
files, [P] **macro**, [P] **preserve**, [P] **scalar**
names, [P] **macro**, [P] **matrix**, [P] **scalar**, [U] **18.7.2 Temporary scalars and matrices**
variables, [P] **macro**, [U] **18.7.1 Temporary variables**
tempvar command, [P] **macro**
tempvar macro extended function, [P] **macro**
termcap(5), [U] **10 Keyboard use**

terminal
obtaining input from, [P] **display**
suppressing output, [P] **quietly**
terminfo(4), [U] **10 Keyboard use**
test command, [SVY] **survey**, [R] **anova postestimation**, [R] **test**, [SEM] **estat stdize**, [SEM] **example 8**, [SEM] **example 9**, [SEM] **example 16**, [SEM] **lrtest**, [SEM] **test**, [SVY] **svy postestimation**, [U] **20.12 Performing hypothesis tests on the coefficients**
test, mi subcommand, [MI] **mi test**
test,
ARCH effect, [R] **regress postestimation time series**
association, [R] **correlate**, [R] **spearman**, [R] **tabulate twoway**, [ST] **epitab**, [SVY] **svy: tabulate twoway**
autocorrelation, *see* autocorrelation test
autoregressive conditional heteroskedasticity, [R] **regress postestimation time series**
binomial probability, [R] **bitest**
bioequivalence, *see* bioequivalence tests
Breitung, [XT] **xtunitroot**
Breusch–Godfrey, *see* Breusch–Godfrey test
Breusch–Pagan, [R] **mvreg**, [R] **sureg**
Breusch–Pagan Lagrange multiplier, *see* Breusch–Pagan Lagrange multiplier test
chi-squared, *see* chi-squared test of independence
chi-squared hypothesis, *see* chi-squared hypothesis test
Chow, *see* Chow test
Cox proportional hazards model, assumption, *see* Cox proportional hazards model, test of assumption
cusum, [R] **cusum**
differences of two means, [SVY] **svy postestimation**
Dickey–Fuller, [TS] **dfgls**, [TS] **dfuller**
Durbin's alternative, *see* Durbin's alternative test
endogeneity, [R] **ivregress postestimation**
Engle's LM, *see* Engle's LM test
equality of
coefficients, [R] **pwcompare**, [R] **test**, [R] **testnl**, [SVY] **svy postestimation**
distributions, *see* distributions, testing equality of
margins, [R] **margins**, [R] **pwcompare**
means, [R] **contrast**, [R] **pwmean**, [R] **sampsi**, [R] **ttest**, [SVY] **svy postestimation**
medians, [R] **ranksum**
proportions, [R] **bitest**, [R] **prtest**, [R] **sampsi**
ROC areas, [R] **roccomp**, [R] **rocreg**
survivor functions, *see* equality test, survivor functions
variances, [R] **sdtest**
equivalence, [R] **pk**, [R] **pkequiv**
exogeneity, *see* endogeneity subentry
Fisher-type, [XT] **xtunitroot**
Fisher's exact, *see* Fisher's exact test

test, *continued*
  goodness-of-fit, [SVY] **estat**, *see* goodness-of-fit
    statistics and tests
  group invariance, *see* group invariance test
  granger causality, [TS] **vargranger**
  Hadri Lagrange multiplier stationarity,
    [XT] **xtunitroot**
  Harris–Tzavalis, [XT] **xtunitroot**
  Hausman specification, *see* Hausman specification
    test
  heterogeneity, *see* heterogeneity test
  heteroskedasticity, [R] **regress postestimation**,
    [R] **sdtest**
  homogeneity, *see* homogeneity test
  hypothesis, *see* hypothesis tests
  Im–Pesaran–Shin, [XT] **xtunitroot**
  independence, [R] **correlate**, [R] **spearman**,
    [R] **tabulate twoway**, [ST] **epitab**,
    [SVY] **svy: tabulate twoway**, *also see* Breusch–
    Pagan subentry
  independence of irrelevant alternatives, *see* IIA
  information matrix, *see* information matrix test
  internal consistency, [R] **alpha**
  interrater agreement, [R] **kappa**
  Kolmogorov–Smirnov, *see* Kolmogorov–Smirnov test
  Kruskal–Wallis, *see* Kruskal–Wallis test
  kurtosis, [R] **regress postestimation**, [R] **sktest**
  Lagrange multiplier, *see* Lagrange-multiplier test
  Levin–Lin–Chu, [XT] **xtunitroot**
  likelihood-ratio, *see* likelihood-ratio test
  linear hypotheses after estimation, *see* linear
    hypothesis test after estimation
  log-rank, *see* log-rank test
  Mantel–Haenszel, *see* Mantel–Haenszel test
  marginal homogeneity, [R] **symmetry**
  margins, [R] **margins**, [R] **pwcompare**
  McNemar's chi-squared test, *see* McNemar's chi-
    squared test
  model coefficients, [R] **lrtest**, [R] **test**, [R] **testnl**,
    [SVY] **svy postestimation**
  model simplification, *see* model simplification tests
  model specification, *see* specification test
  modification indices, *see* modification indices
  nonlinear, [SVY] **svy postestimation**
  nonlinear hypotheses after estimation, *see* nonlinear
    hypothesis test after estimation
  normality, *see* normal distribution and normality
  omitted variables, *see* omitted variables test
  overidentifying restrictions, *see* overidentifying
    restrictions, tests of
  permutation, [R] **permute**
  quadrature, [XT] **quadchk**
  Ramsey, *see* Ramsey test
  random order, [R] **runtest**
  RESET, *see* RESET test
  score, *see* score tests
  serial correlation, [R] **regress postestimation time
    series**

test, *continued*
  serial independence, [R] **runtest**
  Shapiro–Francia, *see* Shapiro–Francia test for
    normality
  Shapiro–Wilk, *see* Shapiro–Wilk test for normality
  skewness, [R] **regress postestimation**, [R] **sktest**
  symmetry, [R] **symmetry**
  Szroeter's, *see* Szroeter's test for heteroskedasticity
  TDT, [R] **symmetry**
  trend, *see* trend test
  unit root, [XT] **xtunitroot**
  variance-comparison, [R] **sdtest**
  Wald, *see* Wald tests
  weak instruments, [R] **ivregress postestimation**
test, sts subcommand, [ST] **sts test**
test-based confidence intervals, [ST] **epitab**
testnl command, [R] **testnl**, [SEM] **estat stdize**,
    [SEM] **testnl**, [SVY] **svy postestimation**
testparm command, [R] **test**, [SEM] **test**, [SVY] **svy
    postestimation**
tests
  after estimation, *see* estimation, tests after
  multivariate, [MV] **mvtest**
  normality, [MV] **mvtest normality**
  variance-comparison, [MV] **mvtest covariances**
testtransform, mi subcommand, [MI] **mi test**
tetrachoric command, [R] **tetrachoric**
tetrachoric correlation, [MV] **Glossary**
text
  adding, [G-3] *added_text_options*
  and textboxes, relationship between, [G-4] *textstyle*
  angle of, [G-4] *anglestyle*
  captions, [G-3] *title_options*
  look of, [G-4] *textboxstyle*, [G-4] *textstyle*
  note, [G-3] *title_options*
  resizing, [G-3] *scale_option*
  running outside of borders,
    [G-3] *added_text_options*
  size of, [G-3] *textbox_options*
  subtitle, [G-3] *title_options*
  title, [G-3] *title_options*
  vertical alignment, [G-4] *alignmentstyle*
text,
  exporting, *see* exporting data
  reading data in, *see* importing data
  saving data in, *see* exporting data
text() option, [G-3] *added_text_options*,
    [G-3] *aspect_option*
text, in graphs, [G-4] *text*
textboxes, [G-3] *textbox_options*
  orientation of, [G-4] *orientationstyle*
textboxstyle, [G-4] **textboxstyle**
textsizestyle, [G-4] **textsizestyle**
textstyle, [G-4] **textstyle**
th() pseudofunction, [D] **datetime**, [D] **functions**
thickness of lines, [G-4] *linewidthstyle*
Thomson scoring, [MV] **factor postestimation**

thrashing, [ST] **Glossary**

three-dimensional graph, [G-2] **graph twoway contour**,
[G-2] **graph twoway contourline**

three-stage least squares, [R] **reg3**

tick,
definition, [G-4] *tickstyle*
suppressing, [G-4] *tickstyle*

ticksetstyle, [G-4] *ticksetstyle*

tickstyle, [G-4] *tickstyle*

ties, [MV] **Glossary**

TIFF, [G-3] *tif_options*

time and date, [M-5] **c( )**, [M-5] **date( )**

time-domain analysis, [TS] **arch**, [TS] **arfima**,
[TS] **arima**, [TS] **Glossary**

time of day, [P] **creturn**

time-series
analysis, [D] **egen**, [R] **regress postestimation time
series**
filter, [TS] **psdensity**, [TS] **ucm**
formats, [D] **format**
functions, [D] **functions**
operators, [TS] **tsset**
operated variable, [M-5] **st_data( )**,
[M-5] **st_tsrevar( )**, [M-6] **Glossary**
plots, [G-2] **graph other**, [G-2] **graph twoway
tsline**

time-span data, [ST] **snapspan**

time stamp, [D] **describe**

time variables and values, [D] **datetime**

time-varying covariates, [ST] **Glossary**

time-varying variance, [TS] **arch**

time-versus-concentration curve, [R] **pk**

timeout1, set subcommand, [R] **netio**, [R] **set**

timeout2, set subcommand, [R] **netio**, [R] **set**

timer
clear command, [P] **timer**
list command, [P] **timer**
off command, [P] **timer**
on command, [P] **timer**

time-series
analysis, [P] **matrix accum**
estimation, [U] **26.16 Models with time-series data**
operators, [U] **13.9 Time-series operators**
unabbreviating varlists, [P] **unab**
varlists, [U] **11.4.4 Time-series varlists**

timing code, [P] **timer**

tin( ) function, [D] **functions**

title, estimates subcommand, [R] **estimates title**

title( ) option, [G-3] *title_options*

titles, [G-3] *title_options*
of axis, [G-3] *axis_title_options*

tlabel( ) option, [G-3] *axis_label_options*

TLI, *see* Tucker–Lewis index

tm( ) pseudofunction, [D] **datetime**, [D] **functions**

tmlabel( ) option, [G-3] *axis_label_options*

TMPDIR Unix environment variable, [P] **macro**

tmtick( ) option, [G-3] *axis_label_options*

tnbreg command, [R] **tnbreg**, [R] **tnbreg
postestimation**

tobit command, [R] **tobit**, [R] **tobit postestimation**

tobit regression, [R] **ivtobit**, [R] **tobit**, [SVY] **svy
estimation**, *also see* intreg command, *also see*
truncreg command
random-effects, [XT] **xttobit**
with endogenous regressors, [SVY] **svy estimation**

.toc filename suffix, [R] **net**

Toeplitz( ) function, [M-5] **Toeplitz( )**

tokenallowhex( ) function, [M-5] **tokenget( )**

tokenallownum( ) function, [M-5] **tokenget( )**

tokenget( ) function, [M-5] **tokenget( )**

tokengetall( ) function, [M-5] **tokenget( )**

tokeninit( ) function, [M-5] **tokenget( )**

tokeninitstata( ) function, [M-5] **tokenget( )**

tokenize command, [P] **tokenize**

tokenoffset( ) function, [M-5] **tokenget( )**

tokenpchars( ) function, [M-5] **tokenget( )**

tokenpeek( ) function, [M-5] **tokenget( )**

tokenqchars( ) function, [M-5] **tokenget( )**

tokenrest( ) function, [M-5] **tokenget( )**

tokens( ) function, [M-5] **tokens( )**

tokenset( ) function, [M-5] **tokenget( )**

tokenwchars( ) function, [M-5] **tokenget( )**

tolerance( ) option, [R] **maximize**

tolerances, [M-1] **tolerance**, [M-5] **solve_tol( )**

top( ) suboption, [G-4] *alignmentstyle*

tostring command, [D] **destring**

total command, [R] **total**, [R] **total postestimation**

total( ), egen function, [D] **egen**

total inertia, [MV] **Glossary**

total principal inertia, [MV] **Glossary**

totals, estimation, [R] **total**

totals, survey data, [SVY] **svy estimation**

toward a target rotation, [MV] **procrustes**, [MV] **rotate**,
[MV] **rotatemat**

tpoisson command, [R] **tpoisson**, [R] **tpoisson
postestimation**

tq( ) pseudofunction, [D] **datetime**, [D] **functions**

trace( ) function, [D] **functions**, [M-5] **trace( )**,
[P] **matrix define**

trace, ml subcommand, [R] **ml**

trace of matrix, [M-5] **trace( )**, [P] **matrix define**

trace option, [R] **maximize**

trace, set subcommand, [P] **creturn**, [P] **trace**,
[R] **set**

traceback log, [M-2] **errors**, [M-5] **error( )**,
[M-6] **Glossary**

tracedepth, set subcommand, [P] **creturn**, [P] **trace**,
[R] **set**

traceexpand, set subcommand, [P] **creturn**,
[P] **trace**, [R] **set**

tracehilite, set subcommand, [P] **creturn**,
[P] **trace**, [R] **set**

traceindent, set subcommand, [P] **creturn**,
[P] **trace**, [R] **set**

tracenumber, set subcommand, [P] **creturn**,
    [P] **trace**, [R] **set**
tracesep, set subcommand, [P] **creturn**, [P] **trace**,
    [R] **set**
tracing iterative maximization process, [R] **maximize**
trademark symbol, [G-4] *text*
training, [U] **3.7 Conferences and training**
transferring data
    copying and pasting, [D] **edit**
    from Stata, [D] **export**, [U] **21.4 Transfer programs**
    into Stata, [D] **import**, [U] **21 Inputting and
        importing data**, [U] **21.4 Transfer programs**
transformation, [MV] **procrustes**
transformations
    to achieve normality, [R] **boxcox**, [R] **ladder**
    to achieve zero skewness, [R] **lnskew0**
transformations,
    log, [R] **lnskew0**
    modulus, [R] **boxcox**
    power, [R] **boxcox**, [R] **lnskew0**
transformed coefficients, [MI] **mi estimate**, [MI] **mi
    estimate using**, [MI] **mi test**
translate command, [R] **translate**
translate logs, [R] **translate**
translation, file, [D] **changeeol**, [D] **filefilter**
translator
    query command, [R] **translate**
    reset command, [R] **translate**
    set command, [R] **translate**
transmap
    define command, [R] **translate**
    query command, [R] **translate**
transmission-disequilibrium test, [R] **symmetry**
transmorphic, [M-2] **declarations**, [M-6] **Glossary**
transpose, [M-6] **Glossary**, *also see* conjugate transpose
    in place, [M-5] **_transpose( )**
    operator, [M-2] **op_transpose**
    without conjugation, [M-5] **transposeonly( )**
_transpose( ) function, [M-5] **_transpose( )**
_transposeonly( ) function, [M-5] **transposeonly( )**
transposeonly( ) function, [M-5] **transposeonly( )**
transposing data, [D] **xpose**
transposing matrices, [P] **matrix define**
transposition, [M-2] **op_transpose**, [M-5] **_transpose( )**,
    [M-5] **transposeonly( )**
treatment effects, [R] **treatreg**
treatment-effects regression, [SVY] **svy estimation**
treatreg command, [R] **treatreg**, [R] **treatreg
    postestimation**
tree, misstable subcommand, [R] **misstable**
trees, [MV] **cluster**, [MV] **cluster dendrogram**
trend, [TS] **Glossary**
trend test, [R] **nptrend**, [R] **symmetry**, [ST] **epitab**,
    [ST] **sts test**
trend, test for, [R] **nptrend**, [R] **symmetry**
triangular matrix, [M-5] **solvelower( )**, [M-6] **Glossary**
trigamma( ) function, [D] **functions**, [M-5] **factorial( )**

trigonometric functions, [D] **functions**, [M-5] **sin( )**
trim( ) function, [D] **functions**
trunc( ) function, [D] **functions**, [M-5] **trunc( )**
truncated
    negative binomial regression, [R] **tnbreg**, [SVY] **svy
        estimation**
    Poisson regression, [R] **tpoisson**, [SVY] **svy
        estimation**
    regression, [R] **truncreg**, [SVY] **svy estimation**
truncating
    real numbers, [D] **functions**
    strings, [D] **functions**
truncation, [ST] **Glossary**, *see* imputation, truncated
    data
truncreg command, [R] **truncreg**, [R] **truncreg
    postestimation**
tsappend command, [TS] **tsappend**
tscale, graph twoway subcommand, [G-2] **graph
    twoway tsline**
tscale( ) option, [G-3] *axis_scale_options*
tsfill command, [TS] **tsfill**
tsfilter, [TS] **tsfilter**
    bk command, [TS] **tsfilter bk**
    bw command, [TS] **tsfilter bw**
    cf command, [TS] **tsfilter cf**
    hp command, [TS] **tsfilter hp**
tsline command, [TS] **tsline**
tsline, graph twoway subcommand, [G-2] **graph
    twoway tsline**
tsnorm macro extended function, [P] **macro**
tsreport command, [TS] **tsreport**
tsrevar command, [TS] **tsrevar**
tsrline command, [TS] **tsline**
tsrline, graph twoway subcommand, [G-2] **graph
    twoway tsline**
tsset command, [MI] **mi XXXset**, [TS] **tsset**
tsset, mi subcommand, [MI] **mi XXXset**
tssmooth, [TS] **tssmooth**
    shwinters command, [TS] **tssmooth shwinters**
    dexponential command, [TS] **tssmooth
        dexponential**
    exponential command, [TS] **tssmooth exponential**
    hwinters command, [TS] **tssmooth hwinters**
    ma command, [TS] **tssmooth ma**
    nl command, [TS] **tssmooth nl**
tsunab command, [P] **unab**
ttail( ) function, [D] **functions**, [M-5] **normal( )**
ttest and ttesti commands, [MV] **hotelling**,
    [R] **ttest**
ttick( ) option, [G-3] *axis_label_options*
ttitle( ) option, [G-3] *axis_title_options*
Tucker–Lewis index, [SEM] **estat gof**, [SEM] **methods
    and formulas**
Tukey's multiple-comparison adjustment, *see* multiple
    comparisons, Tukey's method
tukeyprob( ) function, [D] **functions**, [M-5] **normal( )**
Tukey's Studentized range, [D] **functions**
tuning constant, [R] **rreg**

tutorials, [U] **1.2.2 Example datasets**

tw( ) pseudofunction, [D] **datetime**, [D] **functions**

twithin( ) function, [D] **functions**

Twitter, *see* Stata on Twitter

two-level model, [XT] **Glossary**

two-stage least squares, [R] **gmm**, [R] **ivregress**,
    [R] **nlsur**, [R] **regress**, [SVY] **svy estimation**,
    [XT] **xthtaylor**, [XT] **xtivreg**

two-way
    analysis of variance, [R] **anova**
    multivariate analysis of variance, [MV] **manova**
    scatterplots, [R] **lowess**

type, [M-2] **declarations**, [M-6] **Glossary**

type
    command, [D] **type**
    macro extended function, [P] **macro**
    parameter, [D] **generate**

type,
    set subcommand, [D] **generate**, [R] **set**
    ssc subcommand, [R] **ssc**

type, broad, [M-6] **Glossary**

type I error, [ST] **Glossary**

type I study, [ST] **Glossary**

type II error, [ST] **Glossary**

type II study, [ST] **Glossary**

U

U statistic, [R] **ranksum**

UCM
    model, [TS] **ucm**
    postestimation, [TS] **psdensity**, [TS] **ucm
        postestimation**

ucm command, [TS] **ucm**

unab command, [P] **unab**

unabbreviate
    command names, [P] **unabcmd**
    variable list, [P] **syntax**, [P] **unab**

unabcmd command, [P] **unabcmd**

unaddgroup, ssd subcommand, [SEM] **ssd**

.uname built-in class function, [P] **class**

unary operator, [M-6] **Glossary**

unbalanced data, [XT] **Glossary**

uncompress files, [D] **zipfile**

under observation, [ST] **Glossary**

underlining in syntax diagram, [U] **11 Language syntax**

underscore c( ) function, [D] **functions**

underscore functions, [M-1] **naming**, [M-6] **Glossary**

underscore variables, [U] **13.4 System variables
    (_variables)**

unhold, _estimates subcommand, [P] **_estimates**

uniform prior, [MI] **mi impute mvn**

uniformly distributed random
    number function, [D] **functions**, [M-5] **runiform( )**,
        [R] **set seed**
    variates, [M-5] **runiform( )**

uninstall,
    net subcommand, [R] **net**
    ssc subcommand, [R] **ssc**

uniqrows( ) function, [M-5] **uniqrows( )**

unique options, [G-4] **concept: repeated options**

unique value labels, [D] **labelbook**

unique values,
    counting, [D] **codebook**, [R] **table**, [R] **tabulate
        oneway**
    determining, [D] **inspect**, [D] **labelbook**

uniqueness, [MV] **Glossary**

unit loading, [SEM] **intro 3**

unit-root
    models, [TS] **vec intro**, [TS] **vec**
    process, [TS] **Glossary**
    test, [TS] **dfgls**, [TS] **dfuller**, [TS] **pperron**,
        [TS] **Glossary**, [XT] **xtunitroot**

unit vectors, [M-5] **e( )**

unitary matrix, [M-6] **Glossary**

unitcircle( ) function, [M-5] **unitcircle( )**

univariate
    distributions, displaying, [R] **cumul**, [R] **diagnostic
        plots**, [R] **histogram**, [R] **ladder**, [R] **lv**,
        [R] **stem**
    imputation, *see* imputation, univariate
    kernel density estimation, [R] **kdensity**
    time series, [TS] **arch**, [TS] **arfima**, [TS] **arima**,
        [TS] **newey**, [TS] **prais**, [TS] **ucm**

Unix,
    keyboard use, [U] **10 Keyboard use**
    pause, [P] **sleep**
    specifying filenames, [U] **11.6 Filenaming
        conventions**

_unlink( ) function, [M-5] **unlink( )**

unlink( ) function, [M-5] **unlink( )**

unobserved-components model, *see* UCM

unorder( ) function, [M-5] **sort( )**

unregister, mi subcommand, [MI] **mi set**

unregistered variables, *see* variables, unregistered

unrestricted FMI test, [MI] **mi estimate**, [MI] **mi test**,
    [MI] **Glossary**

unrestricted transformation, [MV] **Glossary**

unstandardized coefficient, [SEM] **Glossary**

unstructured, [SEM] **Glossary**

unzipfile command, [D] **zipfile**

update
    all command, [R] **update**
    command, [R] **update**
    from command, [R] **update**
    query command, [R] **update**

update,
    mi subcommand, [MI] **mi update**, [MI] **noupdate
        option**
    view subcommand, [R] **view**

update_d, view subcommand, [R] **view**

update_interval, set subcommand, [R] **set**,
    [R] **update**

update_prompt, set subcommand, [R] set,
    [R] update
update_query, set subcommand, [R] set, [R] update
UPDATES directory, [P] sysdir, [U] **17.5 Where does
    Stata look for ado-files?**
updates to Stata, [R] adoupdate, [R] net, [R] sj,
    [R] update, [U] **3.5 The Stata Journal**,
    [U] **3.6 Updating and adding features from the
    web**, [U] **17.6 How do I install an addition?**,
    [U] **28 Using the Internet to keep up to date**
upper() function, [D] functions
uppercase, [M-5] strupper( )
uppercase-string function, [D] functions
_uppertriangle() function, [M-5] lowertriangle( )
uppertriangle() function, [M-5] lowertriangle( )
upper-triangular matrix, *see* triangular matrix
use,
    cluster subcommand, [MV] cluster utility
    command, [D] use
    estimates subcommand, [R] estimates save
    graph subcommand, [G-2] graph use
    serset subcommand, [P] serset
uselabel command, [D] labelbook
user interface, [P] dialog programming
user-written additions,
    installing, [R] net, [R] ssc
    searching for, [R] net search, [R] ssc
using,
    cmdlog subcommand, [R] log
    log subcommand, [R] log
using data, [D] sysuse, [D] use, [D] webuse,
    [P] syntax, *also see* importing data
using graphs, [G-2] graph use
utilities, programming, [MV] cluster utility
utility routines, [MI] technical

V

valofexternal() function, [M-5] valofexternal( )
value label macro extended function, [P] macro
value labels, [D] codebook, [D] describe, [D] edit,
    [D] encode, [D] inspect, [D] label, [D] label
    language, [D] labelbook, [D] varmanage,
    [P] macro, [U] **12.6.3 Value labels**,
    [U] **13.10 Label values**
    potential problems in, [D] labelbook
values, label subcommand, [D] label
Vandermonde() function, [M-5] Vandermonde( )
VAR, [TS] dfactor, [TS] sspace, [TS] ucm, [TS] var
    intro, [TS] var, [TS] var svar, [TS] varbasic,
    [TS] Glossary
    postestimation, [TS] fcast compute, [TS] fcast
    graph, [TS] irf, [TS] irf create, [TS] var
    postestimation, [TS] vargranger, [TS] varlmar,
    [TS] varnorm, [TS] varsoc, [TS] varstable,
    [TS] varwle
var command, [TS] var, [TS] var postestimation
varabbrev command, [P] varabbrev
varabbrev, set subcommand, [R] set

varbasic command, [TS] varbasic, [TS] varbasic
    postestimation
vargranger command, [TS] vargranger
variable, [M-2] declarations, [M-5] st_data( ),
    [M-6] Glossary
    abbreviation, [P] varabbrev
    description, [D] describe
    labels, [D] codebook, [D] edit, [D] describe,
        [D] label, [D] label language, [D] notes,
        [D] varmanage, [P] macro, [U] **11.4 varlists**,
        [U] **12.6.2 Variable labels**
    lists, *see* varlist
    identifying panels, [XT] xtset
    types, [P] macro, [U] **11.4 varlists**,
        [U] **12.2.2 Numeric storage types**,
        [U] **12.4.4 String storage types**
    class, [P] class
variable, confirm subcommand, [P] confirm
variable label macro extended function, [P] macro
variable, label subcommand, [D] label
variable types, [D] codebook, [D] data types,
    [D] describe, [I] data types, [SEM] intro 3
variable-naming convention, [M-1] naming
_variables, [U] **11.3 Naming conventions**,
    [U] **13.4 System variables (_variables)**
variables,
    alphabetizing, [D] order
    categorical, *see* categorical data, agreement,
        measures for, *see* categorical data
    changing storage types of, [D] recast
    characteristics of, [P] char, [P] macro,
        [U] **12.8 Characteristics**
    comparing, [D] compare
    copying, [D] clonevar
    creating, [D] varmanage
    creating new, [D] separate
    describing, [D] codebook, [D] notes
    determining storage types of, [D] describe
    displaying contents of, [D] edit, [D] list
    documenting, [D] codebook, [D] labelbook,
        [D] notes
    dropping, [D] drop
    dummy, *see* indicator variables
    factor, *see* factor variables
    filtering, [D] varmanage
    finding, [D] lookfor
    generate, summary, or grouping, [MV] cluster
        generate
    generating, [ST] stgen
    imputed, [MI] mi rename, [MI] mi reset, [MI] mi
        set, [MI] Glossary
    in dataset, maximum number of, [D] memory,
        [U] **6 Managing memory**
    in model, maximum number, [R] matsize
    list values of, [P] levelsof
    listing, [D] codebook, [D] describe, [D] labelbook,
        [D] edit, [D] list
    mapping numeric to string, [D] destring

variables, *continued*
    naming, [D] **rename**, [U] **11.2 Abbreviation rules**,
        [U] **11.3 Naming conventions**
    naming groups of, [D] **rename group**
    ordering, [D] **sort**
    orthogonalize, [R] **orthog**
    passive, [MI] **mi impute**, [MI] **mi passive**, [MI] **mi
        rename**, [MI] **mi reset**, [MI] **mi set**, [MI] **mi
        xeq**, [MI] **Glossary**
    registered, [MI] **mi rename**, [MI] **mi set**,
        [MI] **Glossary**
    regular, [MI] **mi rename**, [MI] **mi set**,
        [MI] **Glossary**
    renaming, *see* renaming variables
    reordering, [D] **order**
    setting properties of, [D] **varmanage**
    sorting, [D] **gsort**, [D] **sort**, [D] **varmanage**
    standardizing, [D] **egen**
    storage types, *see* storage types
    string, *see* string variables
    system, *see* system variables
    tab expansion of, [U] **10.6 Tab expansion of
        variable names**
    temporary, [P] **macro**
    transposing with observations, [D] **xpose**
    unabbreviating, [P] **syntax**, [P] **unab**
    unique values, [D] **codebook**, [D] **duplicates**
    unique values, determining, [D] **duplicates**,
        [D] **inspect**
    unregistered, [MI] **mi rename**, [MI] **mi set**,
        [MI] **Glossary**
    varying and super varying, [MI] **mi passive**,
        [MI] **mi predict**, [MI] **mi set**, [MI] **mi varying**,
        [MI] **Glossary**, [ST] **stvary**
Variables Manager, [D] **varmanage**
variance, [SEM] **intro 3**, *also see* constraints, specifying
    analysis of, [MV] **manova**, [R] **anova**, [R] **loneway**,
        [R] **oneway**
    components, [XT] **Glossary**, *also see* mixed model
    creating dataset of, [D] **collapse**
    creating variable containing, [D] **egen**
    decompositions, *see* FEVD
    displaying, [R] **summarize**, [R] **table**, [R] **tabstat**,
        [R] **tabulate, summarize()**, [XT] **xtsum**, [R] **lv**
    estimation, [SVY] **variance estimation**,
        [SVY] **Glossary**
    estimators, [R] *vce_option*, [XT] *vce_options*
    Huber/White/sandwich estimator, *see* robust,
        Huber/White/sandwich estimator of variance
    inflation factors, [R] **regress postestimation**
    linearized, [SVY] **variance estimation**
    nonconstant, *see* robust, Huber/White/sandwich
        estimator of variance
    stabilizing transformations, [R] **boxcox**
    testing equality of, [R] **sdtest**
variance-comparison test, [R] **sdtest**

variance–covariance matrix of estimators, [P] **ereturn**,
    [P] **matrix get**, [R] **correlate**, [R] **estat**,
    [SEM] **Glossary**, *also see* sem option vce()
variance–covariance matrix, obtaining,
    [U] **20.9 Obtaining the variance–covariance
    matrix**
variance() function, [M-5] **mean( )**
variance() option, *see* sem option variance()
variance-weighted least squares, [R] **vwls**
varimax rotation, [MV] **rotate**, [MV] **rotatemat**,
    [MV] **Glossary**
varkeyboard, set subcommand, [R] **set**
varlist, [P] **syntax**, [U] **11 Language syntax**,
    [U] **11.4 varlists**
    existing, [U] **11.4.1 Lists of existing variables**
    new, [U] **11.4.2 Lists of new variables**
    time series, [U] **11.4.4 Time-series varlists**
varlmar command, [TS] **varlmar**
varmanage command, [D] **varmanage**
varnorm command, [TS] **varnorm**
varsoc command, [TS] **varsoc**
varstable command, [TS] **varstable**
varwle command, [TS] **varwle**
varying
    conditional-correlation model, [TS] **mgarch**,
        [TS] **mgarch vcc**
    estimation sample, [MI] **mi estimate**
    variables, *see* variables, varying and super varying
varying, mi subcommand, [MI] **mi varying**
vcc, mgarch subcommand, [TS] **mgarch vcc**
VCE, *see* variance–covariance matrix of the estimator
vce, estat subcommand, [R] **estat**, [SVY] **estat**
vce() option, [R] *vce_option*, [XT] *vce_options*, *see*
    sem option vce()
vec command, [TS] **vec**, [TS] **vec postestimation**
vec() function, [D] **functions**, [M-5] **vec( )**, [P] **matrix
    define**
vecaccum, matrix subcommand, [P] **matrix accum**
vecdiag() function, [D] **functions**, [P] **matrix define**
vech() function, [M-5] **vec( )**
veclmar command, [TS] **veclmar**
VECM, [TS] **vec intro**, [TS] **vec**, [TS] **Glossary**, *also
    see* multivariate GARCH
    postestimation, [TS] **fcast compute**, [TS] **fcast
        graph**, [TS] **irf**, [TS] **irf create**, [TS] **varsoc**,
        [TS] **vec postestimation**, [TS] **veclmar**,
        [TS] **vecnorm**, [TS] **vecrank**, [TS] **vecstable**
vecnorm command, [TS] **vecnorm**
vecrank command, [TS] **vecrank**
vecstable command, [TS] **vecstable**
vector, [M-2] **declarations**, [M-6] **Glossary**
vector autoregression, *see* VAR
vector autoregressive forecast, [TS] **fcast compute**,
    [TS] **fcast graph**
vector autoregressive (VAR) models, [G-2] **graph other**
vector autoregressive moving-average model,
    [TS] **dfactor**, [TS] **sspace**, [TS] **ucm**
vector error-correction model, *see* VECM

vector norm, [M-5] **norm( )**

vectors, *see* matrices

verifying data, [D] **assert**, [D] **count**,
    [D] **datasignature**, [D] **inspect**, *also see*
    certifying data

verifying mi data are consistent, [MI] **mi update**

version command, [M-2] **version**, [P] **version**,
    [U] **16.1.1 Version**, [U] **18.11.1 Version**

    class programming, [P] **class**

version control, [M-2] **version**, [M-5] **callersversion( )**,
    see version command

version of ado-file, [R] **which**

version of Stata, [M-5] **stataversion( )**, [R] **about**

vertical alignment of text, [G-4] *alignmentstyle*

view

    ado command, [R] **view**

    ado_d command, [R] **view**

    browse command, [R] **view**

    command, [R] **view**

    help command, [R] **view**

    help_d command, [R] **view**

    net command, [R] **view**

    net_d command, [R] **view**

    news command, [R] **view**

    search command, [R] **view**

    search_d command, [R] **view**

    update command, [R] **view**

    update_d command, [R] **view**

    view_d command, [R] **view**

view matrix, [M-5] **isview( )**, [M-5] **st_subview( )**,
    [M-5] **st_view( )**, [M-5] **st_viewvars( )**,
    [M-6] **Glossary**

view source code, [P] **viewsource**

view_d, view subcommand, [R] **view**

viewing previously typed lines, [R] **#review**

viewsource, [M-1] **source**

viewsource command, [P] **viewsource**

vif, estat subcommand, [R] **regress postestimation**

vignettes, [U] **1.2.7 Vignettes**

virtual, [M-2] **class**

virtual memory, [D] **memory**

void

    function, [M-2] **declarations**, [M-6] **Glossary**

    matrix, [M-2] **void**, [M-6] **Glossary**

vwls command, [R] **vwls**, [R] **vwls postestimation**

W

Wald tests, [R] **contrast**, [R] **predictnl**,
    [R] **test**, [R] **testnl**, [SEM] **intro 6**,
    [SEM] **estat eqtest**, [SEM] **estat ginvariant**,
    [SEM] **example 13**, [SEM] **example 22**,
    [SEM] **methods and formulas**, [SEM] **test**,
    [SEM] **testnl**, [SEM] **Glossary**, [SVY] **svy**
    postestimation, [TS] **vargranger**, [TS] **varwle**,
    [U] **20.12 Performing hypothesis tests on the**
    coefficients, [U] **20.12.4 Nonlinear Wald tests**

wardslinkage,
    clustermat subcommand, [MV] **cluster linkage**
    cluster subcommand, [MV] **cluster linkage**

Ward's linkage clustering, [MV] **cluster**,
    [MV] **clustermat**, [MV] **cluster linkage**,
    [MV] **Glossary**

Ward's method clustering, [MV] **cluster**,
    [MV] **clustermat**

warning messages, [M-2] **pragma**

waveragelinkage,
    clustermat subcommand, [MV] **cluster linkage**
    cluster subcommand, [MV] **cluster linkage**

wcorrelation, estat subcommand, [XT] **xtgee**
    postestimation

weak instrument test, [R] **ivregress postestimation**

weakly balanced, [XT] **Glossary**

website,
    stata.com, [U] **3.2.1 The Stata website**
    (www.stata.com)
    stata-press.com, [U] **3.3 Stata Press**

webuse

    query command, [D] **webuse**

    set command, [D] **webuse**

    command, [D] **webuse**

week( ) function, [D] **datetime**, [D] **functions**,
    [M-5] **date( )**

weekly( ) function, [D] **datetime**, [D] **datetime**
    translation, [D] **functions**, [M-5] **date( )**

Weibull distribution, [ST] **streg**

Weibull survival regression, [ST] **streg**

weight, [P] **syntax**

[weight=*exp*] modifier, [U] **11.1.6 weight**,
    [U] **20.22 Weighted estimation**

weighted data, [U] **11.1.6 weight**, [U] **20.22 Weighted**
    estimation, *also see* survey data

weighted least squares, [R] **regress**, [SEM] **methods**
    and formulas, [SEM] **Glossary**

    for grouped data, [R] **glogit**

    generalized linear models, [R] **glm**

    generalized method of moments estimation,
        [R] **gmm**

    instrumental-variables regression, [R] **gmm**,
        [R] **ivregress**

    nonlinear least-squares estimation, [R] **nl**

    nonlinear systems of equations, [R] **nlsur**

    variance, [R] **vwls**

weighted moving average, [TS] **tssmooth**,
    [TS] **tssmooth ma**

weighted-average linkage clustering, [MV] **cluster**,
    [MV] **clustermat**, [MV] **cluster linkage**,
    [MV] **Glossary**

weights, [G-2] **graph twoway scatter**

    probability, [SVY] **survey**, [SVY] **svydescribe**,
    [SVY] **svyset**

    sampling, [SVY] **survey**, [SVY] **svydescribe**,
    [SVY] **svyset**

Welsch distance, [R] **regress postestimation**

which, class, [P] **classutil**

which command, [M-3] **mata which**, [R] **which**,
  [U] **17.3 How can I tell if a command is built
  in or an ado-file?**
which, classutil subcommand, [P] **classutil**
while command, [M-2] **while**, [M-2] **continue**,
  [M-2] **break**, [M-2] **semicolons**, [P] **while**
White/Huber/sandwich estimator of variance, *see* robust,
  Huber/White/sandwich estimator of variance
white noise, [TS] **Glossary**, [XT] **Glossary**
white-noise test, [TS] **wntestb**, [TS] **wntestq**
White's test for heteroskedasticity, [R] **regress
  postestimation**
wide
  data style, [MI] **styles**, [MI] **Glossary**
  technical description, [MI] **technical**
wide, reshape subcommand, [D] **reshape**
width of % *fmt*, [M-5] **fmtwidth( )**
width() option, [G-2] **graph twoway histogram**
Wilcoxon
  rank-sum test, [R] **ranksum**
  signed-ranks test, [R] **signrank**
  test (Wilcoxon–Breslow, Wilcoxon–Gehan,
    Wilcoxon–Mann–Whitney), [ST] **sts test**
wildcard, *see* regexm() string function, *see* regexr()
  string function, *see* regexs() string function, *see*
  strmatch() string function
Wilks'
  lambda, [MV] **canon**, [MV] **manova**, [MV] **mvtest
    means**, [MV] **Glossary**
  likelihood-ratio test, [MV] **canon**, [MV] **manova**,
    [MV] **mvtest means**
window
  fopen command, [P] **window programming**
  fsave command, [P] **window programming**
  manage command, [P] **window programming**
  menu command, [P] **window programming**
  push command, [P] **window programming**
  stopbox command, [P] **window programming**
Windows Metafile, [G-2] **graph export**
Windows metafiles programming, [P] **automation**
Windows, pause, [P] **sleep**
Windows programming, [P] **automation**
Windows,
  filenames, [U] **18.3.11 Constructing Windows
    filenames by using macros**
  keyboard use, [U] **10 Keyboard use**
  specifying filenames, [U] **11.6 Filenaming
    conventions**
winexec command, [D] **shell**
Wishart distribution, [MV] **Glossary**
withdrawal, [ST] **Glossary**
within-cell means and variances, [XT] **xtsum**
within estimators, [XT] **xthtaylor**, [XT] **xtivreg**,
  [XT] **xtreg**, [XT] **xtregar**, [XT] **Glossary**
within matrix, [MV] **Glossary**
within-imputation variability, [MI] **mi estimate**, [MI] **mi
  predict**
WLF, *see* worst linear function

WLS, *see* weighted least squares
wntestb command, [TS] **wntestb**
wntestq command, [TS] **wntestq**
wofd() function, [D] **datetime**, [D] **functions**,
  [M-5] **date( )**
Woolf confidence intervals, [ST] **epitab**
word macro extended function, [P] **macro**
word() function, [D] **functions**
wordcount() function, [D] **functions**
workflow, [MI] **workflow**
worst linear function, [MI] **mi impute mvn**,
  [MI] **Glossary**
wrap() option, *see* sem option wrap()
write, file subcommand, [P] **file**
writing and reading ASCII text and binary files, [P] **file**
writing data, *see* exporting data, *see* saving data
www.stata.com website, [U] **3.2.1 The Stata website
  (www.stata.com)**
www.stata-press.com website, [U] **3.3 Stata Press**

X

xaxis() suboption, [G-3] *axis_choice_options*
X-bar charts, [G-2] **graph other**
xchart command, [R] **qc**
xcommon option, [G-2] **graph combine**
xcorr command, [TS] **xcorr**
xeq, mi subcommand, [MI] **mi xeq**
xi prefix command, [R] **xi**
xlabel() option, [G-3] *axis_label_options*
xline() option, [G-3] *added_line_options*
XML, [D] **xmlsave**
xmlabel() option, [G-3] *axis_label_options*
xmlsave command, [D] **xmlsave**
xmluse command, [D] **xmlsave**
xmtick() option, [G-3] *axis_label_options*
xpose command, [D] **xpose**
xscale() option, [G-3] *axis_scale_options*
xshell command, [D] **shell**
xsize() option, [G-2] **graph display**,
  [G-3] *region_options*
xtabond command, [XT] **xtabond**
xtcloglog command, [XT] **quadchk**, [XT] **xtcloglog**
xtdata command, [XT] **xtdata**
xtdescribe command, [XT] **xtdescribe**
xtdpd command, [XT] **xtdpd**
xtdpdsys command, [XT] **xtdpdsys**
xtfrontier command, [XT] **xtfrontier**
xtgee command, [XT] **xtgee**
xtgls command, [XT] **xtgls**
xthtaylor command, [XT] **xthtaylor**
xtick() option, [G-3] *axis_label_options*
xtile command, [D] **pctile**
xtintreg command, [XT] **quadchk**, [XT] **xtintreg**
xtitle() option, [G-3] *axis_title_options*
xtivreg command, [XT] **xtivreg**
xtline command, [XT] **xtline**

xtlogit command, [XT] **quadchk**, [XT] **xtlogit**
xtmelogit command, [XT] **xtmelogit**
xtmepoisson command, [XT] **xtmepoisson**
xtmixed command, [XT] **xtmixed**
xtnbreg command, [XT] **xtnbreg**
xtpcse command, [XT] **xtpcse**
xtpoisson command, [XT] **quadchk**, [XT] **xtpoisson**
xtprobit command, [XT] **quadchk**, [XT] **xtprobit**
xtrc command, [XT] **xtrc**
xtreg command, [XT] **xtreg**
xtregar command, [XT] **xtregar**
xtset command, [XT] **xtset**
xtset command for mi data, [MI] **mi XXXset**
xtset, mi subcommand, [MI] **mi XXXset**
xtsum command, [XT] **xtsum**
xttab command, [XT] **xttab**
xttest0 command, [XT] **xtreg postestimation**
xttobit command, [XT] **quadchk**, [XT] **xttobit**
xttrans command, [XT] **xttab**
xtunitroot command, [XT] **xtunitroot**
xvarformat() option, [G-3] *advanced_options*
xvarlabel() option, [G-3] *advanced_options*
xxxset, programming, [MI] **technical**

Y

yaxis() suboption, [G-3] *axis_choice_options*
ycommon option, [G-2] **graph combine**
year() function, [D] **datetime**, [D] **functions**,
        [M-5] **date( )**, [U] **24.5 Extracting components
        of dates and times**
yearly() function, [D] **datetime**, [D] **datetime
        translation**, [D] **functions**, [M-5] **date( )**
yh() function, [D] **datetime**, [D] **functions**,
        [M-5] **date( )**
ylabel() option, [G-3] *axis_label_options*
yline() option, [G-3] *added_line_options*
ym() function, [D] **datetime**, [D] **functions**,
        [M-5] **date( )**
ymlabel() option, [G-3] *axis_label_options*
ymtick() option, [G-3] *axis_label_options*
yofd() function, [D] **datetime**, [D] **functions**,
        [M-5] **date( )**
yq() function, [D] **datetime**, [D] **functions**,
        [M-5] **date( )**
yscale() option, [G-3] *axis_scale_options*
ysize() option, [G-2] **graph display**,
        [G-3] *region_options*
ytick() option, [G-3] *axis_label_options*
ytitle() option, [G-3] *axis_title_options*
Yule coefficient similarity measure,
        [MV] *measure_option*
Yule–Walker equations, [TS] **corrgram**, [TS] **Glossary**
yvarformat() option, [G-3] *advanced_options*
yvarlabel() option, [G-3] *advanced_options*
yw() function, [D] **datetime**, [D] **functions**,
        [M-5] **date( )**

Z

Zellner's seemingly unrelated regression, [R] **sureg**,
        [R] **reg3**, [R] **suest**
zero-altered, *see* zero-inflated
zero-inflated
        negative binomial regression, [R] **zinb**, [SVY] **svy
        estimation**
        Poisson regression, [R] **zip**, [SVY] **svy estimation**
zero matrix, [P] **matrix define**
zero-skewness transform, [R] **lnskew0**
zinb command, [R] **zinb**, [R] **zinb postestimation**
zip command, [R] **zip**, [R] **zip postestimation**
zipfile command, [D] **zipfile**
zlabel() option, [G-3] *axis_label_options*
zmlabel() option, [G-3] *axis_label_options*
zmtick() option, [G-3] *axis_label_options*
zscale() option, [G-3] *axis_scale_options*
ztick() option, [G-3] *axis_label_options*
ztitle() option, [G-3] *axis_title_options*